Tropical Peatland Eco-evaluation

Mitsuru Osaki • Nobuyuki Tsuji •
Tsuyoshi Kato • Albertus Sulaiman
Editors

Tropical Peatland Eco-evaluation

Springer

Editors
Mitsuru Osaki
Research Faculty of Agriculture
Hokkaido University
Sapporo, Hokkaido, Japan

Nobuyuki Tsuji
Japan Peatland Society
Yanagawa, Fukuoka, Japan

Tsuyoshi Kato
Sumitomo Forestry Co., LTD.
Chiyoda-ku, Tokyo, Japan

Albertus Sulaiman
Research Center for Climate and Atmosphere
Badan Riset dan Inovasi Nasional (BRIN)
Bandung, West Java, Indonesia

ISBN 978-981-99-6789-6 ISBN 978-981-99-6790-2 (eBook)
https://doi.org/10.1007/978-981-99-6790-2

This Springer imprint is published by the registered company Springer Nature Singapore Pte Ltd.
The registered company address is: 152 Beach Road, #21-01/04 Gateway East, Singapore 189721, Singapore

Paper in this product is recyclable.

Preface

Tropical peatlands are paradoxical ecosystems for plant growth and are based on modern sciences such as plant physiology, soil science, ecology, and hydrology. Pristine tropical peatlands have a "high ground water level (GWL)" that causes a poor oxygen supply to roots (then root cannot absorb nutrients even if rich nutrients are available), and the "extremely poor nutrients in peat soil and peat water" cause serous nutrient deficiencies.

The "Tropical Peatlands Paradox" has been solved in previous books (Springer); first "Tropical Peatland Ecosystem" and second "Tropical Peatland Eco-management." According to this Paradox analysis, we have developed the concepts of "Tri-harmony" and "Tri-lemma" for analysis and management of tropical peatland ecosystems.

Here, *Tropical Peatland Eco-Evaluation* is the third book on Tropical Peatland issues, focusing on the Eco-evaluation of "Tri-harmony/Tri-lemma." In this third book, the eco-evaluation focuses on (1) "Tri-harmony" system evaluation (concept), (2) "Tri-harmony" system analysis (technology: monitoring and sensing, carbon-water modeling, mapping, and disaster prediction), and (3) "Tri-harmony" system management (land surface management (LSM)).

In tropical peatlands, three elements are key in maintaining harmonized ecosystems, water, oxygen, and nutrients, and these three elements' interrelations and interactions are mainly studied in the Tri-harmony/Tri-lemma system. The Tri-harmony/Tri-lemma concept is applied not only to peatlands, wetlands, and mangroves but also to drylands and semiarid lands. Thus, the *Tropical Peatland Eco-Evaluation* book contributes to the analysis of natural capital and nature assets. This book provides good guidance to each country's NDCs (nationally determined contributions) submitted to the UNFCCC (United Nations Framework Convention and Climate Change) by 12 October 2021; 105 out of 114 (92%) enhanced NDCs included nature-based solutions (NbS), of which the countries were located mainly in tropical and subtropical areas. NDCs embody efforts by each country to reduce national emissions and adapt to the impacts of climate change based on NbS. NbS are initiatives that protect, restore, and manage land and ocean ecosystems such as

forests, peatlands, wetlands, savannahs, coral reefs, and mangroves while simultaneously addressing other societal challenges. To achieve NDCs, an evaluation of NbS or natural capital assets is necessary; however, a comprehensive system for eco-evaluation in the tropics has not yet been established. This book challenges the review and proposal of eco-evaluation methods in the tropics ecosystem, focusing mainly on peatland ecosystems and others, including social capital such as credit, bonds, and national accounting.

Sapporo, Hokkaido, Japan Mitsuru Osaki

Contents

Part III Eco-Evaluation Technology in Tropical Peatland

Part I
Eco-Evaluation of Tri-Harmony/Lemma in Tropical Peatland

Chapter 1
Innovative Eco-evaluation System for Tropical Peatlands

Mitsuru Osaki, Nobuyuki Tsuji, and Tsuyoshi Kato

Abstract Tropical peatlands are complex and fragile ecosystems because three key elements, water (low oxygen), carbon, and nutrients, are interrelated, interconnected, and interact. Therefore, once the delicate balance of a tropical peatland ecosystem is disrupted, the ecosystem catastrophically collapses and enters a state that cannot be restored. On the other hand, forest biomass productivity in pristine tropical peatlands is much higher than that in tropical mineral soils, even under a high groundwater level (GWL) (due to low oxygen) and low nutrients, which is similar to the paradoxical scenario, called here the "tropical peatland paradox." Without solving this "tropical peatland paradox," tropical peatland has been developed by applying mainly the "Drainage-based water management" (drainage-based WM), which causes tremendous damage to peatland ecosystems and generates extensive disasters in the global environment.

One of the reasons why the "tropical peatland paradox" cannot be solved is related mainly to the bias and preconception of the temperate/boreal peatland model (a concept based on science and management based on technology [drainage or canal construction without evaluation]). Unfortunately, tropical peatland ecosystems have high GWLs, which ensure high biomass productivity. From observations of pristine peatland forests, we found simple strategies for plants (mainly trees) under high GWL conditions, e.g., the development of aerial roots and mound roots on the peatland surface at or above the water table. These roots support oxygen and nutrient (N_2 fixation and litter decomposition) absorption.

The key elements of plant growth (carbon stock and carbon flow) in tropical peatlands are water, oxygen, and nutrients. We define triharmony management and trilemma management as the respective appropriate and poor management of water–

M. Osaki (✉)
Hokkaido University, Sapporo, Japan
e-mail: mosaki@agr.hokudai.ac.jp

N. Tsuji
Japan Peatland Society (JPS), Fukuoka, Japan

T. Kato
Forest Carbon Sink Business Department, Sumitomo Forestry Co., Ltd., Tokyo, Japan

M. Osaki et al. (eds.), *Tropical Peatland Eco-evaluation*,
https://doi.org/10.1007/978-981-99-6790-2_1

oxygen–nutrients. "Stock-based water management" (stock-based WM) and "drainage-based water management" (drainage-based WM) correspond to triharmony management and trilemma management, respectively.

It is concluded that the "tropical peatland paradox" answer is "aerial root and mound root" formation, even under high GWLs in peatland. Based on this "tropical peatland paradox" answer, we create and propose an innovative eco-management system in tropical peatlands called the "Stock-based water management" (stock-based WM) and land surface management (LSM), which for the first time makes it possible to rehabilitate, restore, and regenerate disturbed tropical peatland ecosystems. However, stock-based WM and LSM technologies must be applied under the appropriate operation of the integrated measuring, reporting, and verifying (iMRV) system based on iEOS (informatics based on an equator/Earth observation system).

However, as tropical peatlands are very sensitive, fragile, and complex ecosystems, we must develop a new science discipline to understand the overall mechanisms of peatland ecosystems composed of interacting, interrelated, and interconnected elements and systems, which is called "transdisciplinary science" based on the "triharmony/trilemma concept." Finally, we reach the universal concept of "envelopment mechanisms" based on material (re)cycling, symbiosis, and coexistence, not on "development mechanisms," which can be applied to analyze ecosystems, manage ecosystems, and evaluate ecosystems such as peatlands/wetlands as well as arid/semiarid/saline land and arable land.

Keywords Tropical peatland paradox · Triharmony/trilemma system · Envelopment/development mechanism · Land surface management (LSM) · Integrated measuring, reporting and verifying (iMRV) · Informatics based on an Equator Observation System (iEOS) · Drainage-based/Stock-based Water Management (WM)

1.1 Introduction

Modern science and technology related to tropical peatlands have unfortunately failed (1) to understand these unique and mysterious *ecosystems* (extremely high plant productivity in pristine peatlands, even under a poor supply of nutrients and oxygen), (2) to establish an *eco-management system* (on the other hand, tropical peatlands have become completely degraded, disrupted, and destroyed by inappropriate management, such as water drainage), and (3) to establish an *eco-evaluation system* (especially an integrated measuring, reporting, and verifying (iMRV) system, which is still immature).

To address these gaps in knowledge (science) and management (technology) for tropical peatlands, this chapter provides a guide to revive and develop an innovative (1) concept of peatlands (optimization of water–oxygen–nutrient balance), (2) eco-management method (introduction of land surface management (LSM) technology), and (3) eco-evaluation system (establishment of informatics of iMRV system).

In tropical peatlands, the three key elements, water (oxygen), carbon, and nutrients, are interrelated, interconnected, and interact. When these elements are in balance, a triharmony state occurs. Conversely, a trilemma state occurs when these elements are out of balance. Here, we focus on water, carbon, and nutrient assets in the context of triharmony/trilemma.

As tropical peatlands are composed of complex ecosystems, they are easily destroyed by inappropriate development. In this chapter, we aim to introduce an eco-evaluation system that will contribute to stopping peatland destruction and accelerating the regeneration and restoration of disturbed peatlands.

1.2 Tropical Peatland Paradoxes

Tropical peatlands are unique ecosystems; however, conventional existing scientific concepts, information, knowledge, and technology have largely been concentrated on temperate/boreal peatlands. Thus, unfortunately, once temperate/boreal peatland knowledge/management is applied to tropical peatlands, it misleads tropical peatland science and management, causing catastrophic disasters. We find that tropical peatlands are mysterious "paradoxical ecosystems" compared with temperate/boreal peatlands. Here, "paradox" means bias, prejudice, preconception, and preoccupation.

1.2.1 Tree Growth Paradox

Peat forms in areas with (1) a high precipitation rate (water flow) and (2) high groundwater levels (GWLs) (water stock), which result in naturally low oxygen (physically low O_2 solubilization in water) and nutrient concentrations (extremely low application from rain and rivers [because of the dome structure]). Paradoxically, however, biomass productivity (e.g., tree growth) is much higher in tropical peat soil and sandy podzolic soil with sallow peat than in tropical mineral soils. For example, wood production rates are approximately 5 t/ha/year in mineral soils, while they are approximately 10–15 t/ha/year in peat soils and approximately 12–20 t/ha/year in sandy podzolic soil with sallow peat (Fig. 1.1) (Osaki et al. 2021a). Thus, the two peatland forest types have 1.5–2 times higher biomass production rates than those of the mixed dipterocarp forest. The turnover rate of the number of individuals was reported to be higher in the two peatland forest types than in the mixed dipterocarp forest with mineral soil. In the two peatland forest types, while oligotrophic conditions result in high levels of canopy tree mortality, there is sufficient light and nutrient supply to support replacement individuals, resulting in high biomass production rates (Miyamoto et al. 2016; Osaki et al. 2021a).

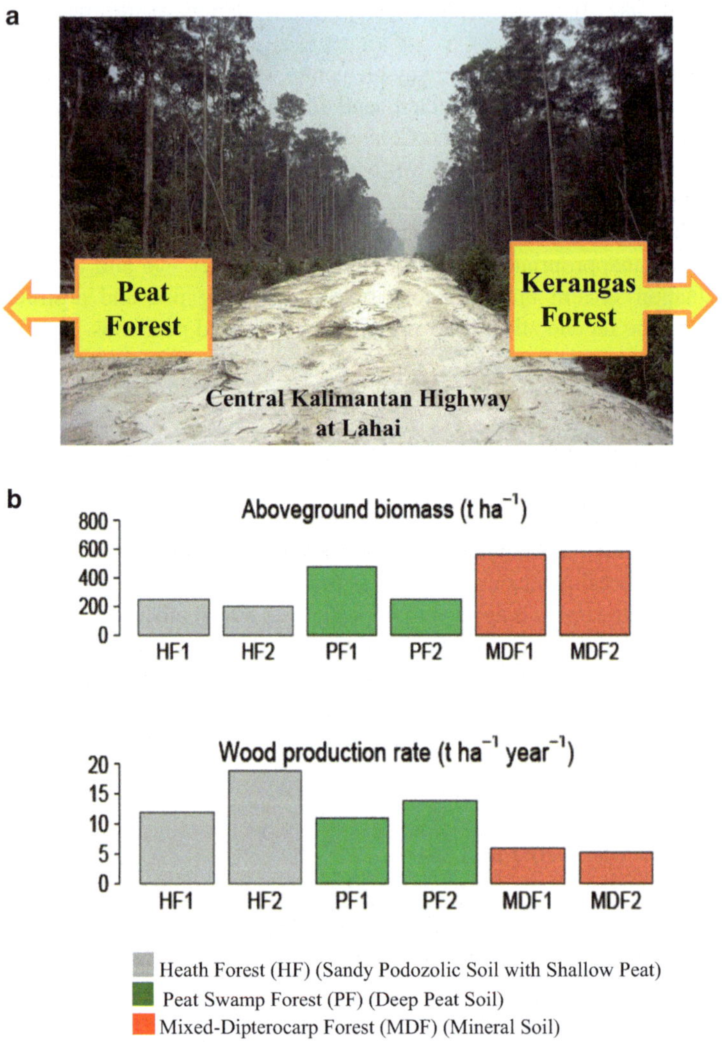

Fig. 1.1 (**a**) Majestic forest in pristine peatlands. (**a**) Left side: pristine peat forest (deep peat), right side: pristine kerangas forest (shallow peat and sandy podzolic subsoil), (**b**) aboveground biomass and wood production rate of the heath forest (HF), peat swamp forest (PF), and mixed dipterocarp forest (MDF). Modified from Osaki et al. (2021a). Heath Forest (HF) (Sandy Podozolic Soil with Shallow Peat). Peat Swamp Forest (PF) (Deep Peat Soil). Mixed-Dipterocarp Forest (MDF) (Mineral Soil)

Thus, tropical peatlands are paradoxical ecosystems in terms of plant growth but nevertheless have high GWLs. Pristine tropical peatlands have high GWLs that result in a low oxygen supply to roots (therefore, roots cannot absorb nutrients even if high amounts of nutrients are available), and extremely poor nutrient

Fig. 1.2 Aerial roots and mound roots for oxygen and nutrient absorption in a native peatland ecosystem. Mound roots exist in mounds and have several morphological types, such as lateral roots, aerial roots, and mangrove-like roots. Modified from Osaki et al. (2021a)

concentrations in peat soil and peat water cause serious nutrient deficiencies. Therefore, the tropical peatland paradox is defined as "biomass productivity is high even under extremely poor nutrient conditions." In these terms of the tropical peatlands paradox, the "**tree growth paradox** "has already been addressed in previous books (Springer): (1) "Tropical Peatland Ecosystem" eds. by "Tropical Peatland Ecosystems" (2016) eds. by M. Osaki and N. Tsuji and (2) "Tropical Peatland Eco-management" (2021) eds. by Osaki M, Tsuji N, Foead N, and Rieley J. A summary of this information is as follows.

Oxygen supply and absorption: As oxygen solubility in water is extremely low, plant roots in water cannot absorb nutrients because the nutrient absorption process depends on the energy supplied by O_2-dependent respiration. Therefore, aeration is an important aspect of hydro-culture systems, even in rice crops adapted to aquatic conditions. Because nutrient absorption depends on active absorption mechanisms, oxygen is essential to the production of energy in the roots (Osaki et al. 2023b).

Two strategies for oxygen absorption have been observed in pristine peatland forests (Fig. 1.2):

1. Aerial root formation: oxygen absorption from the air
2. Mound root formation: oxygen absorption from the air and nutrient absorption from the litter piled on the mound

Nutrient supply and absorption: Nutrient stocks and supplies are extremely low in peat soil and peat water for the following reasons:

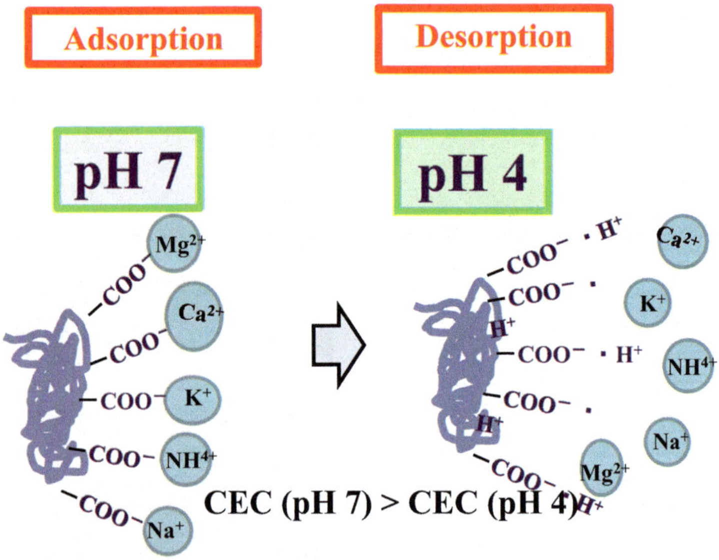

Fig. 1.3 Cation adsorption and desorption mechanisms depending on pH in soil organic matter (SOM). Cation leaching mechanisms in peat (pH 3.0–3.5) because of the cation desorption trait of peat at low pH (less than 4.0). Modified from Osaki et al. (2021a)

1. Low cation absorption capacity of peat materials.

   As peat material is composed of carbon fiber that is made mainly of lignin, peat material contains very few nutrients and desorbs nutrient ions, especially under low pH conditions. Peat interacts with water molecules by physical (van der Waals) force, which, together with bulk density, determines the soil water holding capacity. Thus, the main functions of peat are to support plants and provide water retention in peatlands.

   The cation absorption capacity of organic matter (peat soil) is extremely low at a low pH (less than 4.0) (Fig. 1.3). Thus, even if fertilizer is applied to peat soil, all nutrients, especially K^+, leach immediately through desorption mechanisms. As K^+ is always in ionic form (does not form compounds), it leaches easily. As other nutrients are absorbed to some extent by microorganisms and form chemical compounds, the chemical compounds are retained and decompose slowly in peat (Osaki et al. 2021a, 2023b).

2. Poor nutrient supply.

   As peatland water is supplied mainly by rain, the nutrient supply is very low, and nutrients leach gradually due to acidic rain (pH 5.6).

   Rivers supply clay and water, and the sea supplies nutrients and seawater through tidal effects; however, these effects are small for the water and nutrient supply because of the peat dome topography in peatland areas. A peat dome is an almost flat structure with a height of 10 m at the top of the dome and a distance of 10 km from ridge to top (10 m/10 km inclination); however, it is important to prevent water flow from the bottom to the top of a peat dome. Thus, as the water supply in dome-type peatlands depends only on rainfall, the nutrient supply is extremely limited over the long term (Osaki et al. 2021a, 2023b).

3. Adaptation to nutrient limitation.

   As peatlands are extremely nutrient poor, nutrient cycling is the key to nutrient (mainly P, Ca, Mg, and micronutrients) utilization by plants through the following mechanisms:

   (a) Recycling (litter supply and decomposition)
   (b) Retranslocation (old organs to new organs)
   (c) Reuse (metabolite reuse in cells)

      As recycling (litter supply and decomposition) depends on microorganism activity in the surface layer of a peatland, symbiotic systems on the land surface are important and essential (Osaki et al. 2021a, 2023b).

4. Nutrient cycling disturbance.

   Nutrient cycling is severely disturbed by (1) intense drainage that results in effects such as a low water table and low soil moisture and (2) frequent burning that results in effects such as serious microflora loss, nutrient leaching, low water-holding capacity (water loss), land surface drying due to direct solar radiation and fern/grass vegetation, and flooding in the rainy season (due to peat loss) (Osaki et al. 2021a).

   Thus, peatlands have a high GWL (limited oxygen) and contain poor nutrients in peat soil and water; however, their biomass productivity is very high compared with that of tropical mineral soil forests, which is called **the biomass productivity paradox**.

1.2.2 Tree Function Paradox

According to Osaki et al. (2022a), forests (trees) have photosynthetic ability: $CO_2 + H_2O \rightarrow CH_2O^* + O_2$ by sun radiation energy (hv), where CH_2O^* is the abbreviation for carbohydrate ($C_6H_{12}O_6$). This carbon assimilation appears to be a very simple process. However, in the process of photosynthesis, water emitted from leaves (evapotranspiration) contributes to the water cycle and cools air by latent heat removal. Thus, the photosynthetic process is linked to carbon–water metabolism. The most basic functions of forests include energy fixation (from the sun), CO_2 fixation (from the air), the water cycle, cooling, and green spectrum reflection (reduction of solar energy). We can therefore say that forests are a bioapparatus that store energy (a "biotic-battery"), remove CO_2 from air (biotic-CO_2 capture and storage, CCS), contribute to the water cycle (biotic water-pump), cool air temperature (biotic air conditioner), and reduce solar energy (biotic spectrum reflector). Thus, forests have mulching functions, simply via their leaves through photosynthesis. Trees and forests, due to their leaf function, perform many critical services:

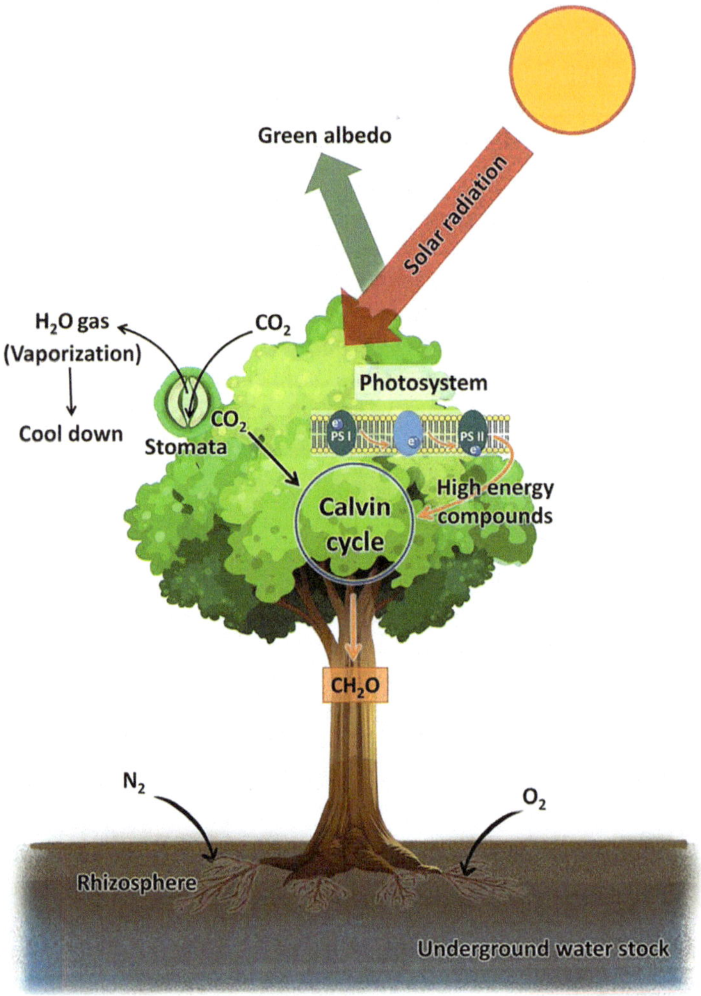

Fig. 1.4 Big Tree Model on CO_2 assimilation, green albedo, and evapotranspiration (latent heat removal) by solar energy, obtained from Kato et al. (2023)

first, CO_2 removal from the air and the conversion of solar energy stock into carbohydrates; second, green spectrum refraction (i.e., excess energy scattering into space, called green albedo); and third, air temperature cooling via latent heat removal through evapotranspiration (via porous structures called stomata on the leaf surface) (Fig. 1.4).

We encounter a new paradox called the "forest function paradox," which is composed of two "paradoxes": (1) the "photosynthetic process paradox" (physiological paradox) and (2) the "natural capital paradox" (economical paradox). First, forests have the highest potential for solar energy utilization and environment *amelioration*, followed by carbohydrate production (solar energy stock), green albedo, and latent heat removal, which is discussed in the above sentence. Second, forests also have the highest potential for natural capital, which is concluded from a basic study on natural capital analysis in Shimokawa, northern Japan, where towns have been orienting and conducting cyclic forest management.

In addition to key functions, i.e., carbon assimilation, green albedo, and latent heat removal, trees or forests have various functions, such as water stocks and flows, nutrient stocks and flows, and protection against soil and land erosion. Plants absorb water and nutrients through their roots, processes that are regulated by leaf function. For example, the water-pump function of the stomata driven by solar energy and carbohydrate production drives root functions and interactions with microorganisms in the rhizosphere (10 μm around the roots), indicating that forests have larger-scale transdisciplinary functions. These include interconnections among ecosystems, modulation of the water–carbon cycle, soil fertility, land protection, and biodiversity conservation within both the biosphere and geosphere. In summary, plants are the basis of solar energy, CO_2, water, and nutrient cycles and form the main currency of natural capital (Osaki et al. 2021a, 2022a; Osaki and Kasuga 2022).

According to actual measurements of the heat balance in peatlands, the heat balance sheet is totally different between pristine forests and nonforests (grass and shrub). Latent heat and sensible heat are dominant in forests and nonforests, respectively (Fig. 1.5a). Thus, forests cool the air by removing latent heat (evapotranspiration) and reflecting green light (Fig. 1.5b).

An essential role of forests is the assimilation of carbon (CO_2 removal from air); however, the cooling-down function of forests unfortunately attracts less attention, even in the discussion of global warming. Industrial technologies cannot perform all of these roles. Solar cells, for example, produce electric energy but cannot cool air (in fact, they may make it hotter) or miserably affect the water cycle (they may have a drying effect) and do not protect against land erosion.

What is the point of the "forest function paradox", which is relevant to peatland forests as well as all other types of land forests? We know the photosynthetic function related to carbon assimilation, green albedo in the process of photosystems I and II, and latent heat removal in the process of evapotranspiration through stomata. However, modern plant and forest science have failed to construct a bioapparatus (forest) model based on forests for mitigation of the earth system, called " Nature-based Solutions (NbS)," as opposed to an industrial apparatus model, called "industrial Technology based Solutions (iTbS)"; solar panels, CCS, air conditioner (heat island), water dams, or water civil engineering, so on. NbSs are cheaper and more effective than ItbSs. Generally, the forest natural capital (NbS) cost is 100–1000 (or more) times cheaper than the industrial technology capital (iTbS) cost (Hojo et al. 2021; Osaki and Kasuga 2022). The industrial apparatus model (iTbS) could invite secondary or chain-reactive environmental disasters, materials mining and pollution, long transportation (energy consumption), manufacturing processes (energy and pollution), setting (destroying forest or land), and waste disposal treatment (energy and pollution), which produce a typical trilemma structure among the ecology, the environment and the economy, indicating

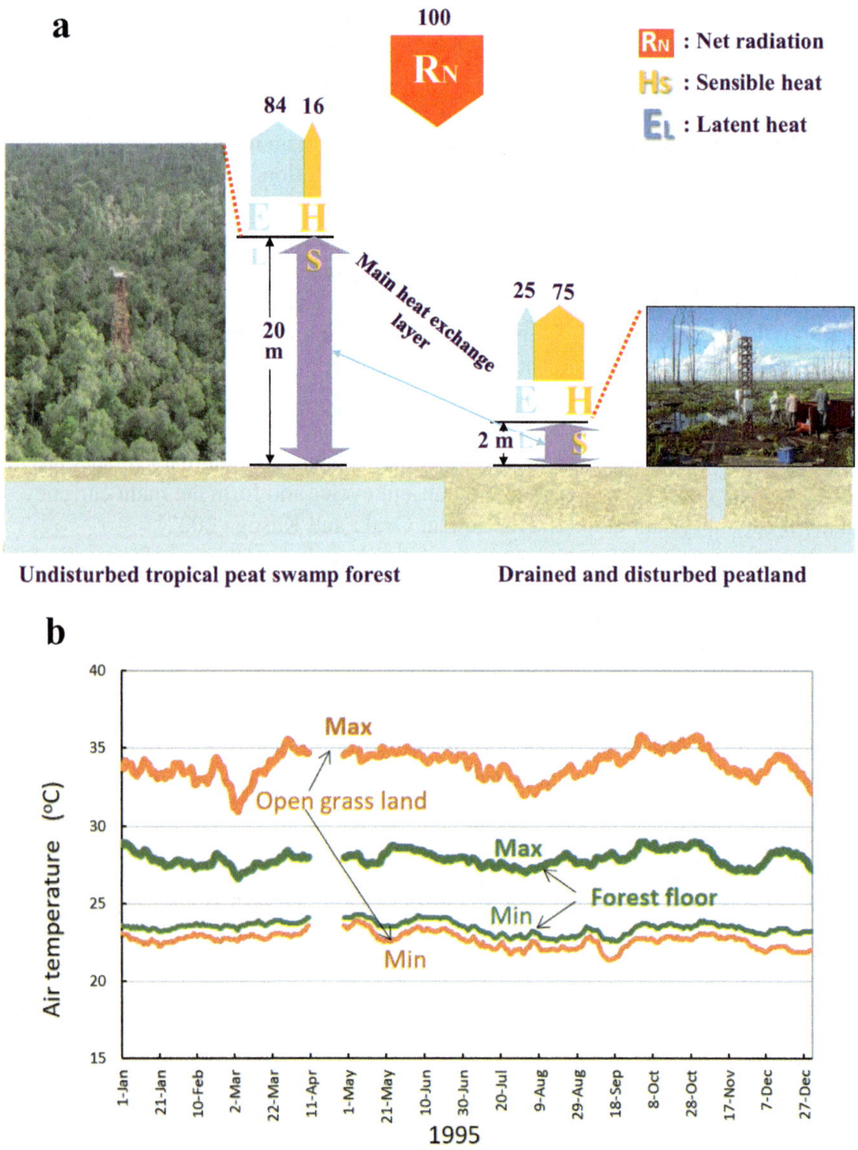

Fig. 1.5 Effect of human impacts on the heat balance of tropical peatlands. a: heat balance in the forest and grassland at flowering, b: maximum and minimum air temperature in the forest and open grassland. Obtained from Osaki et al. (2022a)

that modern science disciplines such as "inventory-based disciplines" (including single, dual, and mulch disciplines) cannot access the design of a balanced ecological, environmental, and economical ecosystem (such as the triharmony system).

1.2.3 Greenhouse Gas Emission Paradox

Greenhouse gas (GHG) emissions (mainly CO_2 and CH_4) are believed to be extremely high in tropical peatlands because (1) surface peat is exposed to oxygen and (2) peatlands have a low redox potential (Eh). However, GHG emissions are not as high from pristine tropical peatlands around the Maritime Continent (MC) for the following reasons:

Redox potential (Eh) in pristine peatland: The Eh of a peat surface clearly responds to the inundation of the peat surface, and the Eh shows large fluctuations according to the water level (Fig. 1.6). The Eh of a high mound (>ca. 40 cm in relative elevation, never inundated) was reported to be constantly oxic (Eh = ca. 600 mV), and the Eh at a low-elevation site on the peat surface (constantly inundated) was reported to be anoxic (approximately 0 mV) (Haraguchi and Yabe 2002).

In a peat dome of a pristine peatland forest, Eh was almost constant (ca. 200–400 mV) from the surface to the bottom of the peat (max. 8 m in depth), with a minimum (ca. 50–200 mV) at an approximately 2 m depth (Haraguchi et al. 2000). However, it is interesting that the Eh increased from a 2 m depth to the peat bottom because of the oxygen supply from the groundwater flow at the mineral horizon under the peat in combination with the low amount of oxygen consumption within the peat layer. Thus, the vertical distribution of the Eh in the pristine peat layer was >0 mV, implying the *low methane emission potential* of pristine peatland forests (Osaki et al. 2021a, 2023a).

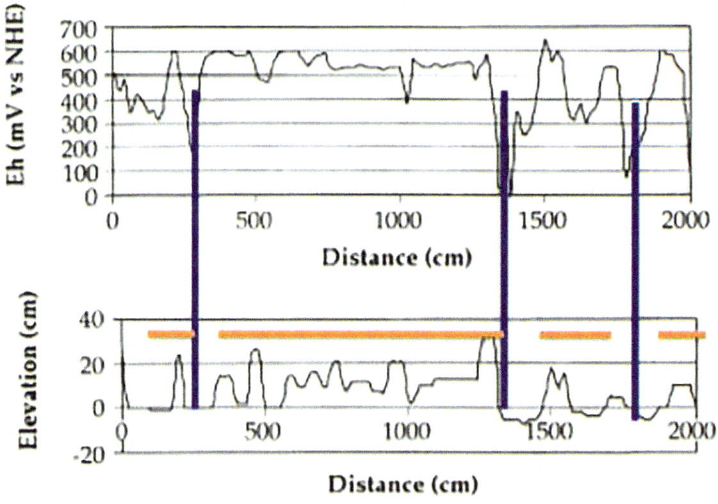

Fig. 1.6 Redox potential (Eh) in mound and nonmound areas of a tropical pristine peatland. A case study in Lahei district, Central Kalimantan, Indonesia. Brown line: mound area, blue line: nonmound area. Obtained from Osaki et al. (2021a)

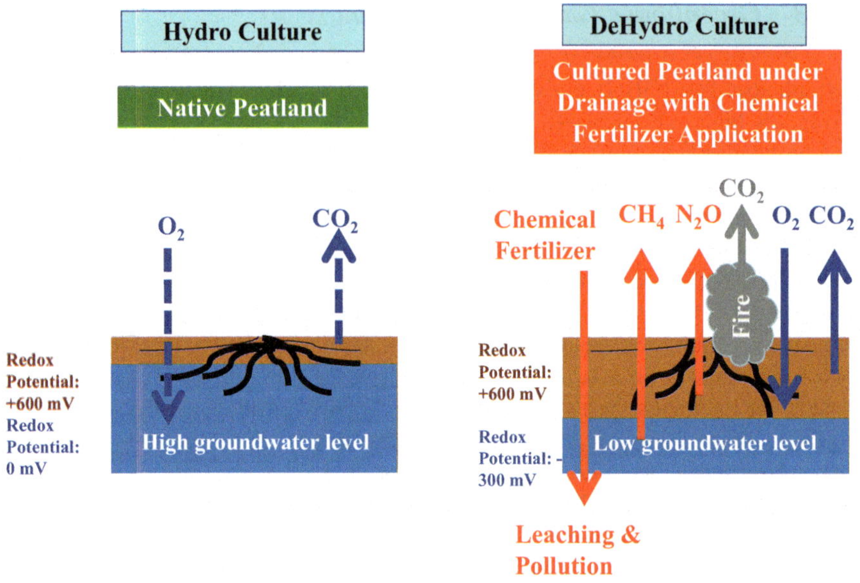

Fig. 1.7 Model of greenhouse gas (GHG) emissions from a pristine peatland and a cultivated peatland with drainage and heavy chemical fertilizer application. Modified frosm Osaki et al. (2023a)

The worst and highest GHG emissions: The GHG emissions from pristine tropical peatlands with high GWLs can be explained by their Eh (approximately 0 mV) and nutrients (Fig. 1.7, left) (Osaki et al. 2021a, 2023a) as follow:

1. CO_2 emissions are low because of the low Eh (approximately 0 mV) and the insufficient oxygen and nutrients to support microorganism activity.
2. CH_4 emissions are also low because the Eh (approximately 0 mV) is too high for methanogenesis.
3. N_2O emissions are low because of the low Eh (approximately 0 mV) and the low nitrate concentration.

Thus, a pristine tropical peatland with a high GWL will have very low GHG emissions; this scenario is quite different from that of boreal peat ecosystems, where CH_4 emissions are especially high (Couwenberg 2009; Lai 2009). Pristine tropical peatlands are *excellent ecosystems, acting as carbon and water reservoirs*, with small amounts of CO_2, CH_4, and N_2O emissions.

However, if tropical peatlands are developed, water drainage is installed, and chemical fertilizers are applied, then a large amount of GHGs could be emitted from these peatlands, such as oil palm plantations under normal management, through (1) heavy water drainage (supplying O_2 to the peat bottom until the lower water table) and (2) heavy nutrient application (increasing microorganism activity, inducing peat material decomposition, and methanogenesis acceleration) (Fig. 1.7, right).

1. CO_2 emissions increase because of the high Eh (higher than 300 mV) at which sufficient oxygen is present for microorganism oxidation activity.
2. CH_4 emissions increase because the Eh (approximately −300 mV) is low enough for methanogenesis.
3. N_2O emissions increase because of the high Eh (more than 0 mV) near the land surface.

Thus, developed tropical peatlands with drainage (low GWLs) and fertilizer application emit extremely high amounts of GHGs; this system represents *the worst cultivation system* among global ecosystem management systems (Osaki et al. 2021a, 2023a).

1.2.4 Rice Growth Paradox

Based on Association of Japanese Agricultural Scientific Societies (AJASS 1975), rice cultivation in Indonesia is classified according to the season and water utilization as follows. Padi Sawah Rendengan (rainy season crops are grown in wetland rice fields), Padi Gogo Rancah (in the rainfed Sawah when water is not available at the beginning of the rainy season, rice is sown directly on dry soil (dry sowing) and later submerged in rain water), Padi Gadu (dry season crops are grown in the irrigated Sawah), Padi Gogo (rice is grown in dry fields without dikes), Padi Ladang (shifting upland rice cultivation occurs in hillside areas, particularly in Sumatera and Kalimantan), and Padi Pasang Surut dan Rawa (rice is grown in tidal swamps and swampy areas, particularly in Sumatera and Kalimantan). Indonesia has a long history and experience in rice cultivation; thus, traditionally, Indonesian people use almost all land types (including seasons) for rice cultivation. However, they never use peatlands (peat swamps) for rice culture. We observed Padi Ladang cultivation around large river dikes (mineral or organic soils) in peatland areas on a map of Central Kalimantan, which showed that rice cultivation occurs only near mineral or organic soil. However, sometimes mineral or organic soil areas are mapped as peatlands, which causes serious confusion. Therefore, we recommend making a more detailed map identifying the area between pure peat soil and mineral soil.

The Lamunti site is a new transmigration area that is included in the One Million Hectare Peatland Development Project and is located on the western side of the Kapuas River, where canals were constructed for drainage, not for irrigation (degrading peatlands yearly) (Osaki et al. 1999). At this site, farmers have not been able to obtain the expected rice yields. For example, in the first cropping year, a farmer only obtained a 10% harvest (a 90% loss from rats, grasshoppers, moles, birds, and wild pigs) on 1.0 ha of land. In the second year, he obtained only 50 kg/ha (one hundred times lower than the normal yield). Soil fertility problems have also contributed to very low productivity because the pH of the A and B soil horizons is 4.7 and 4.4, respectively, causing Al toxicity. This low pH is typical of acid sulfate soil (subsoil beneath peat) after peat decomposes.

In another case study, we surveyed a Balinese farming site in Anjir Basarang, where good rice culture skills based on Bali rice culture were used (Osaki et al. 1999). A farmer we visited, Mr. I Nyoman Tegig, came to this area in 1960 when he was 17 years old. Based on the information from this farmer, the land in the area was formerly covered by a 40- to 100-cm-thick peat layer. From 1965 to 1968, the Balinese thought that peat was not soil; thus, the peat layer in this area was heavily burnt prior to the cultivation of rice. Afterward, from 1968 to 1976, they continued to burn the peat, as did the local people in the surrounding area. In 1976, the peat layer disappeared. Since 1990, they have not been able to cultivate rice; thus, they started to grow vegetables and pineapple because of the highly acidic soil. The rice yield, using local varieties, from 1965 to 1976 was 1.8 ton/ha. Then, the yield continuously decreased and reached a total of 0.9 ton/ha in 1990. Since the area was first cultivated, farmers have never used fertilizer. They only apply ash derived from the burning of shrubs and weeds because local people cultivate rice using these slash and burning methods. Based on our measurements, the top 40 cm of the soil planted with pineapple is an A horizon underlying a clay sublayer. The GWL at this site is 60 cm below the soil surface, and the pyrite layer is 95 cm below the soil surface. Fresh soil pH based on water extraction (1:5) is 3.8 in the A horizon (0–40 cm), 3.2 in the B horizon (40–95 cm), and 3.2 in the pyrite layer (deeper than 95 cm).

We never observed good rice growth and yields in tropical peatlands or even in temperate or boreal peatlands. Thus, the **rice culture paradox** is based on the following: (1) rice can grow in wetlands; thus, there is an expectation that rice will grow in peatlands, based on misunderstanding, wrong information, and prejudice; (2) peatland mapping indicates that rice can grow in organic or mineral soils; however, sometimes these areas are mapped as peatland areas. As rice is a crop with high nutrient requirements, it is impossible to cultivate rice in tropical peatlands and even in temperate peatlands.

1.2.5 Subsoil Paradox

Moreover, once peat becomes degraded, even poorer quality soils appear, including acid sulfate soil (extremely low pH of less than 3), sandy podzolic soil (extremely poor nutrients), and saline soil in mangroves close to peat soils (Fig. 1.8) (Osaki et al. 2021a).

Anderson (1964) emphasized that mangrove-based peatlands developed at an early development stage. However, initial peat formation involves not only mangrove vegetation but also freshwater vegetation upstream (Furukawa and Supiandi 1986), indicating that topography is also important in mangrove formation. The subsoil beneath peat soil is different along the vertical profile (from mountains to the sea) of the dome structure. A common pattern is (1) sandy podzolic subsoils upstream, (2) alluvial subsoils, and (3) (potential) acidic sulfate subsoils near the

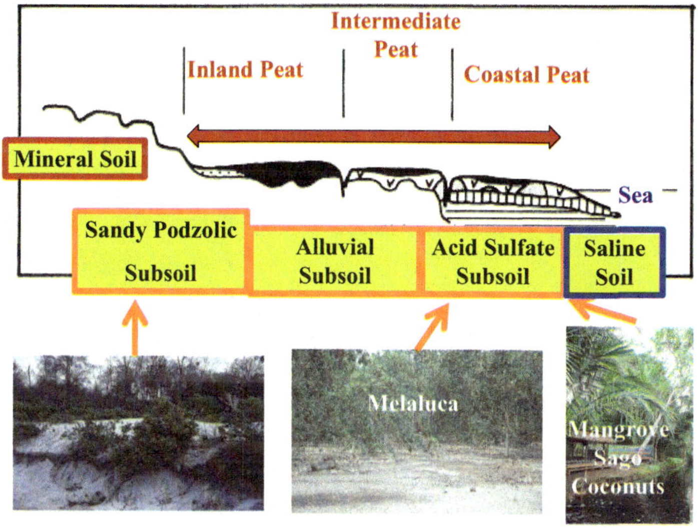

Fig. 1.8 Vertical profile (from the mountains to the sea) of a peatland. Obtained from Osaki et al. (2021a)

sea (Fig. 1.8). The peat developed from these subsoils is typically (1) inland peat, (2) intermediate peat, and (3) coastal peat. After peat degradation, sandy podzolic soil and acid sulfate soil appear. These soils are considered problematic soils because they are extremely nutrient-poor and very acidic (pH less than 3.0) due to the presence of sulfate (Osaki et al. 2021a).

The lowest quality soils in the world, acidic sulfate soil, sandy podzolic soil, and saline soil (close to mangrove areas), exist below (or around) peat soil. However, even when peat covers these low-quality soils, plant growth is maintained, which indicates that peat soil insulates subsoil problems. Thus, the peat soil layer recovers and compensates for subsoil deficits. We call this scenario the **subsoil paradox** because most developers of peatland try to remove and destroy the peat layer by drainage, tractors, or fire, based on the belief that peat is problematic.

1.2.6 Land Surface Paradox

In Sect. 1.2.5, we discussed the "subsoil paradox." As the subsoil in pristine peatland is covered by a peat layer, it is normal that subsoil problems are generally masked in pristine peatland. However, the wrong scientific concept of tropical peatland is accelerated to develop with the wrong civil-engineering technologies, mainly water drainage and land surface disturbance or destruction. In pristine peatlands, the most important dominant factors are primary high water levels and secondary land surface conditions. Thus, the incorrect scientific concept of tropical peatlands

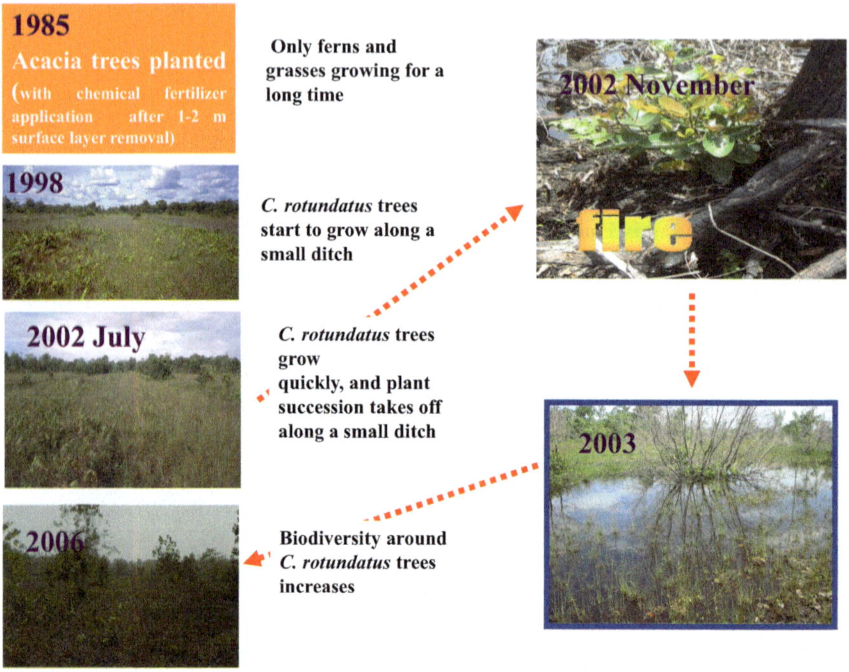

Fig. 1.9 Vegetation succession after peat mining (1–2 m surface layer was removed in the mined plot) in *Kalampangan*, Central Kalimantan

causes two types of biases (called the paradox) in peatland development: (1) peat water drainage and (2) peatland surface disturbance. We observed many paradoxes based on peatland surface disturbance as follows.

Peatland surface disturbance: A peat mining project was performed in Kalampangan, central Kalimantan, which was planned to use peat materials as biofuel. Then, the 1–2 m top peat layer was removed. However, this project failed because the peat material quality as a biofuel was very low. After the failure of the peat mining project, the disturbed peatland was rehabilitated by planting *Acacia* with chemical fertilizer. However, the *Acacia* plants could not be cultivated; they survived for a long period and then died. After this project was decarded, we started to observe the natural regeneration process of the disturbed ecosystem (Fig. 1.9).

After the *Acacia* was planted in 1985, the disturbed land was quickly covered by *Xyris* sp. and fern sp. In approximately 1998, we found that *C. rotundatus* trees (pioneer trees in this area, especially after fire) started to grow along the small ditch side, not in the grass-covered area. In 2002, the *C. rotundatus* trees reached approximately 4–5 m in height, and we observed aerial roots around the *C. rotundatus* stand and seedling germination around the aerial roots. As *C. rotundatus* seeds are spread by wind, we found many seeds in the *Xyris* sp. grassland and area of the fern sp.; however, these seeds never germinated. It is interesting that the *C. rotundatus* seeds germinated around the aerial root area of the

Fig. 1.10 Aerial root, mucigel (rhizosphere), and N_2 fixation by *Combretcarpus rotundatus* (local name: Tumih or Perepat) in the mined plot area

mother tree, not in the *Xyris* sp. grassland and area of the fern sp., indicating that root exudates or microbial symbiosis is important for the germination of *C. rotundatus* seeds (Fig. 1.10). In September 2002, peat fires occurred; however, it was observed that *C. rotundatus* started to regenerate quickly. In 2006, it was surprising that biodiversity around *C. rotundatus* began to occur with various plant species (Fig. 1.9). It is hypothesized that once the root–microbe symbiosis system at the land surface improves, plant (mainly tree) growth increases (coexistence with microbes) and biodiversity increases (cohabitation with various plant species).

In another case study, regeneration ability was dependent on land surface conditions. If fire damage was not so strong (fire damage occurred only once, Fig. 1.11, left), native vegetation (*C. rotundatus*) recovery was quick; however, if fire damage was strong (fire damage occurred twice, Fig. 1.11, left), native vegetation (*C. rotundatus*) recovery was not observed, but grass and fern growth occurred. Thus, we concluded that LSM is of secondary importance after the primary importance of GWL management.

Fig. 1.11 Effect of fire on vegetation recovery in *Kalampangan*, Central Kalimantan, observed in 2006

1.2.6.1 Sago Palm Growth

Sago palm (*Metroxylon sagu* Rottb.) is the most promising starch-producing palm species in peatland, accumulating approximately 200 kg starch or more per plant or in the trunk, which corresponds to 10 times higher than rice starch production in our estimation (8–10 years for starch production, then average starch production per year) (Tampubolon et al. 2021; Ehara et al. 2021). Sago palm can grow in peatlands, brackish water areas, acid sulfate areas (but normally in wetlands, not dryland areas), and wet mineral land areas, indicating that sago palm has a tolerance to submergence, salinity, and acidity. Another unique point of sago palm is its high N_2-fixing ability, which was estimated as 90% from N_2 fixing at the latter stage (13 years old) by the $\delta^{15}N$ analytical method (Lina et al. 2008). Thus, sago palm can grow and assimilate high amounts of starch under various adverse soil conditions. This sago palm adaptation ability is supported by its N_2-fixing ability because nitrogen is a major limiting factor in these adverse soils (poor nitrogen conditions).

Sago palm roots develop "supporting roots" under water level conditions and then "lateral roots" for nutrient absorption (such as mound roots) above the water level (Fig. 1.12). "Aerial roots" develop from ground base parts, and another type of "aerial root" vigorously develops from the trunk. We hypothesize that these aerial roots contribute to N_2 fixation (symbiosis with N_2-fixing bacteria). Aerial roots form under nitrogen-deficient conditions more than under oxygen-deficient conditions (Fig. 1.13 below). Water is not the main factor in the formation of aerial roots, and nutrients (nitrogen) are key to aerial root formation in the case of sago palm. In the field experiment (Fig. 1.13 above), sago palm growth was better in the no fertilizer application plot than in the fertilizer application plot. In the no fertilizer application plot, we observed many aerial roots, even at a low GWL, showing that the oxygen supply was not limiting.

Thus, sago palm is similar to a super plant, producing high amounts of starch, fixing N_2, and tolerating adverse soil conditions. However, we also observed that even with sago palm, it is difficult to grow if the land surface disappears (Fig. 1.14).

Fig. 1.12 Sago root system

At the Sirindhorn Peat Swamp Forest Nature Research and Study Centre, Thailand, we carried out a rehabilitation program for peatland after fire damage. Once peatland was severely damaged by fire, vegetation recovery (including sago palm) was limited (Fig. 1.14, right). However, in areas with mild fire damage, vegetation recovery occurred rapidly (Fig. 1.14, left).

1.2.7 Peatland Ecosystem Paradox

We can point out many paradoxes in tropical peatland described in the above subsections. What is the key paradox in tropical peatland?

To accelerate carbon sequestration, a high GWL must be maintained; however, as (1) oxygen does not solubilize in water and (2) nutrients leach into peat water (because of the low pH in water and nutrient desorption mechanisms of peat organic

Fig. 1.13 Growth and aerial root formation of sago palm with and without chemical fertilizer in Sarawak, Malaysia

Fig. 1.14 Rehabilitation using sago palm and trees after burning at Sirindhorn Peat Swamp Forest Nature Research and Study Centre (Pa Pru to Daeng), Thailand. Left: mild fire damage area, Right: severe fire damage area

matter under low pH), plant growth (biomass productivity) decreases. To improve plant growth, drainage strategies are normally adopted; however, peatland systems are quickly destroyed. This is called the "peatland ecosystem paradox."

How can this "peatland ecosystem paradox" be solved? We find a key plant strategy at high GWLs, for which plants adopt activation on "land (peat) surfaces."

Thus, plants develop a specific ability for compatibility between "high water" and "high productivity."

Consequently, we propose a new evaluation system, called "eco-evaluation", to solve the "ecological paradox", establish "eco-management" technology, and create a new concept of "eco-evaluation", which has been partly discussed and reported in Springer Books ("Tropical Peatland Ecosystems," 2015, and "Tropical Peatland Eco-management," 2021) and research outreach/research future articles (see Acknowledgments). To weave elements (inventory-based disciplinary science and technology) into an innovated evaluation system (transdisciplinary science and technology), we directly cite many parts of articles, written only by our team, from these books and articles of research outreach and research future.

1.3 Concept of Eco-evaluation

Peatland ecosystems are composed of several elements, especially water, oxygen, and nutrients, which change dynamically, affecting ecological and environmental conditions. In Sect. 1.2, tropical peatland paradoxes are pointed out, indicating that tropical peatland studies require both *element analysis* (mainly inventory analysis based on single, dual, and mulch disciplines) and *element integration* (mainly an interrelated and interconnected cause-and-result/mutual interaction based on trans-disciplinary analysis). Osaki's team developed the eco-evaluation concept in nature from (1) disciplinary aspects and (2) the trilemma model/triharmony model, which is reported in Osaki et al. (2022a,b,c,d, 2023a,b) and Kato et al. (2023). We follow these articles and provide more detail in the following discussion.

Scientific disciplines are not all equal. Some are inherently focused on individual issues, while others call on aspects from two or more subdisciplines (known as dual disciplines or mulch disciplines, referred to here as "inventory-based disciplines"). Within the earth sciences, tools such as the Earth Simulator are considered single-disciplinary approaches: they consist of single-element (mainly gases in the atmosphere) models. Dual-disciplinary approaches include the UN's Sustainable Development Goals (SDGs), which list elements of natural/environmental science (e.g., geochemical fluxes, biodiversity, water, air, and soil) and social science (e.g., population, socioeconomic measures, political environment, culture, and technology).

In the case of "multidisciplinary science," planetary boundaries (thresholds within which humanity can survive, develop, and thrive for generations to come) are considered in all nine indicators, i.e., (1) biosphere integrity, (2) climate change, (3) stratospheric ozone depletion, (4) ocean acidification, (5) biogeochemical flows, (6) freshwater use, (7) land-system change, (8) atmospheric aerosol loading, and (9) novel entities.

Finally, "transdisciplinary science" is a fusion approach: feedback systems, self-organization, mutual interactions, cause-and-effect sequences, recycling systems, and multiplicity. Osaki's team has been developing the triharmony/trilemma model, applying transdisciplinary science (Osaki et al. 2022a,b,c,d).

1.3.1 Transdisciplinary Science of Earth's System

Earth's system faces major and unprecedented crises or catastrophes, such as catastrophic events and earth disasters, including global warming, climate change, and planetary–boundary issues, which have been occurring incidentally during the 20th and 21st centuries (Anthropocene period). In recent decades, particularly the Global North has proposed "industrial Technology-based Solution (iTbS) to provide mainly renewable energy to address climate and environmental issues. However, most of these approaches are expensive and ineffective and have negative effects on the environment. To construct solar panels (energy assimilation) and batteries (stock energy), rare metals have become key limiting materials, which are very rare earth resources. These rare earth resources are very expensive and may become another serious pollutant to the Earth system, e.g., solar panels are not environmentally friendly and may be more destructive to the ecosystem because of land surface disturbance.

Osaki's team is developing new approaches based on natural capital (the assets provided by natural systems), among which plants, especially forests, play a key role. Their approach – based on NbSs – is for Earth's systems to be regenerated and become self-sustaining through a process of homeostasis (Osaki et al. 2022a,b,c,d).

1.3.2 Trilemma Model

To introduce transdisciplinary science and technology into ecosystem analysis and management, Osaki's team (2022b) has constructed and proposed the triharmony/ trilemma scheme.

Prior to the industrial revolution, humans were entirely dependent on solar energy and coexisted with nature. With the introduction of fossil fuels, this balance changed, and the human sphere began to interact with natural spheres. In the 19th century, a three-way segregation was observed: the environment, ecology, and resource spheres were divided with the human dimension between them but with minimum interaction (Fig. 1.15a). Over the course of the 20th century, as economic activities began to impact natural spheres more severely, mutual conflicts (or "lemma") arose (Fig. 1.15b). There was a shift to a trilemma state. In a trilemma state, the environment, ecology, and resource spheres overlap, taking the form of a three-way Venn diagram (Fig. 1.15c) (Osaki et al. 2022b), showing that between the ecology and environmental spheres is ecological destruction; between the environment and resource spheres is environmental pollution; and between the ecology and resource

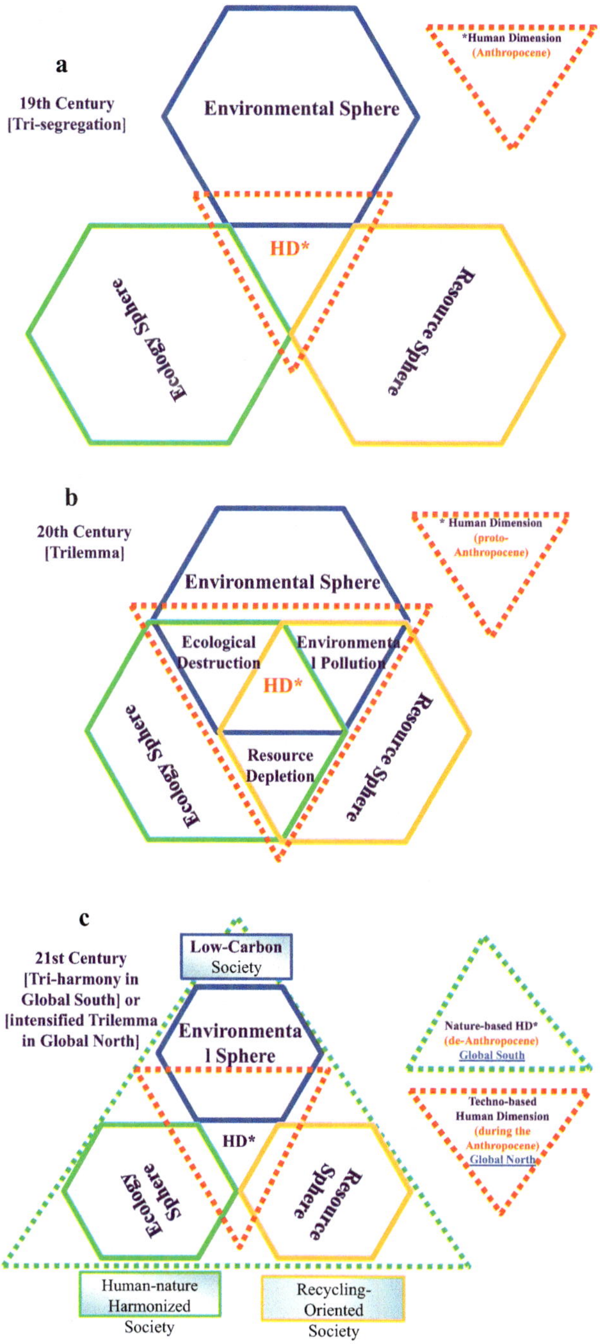

Fig. 1.15 Mutual interactions among the environment-sphere, ecology-sphere and resource-sphere in the nineteenth century (**a**), twentieth century (**b**), and twenty-first century (**c**). The human dimension (HD) is the core of the 3 spheres. The mutual interaction state among the 3 spheres was referred to as tri-segregation in the nineteenth century (**a**), trilemma in the twentieth century (**b**),

spheres is resource depletion. Finally, at the center of all three spheres lies the human dimension.

During the first decades of the 21st century, the trilemma model branched into two zones: the Global North insisted on industrial Technology-based Solutions (ITbSs iTbSs), known as "trilemma," and the global South (however, Global Central is more appropriate terminology for the equatorial belt and tropical/subtropical zones) depended on NbSs, known as "triharmony."

The work in this chapter and book underlines the importance of NbSs for regenerating Earth's systems, especially tropical peatland regeneration.

1.3.3 Triharmony Model

Successful examples of the triharmony model can be found across the Global South (or Global Central) and occasionally in the Global North (Osaki et al. 2022b).

In the triharmony model, Osaki (2014) paid attention to the "satoyama system" concept (or "forest farm," where "sato" denotes society and "yama" denotes nature in Japanese), in which secondary natural areas support the welfare of residents as well as protect biodiversity through human intervention. The satoyama system is largely confined to the Asian Fertile Crescent or Asian monsoon region, in which abundant water, fertile soils, and high biodiversity interact and reinforce each other within an organic farming and agroforestry system developed over a long period (Osaki et al. 2022b) (Fig. 1.16). The most critical feature is the role of the water cycle.

The satoyama system operates on a variety of scales. On a mega scale, Osaki (2014) mentioned that, e.g., the Himalaya and Tibetan Plateau can be seen as the "yama," with glaciers and snow providing a massive reservoir of water. This reservoir ultimately feeds rivers and precipitation across East, South, and Southeast Asia, supporting 40% of the world's population (the megascale "sato") (Fig. 1.16). Thus, the Asian Fertile Crescent is defined as Himalaya and Tibetan Plateau Geography and the Asian monsoon, which is called Mega-satoyama here. The foot of this mega-satoyama covers (1) the river mouth (fertile soil) of a large river originating from the Himalaya and Tibetan Plateau and (2) the maritime continents (the carbon and water stocks and flows, and the high biodiversity). The MC is a large subsystem of satoyama, which is composed of peatland, mangroves, seashores, and coral, surrounding the insular parts of Southeast Asia and the Indonesian archipelago.

Around the Mega-satoyama, the triharmony system has been maintained for a long time. A critical aspect of triharmony depends on solar energy to generate forms

Fig. 1.15 (continued) triharmony in the global South and/or intensified trilemma in the global North in the twenty-first century (**c**). Obtained from Osaki et al. (2022b)

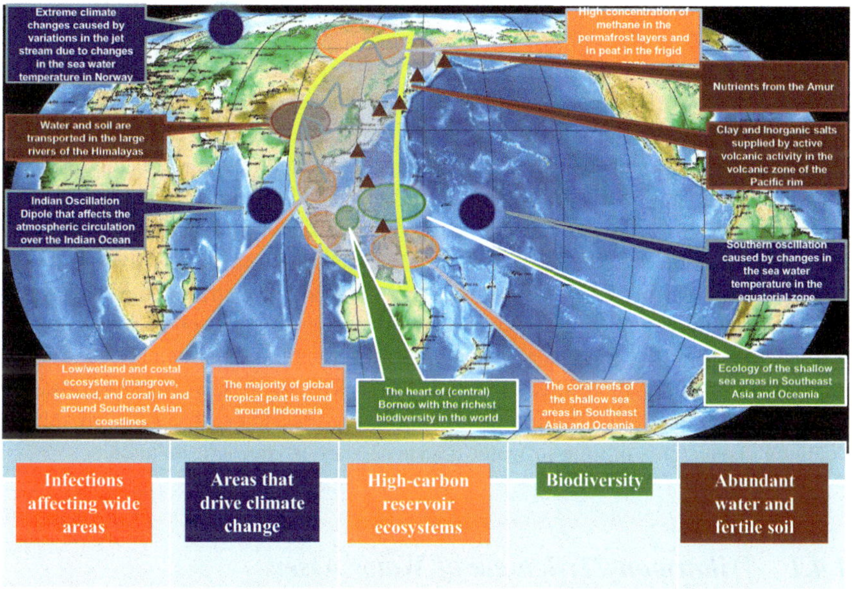

Fig. 1.16 Asian Fertile Crescent. Abundant water, fertile soils, and high biodiversity interact and reinforce each other within an organic farming and agroforestry system. The Maritime Continent is the center of the Asian Fertile Crescent. Obtained from Osaki et al. (2022b)

that can sustain life, which society called the "harvesting sun" society, based on nature. However, human exploitation and the consumption of natural resources (fossil fuels in particular) lead to positive carbon and enhanced climate change, which is called the trilemma in the social system called the "harvested sun" society (Osaki et al. 2022b).

1.4 Eco-evaluation based on "Triharmony/Trilemma"

According to the analysis of tropical peatland paradoxes (see Sect. 1.2), we have been developing the concept of triharmony and trilemma analysis and management, described in Sect. 1.3 (Disciplines on Eco-evaluation), which are applied here to the eco-management and eco-evaluation of tropical peatlands.

In tropical peatlands, three physical, chemical, and biological elements – water, oxygen, and nutrients – are the key to harmonized ecosystems, and these three elements' interrelations and interactions are mainly studied in the triharmony/trilemma system. As the triharmony/trilemma concept has been applied to not only peatlands, wetlands, and mangroves but also drylands and semiarid lands, the "tropical peatland ecoevaluation" concept has, therefore, been extended to the analysis, management and evaluation of natural capital and natural assets of Earth ecosystems. Therefore, this peatland book is a good reference for Earth's regeneration.

"Nationally determined contributions" (NDCs) were submitted to the United Nations Framework Convention on Climate Change (UNFCCC). NDCs embody efforts by each country to reduce national emissions and adapt to the impacts of climate change. Among NDCs, NbSs (nature-based solutions) are selected in many tropical/subtropical countries (Global Central in our definition). NbS are initiatives that protect, restore, and manage the health of terrestrial and coastal ecosystems such as peatlands, wetlands, drylands, and coastal/seashore areas (mangroves and coral reefs). Thus, to achieve NDCs in tropical/subtropical countries, an evaluation of NbSs or natural capital assets is necessary; however, a comprehensive system for eco-evaluation in the tropics/subtropics has not yet been established (Osaki et al. 2022b). This book, therefore, challenges the existing reviews and proposals of eco-evaluation methods in tropical/subtropical ecosystems, focusing mainly on peatland ecosystems unique in water (oxygen)–carbon–nutrient balance.

1.4.1 Triharmony/Trilemma of Water Assets

In tropical peatlands, three key elements, water, carbon, and nutrients, are interrelated, interconnected, and interact. However, as water is an outstanding key element in peatland, it is difficult to sustain peatland without appropriate water management (stock and flow), following rapid peat degradation, reduced carbon assimilation, and nutrient loss; we define a trilemma state when the water element is out of balance.

In contrast, to achieve triharmony in peatland, Kato's team developed an innovative water management system known as "stock-based water management" (stock-based WM). Offering a blueprint for achieving high levels of water retention within peatland ecosystems, stock-based WM contributes to reducing peat degradation through GHG emissions and peat fires, enhances carbon assimilation through biomass production, and conserves nutrient cycling and biodiversity (Kato et al. 2021, 2023).

However, there are superstitious fears that commercial crops in agriculture and plantations are generally susceptible to high soil moisture and GWLs, which we point out as "tropical peatland paradoxes" in Sect. 1.2. Due to these misunderstandings of the water function in peatlands, conventional water management schemes called drainage-based WM are unfortunately predominant in tropical peatland areas (Kato et al. 2021, 2023).

Recently, Osaki's team found that the Indian Ocean Dipole (IOD, the irregular changes in sea surface temperatures in the Indian Ocean) event, not the El Niño event, causes prolonged drought in Indonesia's tropical peatlands around the MCs. Osaki's team study, which continuously measured GWLs for 27 years in tropical peatlands, elucidated the mechanisms of abnormal climate and severe drought around the MCs, which are affected directly by IOD and indirectly by El Niño (Sulaiman et al. 2023). During the 27 years, 2019 was a typical year for water assets in the tropical peatlands because only a serious IOD event occurred, without an El Niño event. In 2019, the IOD was characterized by a decrease in sea surface temperature (SST); consequently, evaporation from the sea decreased, precipitation around the MCs decreased, GWLs dropped, and peatland areas and surrounding areas suffered from fire, haze, and PM10 (causing worsening of people's health).

Thus, a trilemma state is caused mainly by an imbalance in water assets because of both climate abnormalities and human activities. A new water management system, known as triharmony, is therefore urgently needed to address this issue.

Triharmony based on large-scale water management in tropical peatlands: In tropical peatlands, three key elements, water (oxygen), carbon, and nutrients, are interrelated, interconnected, and interact. When these elements are in balance, a triharmony state arises. Therefore, in terms of water, to innovate tropical peatland management systems, stock-based water management (stock-based WM) systems have been developed as ecomanagement principles applied to peatland water resources.

According to Kato et al. (2021, 2023), in large-scale tropical peatland ecosystems, four components are considered: (1) water management and land-use planning, (2) biodiversity conservation planning, (3) new technologies to preserve peatland surfaces, and (4) evaluation systems. The goals of ecomanagement in large-scale tropical peatlands include (1) establishing a carbon-negative system, including mitigation and adaptation strategies to combat global climate change; (2) addressing biodiversity conservation at the landscape level; and (3) providing socioeconomic security to local communities. To meet these goals, Kato's team encourages and recommends preserving and maintaining the functions of the original peatland ecosystem (Kato et al. 2021, 2023). To construct the ideal stock-based WM system, the interconnected water resources and processes within the ecosystem require three main tasks: (1) inventory of water stocks and reservoirs, (2) inventory of irrigation usage, and (3) inventory of water cycling processes that support (1) and (2). However, there is an extensive body of literature on agricultural water management, and most studies have unfortunately focused on drainage-based WM systems in tropical peatland management. Conversely, peatland water management, called stock-based WM, has received relatively little attention; the few studies that have been conducted are superficial, with no in-depth analysis of management approaches or optimal practices. Recently, as IOD events in MC areas have become serious, drainage-based WM systems have revealed major flaws in tropical peatland management. The drainage-based WM system is partially complemented by canal blocking; however, it is not effective in dry seasons, especially during IOD events. There is therefore a vital need to fully reconsider these water management systems (Kato et al. 2021, 2023) (Fig. 1.17).

Drainage-based water management (drainage-based WM): In tropical peatlands, conventional drainage-based WM is designed to drain water from the ecosystem as quickly as possible. A central canal is constructed down the slope of the peat dome, with secondary canals branching off along contour lines. Water then drains into the secondary canals, which in turn drain into the primary canal. Thus, by

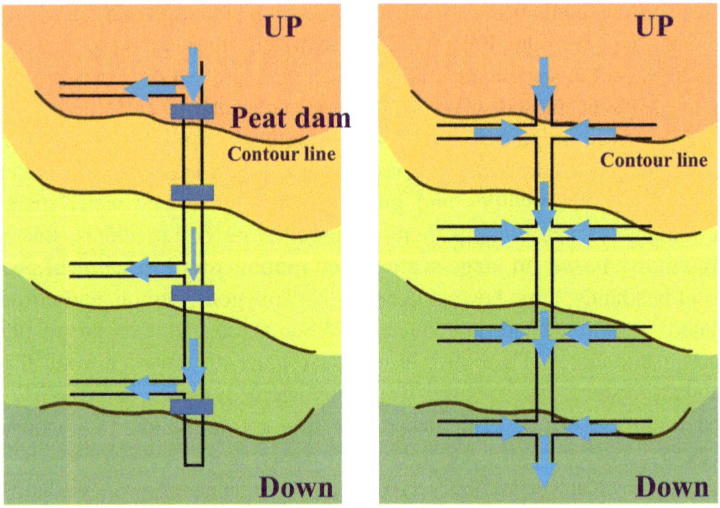

Fig. 1.17 Water management system in tropical peatlands. Modified from Kato et al. (2023). Left: "stock-based water management" (stock-based WM). Right: "drainage-based water management" (drainage-based WM)

draining the water from peatland, a naturally high-water environment results in eventual degradation, resulting in a serious trilemma state (Kato et al. 2021, 2023).

Large-scale oil palm plantation producers and farmers dominate the majority of peatland agriculture in tropical regions and are reluctant to expend additional investment to update their irrigation systems, particularly when they are afraid of the impact on crop growth reduction because of misunderstanding about potential productivity in peatland. As water is the most important asset in peatland, if water is drained from peatland, peatland loses the most important resource and asset. The simplest way to avoid this peatland destruction is by constructing peat dams, i.e., not blocking the main canal at each contour level, allowing water to be retained in the system for a longer period. Moreover, with the IOD impact around MCs reducing the water supply, a stock-based WM system must be adopted to maintain the balance between water inputs, outputs, and storage. However, the implementation of any such system is difficult without information on macro- and microtopography, peat thickness, groundwater, and variations in climate variables such as precipitation and temperature. The innovative stock-based WM system developed by Kato's team refers to the blueprint of traditional paddy-field irrigation and is designed to achieve a high level of water retention within the ecosystem. This approach has four main principles: (1) achieve a stable GWL across the year, (2) prevent peat fires,

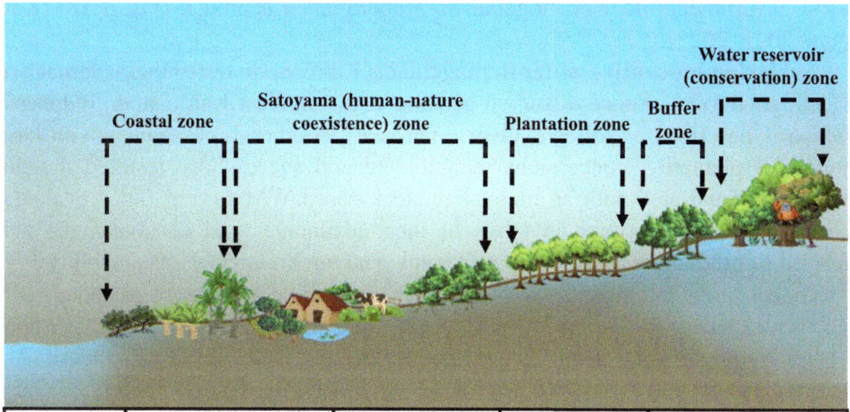

	GWL: 0-40 cm	GWL: 20-40 cm	GWL: 0-20 cm	GWL: around 0 cm
- Mangrove - Sago - Coconut	- Sago - Coconut - *Vitex sp.* - Tengkawang (*Shorea macrophylla*)	- *Acacia crassicarpa*	- *Shorea pachypylla* - *Dyera costulata* - Etc.	- *Diospyros sp.* - *Tristaniopsis obovate* - *Ficus annulate* - *Baccaurea macrocarpa* - Etc.

Fig. 1.18 Triharmony systems achieve good management of water, carbon, and nutrient assets and can be applied to all ecosystems, including mangroves, as shown here. Obtained from Kato et al. (2023)

(3) encourage tree growth, and (4) conserve the native ecosystem as best as possible (Fig. 1.17) (Kato et al. 2021, 2023).

Stock-based water management (stock-based WM): Kato's team tested the stock-based WM system in West Kalimantan, Indonesia, by applying it once peatland areas were drained and damaged by commercial agriculture, slash-and-burn agriculture, and illegal logging. They found that a GWL of 0.4 m below the surface was appropriate for the healthy growth of acacia trees. Then, Kato's team designed a stock-based WM with a higher GWL within 0.4 m.

Peat dams should be used to block the main canal, with the interval between them determined by a 0.5–0.6 m drop in elevation. Spillways, designed based on the maximum expected rainfall, provide a protective mechanism in the case of extreme precipitation by releasing excess water from the site (Kato et al. 2021, 2023). With this canal network design, the GWL is tightly controlled and water is retained long-term due to the retention system; in addition, this system provides a firebreak and ensures tree growth. This design, in turn, helps to support biodiversity corridors. Ultimately, however, successful protection of both water resources and biodiversity requires coordinated management at a large scale; that is, it requires multiple stakeholders (including local communities, smallholders, industrial farms, and local governments) to work together in a coordinated approach (Fig. 1.18) (Kato et al. 2021, 2023).

Achieving innovative water management: Kato's team tested their approach on stock-based WM in once disturbed peatland areas of West Kalimantan, Indonesia, which is described as follows. Over a ten-year period, two companies with large industrial plantation concessions rehabilitated and successfully restored the disturbed peatland area with an innovative stock-based WM system. They evaluated stock-based WM systems in terms of their advantages and disadvantages with respect to productivity, economic efficiency, and environmental sustainability.

Based on their findings, Kato's team suggested that triharmony peatland management adheres to four principles to ensure sustainable peatland management: (1) preservation of intact forests (i.e., no clearing of forests for development); (2) a harmonious balance between economic, environmental, and social factors; (3) business models that emphasize sustainability; and (4) the use of low-cost, simple technologies that are easy to maintain (Kato et al. 2021, 2023). The stock-based WM system can help to support each of these principles. The simplest and quickest way to create such a system is to construct peat dams across existing drainage canals to form a protected environmental buffer zone around the upper reservoir layer that supports water storage and biodiversity (Fig. 1.18) (Kato et al. 2021, 2023).

Satoyama model on peatland management: Using the SATO-YAMA philosophy (Osaki 2014), we designed an integrated tropical peatland management (ITPM) system in which the sustainability of the conservation of forest and peatland in Central Kalimantan depends upon the economic condition of the community (SATO) living around the forest on the peatland (YAMA) (Fig. 1.19).

This SATO-YAMA model is based on AeroHydro Culture methodology (see Chap. 5, and Wetadewi et al. 2021). As the key points of AeroHydro Culture are a high GWL and the symbiotic system, we recommend growing mainly perennial plants, fruits, and cash crops around houses, and we never recommend growing grain crops (rice, maize, soybean) because of their extremely high nutrient requirements (if high amounts of nutrients are applied, land degradation is accelerated, and GHG emissions increase).

Fruits (horticultural crops): (1) mainly banana, pineapple, jackfruit, rambutan, and citrus, (2) secondary carica (papaya), joint fir (mlinjo), guajava (jambu air), cashew nut (jambu mete), star fruit (belimbing), hoplum (kedondong), soursop (nangka belanda), breadfruit (keluwih), mangosteen (manggis), chicu (sawo), horse mango (bacang), avocado, pomegranate (delima), and "duku"

Vegetables (horticultural crops): (1) mainly chili pepper (cabe rawit), sweet pepper (cabe merah), eggplant (terong), tomato, cucumber (mentimun), shallot (bawang merah), leafy shallot (bawang daun), (2) water melon (semangka), "buncis," "kangkung" (Ipomoea), cabbage (kubis), Chinese cabbage (petsai), leaf mustard (sawi), lettuce (selada), celery (seledri), spinach (bayam), okra, luffas (belustru), raddish (lobak), asparagus, and bitter gourds

Cash Crops: sago, coconut, oil palm, coffee, cacao, sugarcane, pepper (lada)

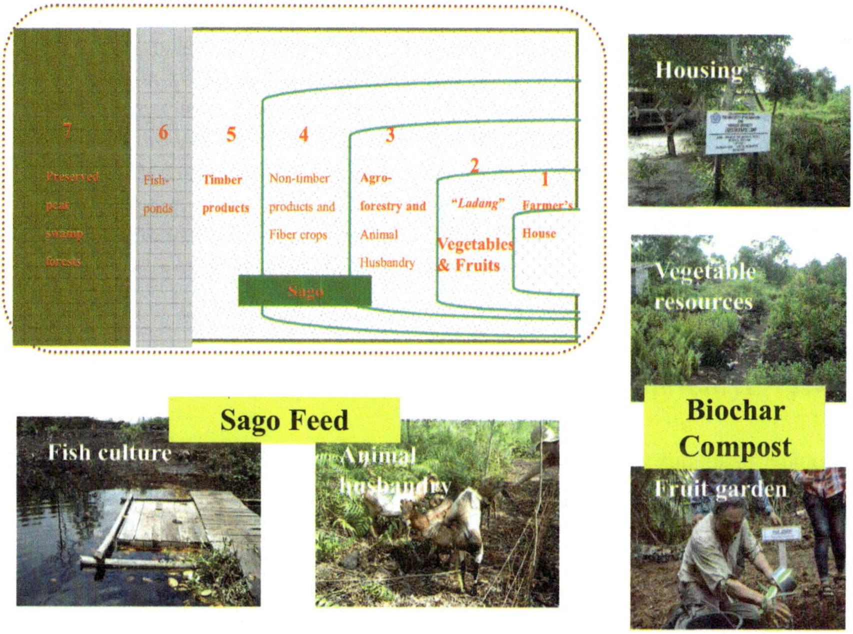

Fig. 1.19 Satoyama (agrarian reform scheme in Indonesia) Model in disturbed peatland, based on AeroHydro Culture

Trees: rubber, Jelutung (ratex), candlenut tree (kemiri), ramie, petai (Parkia peciose), "jengkol" (*Archidendron pauciflorum*), and "kemiri" (candlenut tree)

Fiber crops: Kenaf, Jute, Ramie (Boehmeria Nivea L. GAUD), Abaca (Manila Hemp) (*Musa textilis* L. NÉE), kapok, agave (sisal), "lamtoro" (Fabaceae), pandanus, banana shoot, and coconut fiber

Goats are an important component of the agricultural system in Kalimantan. Goats contribute to income and food sources for people in rural areas. However, the goat population in the fired peatland in Kalimantan is still low compared with that in other regions. Eighteen species of plants were identified as fodder, of which goats preferred to feed on 15 species. Six fodder species are native to Kalimantan. The most preferred fodders are Sasendok, Kalakai, and Pakis Halus. Kid goat production is low (1.3 heads/year), and the growth performance of young goats is 43 g/day. The mortality rate is as high as 72%, and a high level of mortality occurs during the first few days after birth. The supply of mineral blocks containing a mixture of magnesium, phosphorus, and calcium can increase the newborn survival rate. The population is high in the third year because of the number of females who produce more kids. On the other hand, fodder production dramatically decreases in the third year, then we need to introduce the AeroHydro culture system even in grass growing. Grazing by goats slightly improves the pH of acidic soil (Jayadi and Hata 2009).

A fishery reserve is expected to sustain the fish catch based on the natural population in adjacent areas and to conserve fish biodiversity. There were 199,755 hectares of inland fishery reserves recorded in Indonesia in 1986, but most of them were badly managed and developed without a clear theoretical background (Hartoto and Iwakuma 2002). Unfortunately, we still have limited data on how peatland degradation and development affect fishery reserves and fish biodiversity.

Fish culture is important in tropical peatlands. However, its culture is very limited to "fresh water ponds" and "cages" because of poor water nourishment. One idea in our Satoyama system is sago starch feeding of fish in canal brooked ponds.

1.4.2 Triharmony/Trilemma of Carbon Assets

Tropical peatlands are globally important reservoirs of water and carbon, and the cycles of these two elements are intricately connected. Osaki's team (Hirose et al. 2016; Tsuji et al. 2021; Osaki et al. 2023a) developed a model for carbon emissions mapping from tropical peatlands. Our integrated iMRV system measures dynamic changes in water and carbon cycles, biomass and biodiversity, forest degradation, and the impacts of climate change and human activities. A better understanding of these ecosystems by applying iMRV will help eco-management and eco-evaluation in long-term practices in tropical peatlands because the water cycle disturbance has led to increased carbon emissions, which are stimulated by oxidation via either fires or microbial activity (known as "hot fire" and "cold fire" oxidation, respectively).

Oxygen is limited in peatland water because it cannot dissolve in the water based on physical constants. Thus, peatlands have a high water content, which results in low nutrients (low pH due to abundant humic acid, contributing to nutrient leaching) and low oxygen conditions (root activity reduction, reduced root metabolism due to a reduction of oxygen-dependent respiration). Thus, high-water ecosystems are characterized as oxygen-deficient ecosystems, which strongly affect Eh (redox potential) or oxidation-reduction *potential* (ORP), which mainly regulates GHG emissions from tropical peatlands. Unfortunately, when discussing and measuring GHG emissions, Eh (redox potential) information is missing in many cases. As oxygen and Eh information is missing in many cases, tropical peatlands have become mysterious ecosystems, leading to the "tropical peatland paradox," which misleads peatland destruction for development.

In contrast, it is generally believed that tropical peatlands emit high amounts of GHGs; however, normally in pristine tropical peatlands, CO_2 emissions at high GWLs are minimized because the low oxygen concentration and limited nutrients restrict microbial activity. Methane (CH_4) emissions are also minimized because the Eh is relatively high (not absolutely low) for methanogenesis (anaerobic organic degradation) (Fig. 1.5 left). In general, it is incorrect to assume that GHG (mainly CO_2 and CH_4) emissions are extremely high in tropical peatlands; GHG emissions are normally low in tropical peatlands, even in plantations, if high GWLs are retained using stock-based WM systems and no chemical fertilizer application (Osaki et al. 2021a, 2023a).

However, when peatland is drained by applying a drainage-based WM system and is dosed heavily with chemical fertilizer, CO_2 emissions increase considerably due to peat material oxidation, and CH_4 emissions increase drastically because of the absolutely low Eh due to O_2 consumption by microorganisms. In drainage-based WM systems with heavy nitrogen fertilizer application, N_2O emissions also increase (Fig. 1.7, right) (Osaki et al. 2021a, 2023a).

Thus, tropical peatland ecosystems have become the worst model of management (eco-management) among various earth ecosystems when drainage-based WM systems and high amounts of chemical fertilizer are applied, especially in oil palm plantations in tropical peatlands.

In a summary of water–carbon assets, tropical peatlands are globally important reservoirs of water and carbon, and the cycles of these two elements are intricately connected. Water is found within two main ecosystem reservoirs: groundwater and soil water. Levels of water storage are controlled by the balance between evapotranspiration (evaporation and transpiration of water from land and leaves, respectively) and precipitation (the transfer of water from the atmosphere to the ground) (Osaki et al. 2021a, 2023a). According to the explanation provided by Osaki et al. (2023a), these processes are not static and are increasingly variable owing to climate change. For example, in the tropical peatlands of Southeast Asia, phenomena such as the IOD (the irregular changes in sea-surface temperatures in the Indian Ocean area) cause prolonged drought on the MC in Southeast Asia, especially in Indonesia's tropical peatlands. Importantly, a recently published study shows that positive IOD events – not El Niño – act as an indicator of lower GWLs and cause peat fires on the Indonesian MC (IMC) (Sulaiman et al. 2023). In addition, land use, land-use change, and forestry (LULUCF) all have profound impacts on water cycling and consequently carbon cycling.

Deforestation for agriculture and/or plantations not only results in biodiversity losses but also changes the balance between evapotranspiration and precipitation, usually resulting in the drying of peatlands, which cause air temperature to increase. The long-term consequences of this management system are that the peat soil dries (causing an increased fire risk), biomass is reduced (including reduced agricultural productivity), and carbon is released into the atmosphere, which induce a devastating trilemma state (Osaki et al. 2022b).

Carbon sequestration in peat soil: Among Earth's main carbon reservoirs, increases in atmospheric and oceanic carbon are problematic and are already causing significant climatic and environmental issues; storing carbon as fossil fuels takes millions of years; and CCS remains technologically immature and prohibitively expensive. This leaves a final significant carbon reservoir in soil and biomass that can be exploited in land ecosystems (Osaki et al. 2021a, 2023a). The goal for global soil carbon storage, as agreed upon at the United Nations Climate Change Conference in 2021 (COP21) (Paris, 2015), is to increase soil organic matter in the top 1 m of soil by 0.4%/year (called the four per mille initiative); it is estimated that meeting this goal would offset 20–35% of global GHG emissions. According to Osaki's team report (Osaki et al. 2023a), as a global average, 161 t/ha of soil organic carbon (SOC) is estimated to exist; however, the carbon contents of soils vary among the different climate zones. Global maps of the soil carbon content clearly show that arid and semiarid areas support carbon-poor soils (10–30 t C/ha), while the boreal forests of Russia, Scandinavia, and northern North America, which all support rich areas of peatland, are the most carbon-rich areas (120–150 t of C/ha). However, peatlands are also found in tropical areas along the equator, particularly in Southeast Asia, the Congo Basin of Africa, and the Amazon area of South America. After boreal peatlands, these tropical peatland areas are the next richest reservoir of soil-bound carbon (60–150 t of C/ha) (Osaki et al. 2023a).

It is concluded that peatland, especially tropical peatland, is the dominant ecosystem as a carbon reservoir following water reservoirs, which is called the triharmony model (defined as AeroHydro Culture including nutrient cycling, see Sect. 1.5.1 "AeroHydro Culture") based on a stock-based WM system.

Carbon flux and CH_4 emission mechanisms: Carbon flux mapping is critical for managing carbon resources. Various methods and models have been developed to estimate peatland carbon fluxes. Satellite technology enables estimates of carbon fluxes at relatively low costs and high spatiotemporal resolutions. Osaki's team (Osaki et al. 2023a) harnessed this technology to develop a new estimation model, which is described as follows. Osaki's team used data from a National Aeronautics and Space Administration satellite, the Soil Moisture Active Passive satellite, to observe strong correlations among GWLs, soil peat subsidence, and carbon emissions. For every 0.1 m/year decrease in the GWL, there is an increase in carbon emissions of 89 $gC/m^2/year$. Consequently, emissions are higher during the dry season or during droughts (e.g., those caused by IOD) (Sulaiman et al. 2023). Methane (CH_4) is the second most important GHG in terms of global climate change. Wetlands are major contributors to natural CH_4 emissions, accounting for 20% of total global emissions (including natural and anthropogenic emissions), with critical emission zones concentrated in the tropics and boreal North.

Effect of redox potential on carbon flux and CH_4 emissions: Osaki's team (Osaki et al. 2021a, 2023a) mentioned that it is clear that a stable, high-water environment in tropical peatland is conducive to minimizing CO_2 and CH_4 emissions from peat. Even at high water levels, CH_4 emissions are not very high because Eh is not absolutely lower (−300 mV) due to poor nutrient conditions, which restrict microorganism activity, allowing O_2 to diffuse deep into the peat layer. Conversely, in oil palm plantations, when water is drained (drainage-based WM), inducing a low GWL, and a large amount of fertilizer (especially nitrogen fertilizer) is applied, (1) CO_2 emissions increase by oxidation (fire and microorganism degradation), (2) CH_4 emissions increase from the lowest Eh (−300 mV) value due to microorganism activity, and (3) N_2O emissions increase by oxidation (Osaki et al. 2021a, 2023a) (Fig. 1.7).

Thus, CO_2 and CH_4 emissions are closely related to Eh (redox potential); however, there are no remote sensing technologies for Eh evaluation in ecosystems. Hirano et al. (2012) found that net ecosystem exchange (NEE) correlated with the GWL within similar ecotypes in peatland (pristine forest, drained forest, and

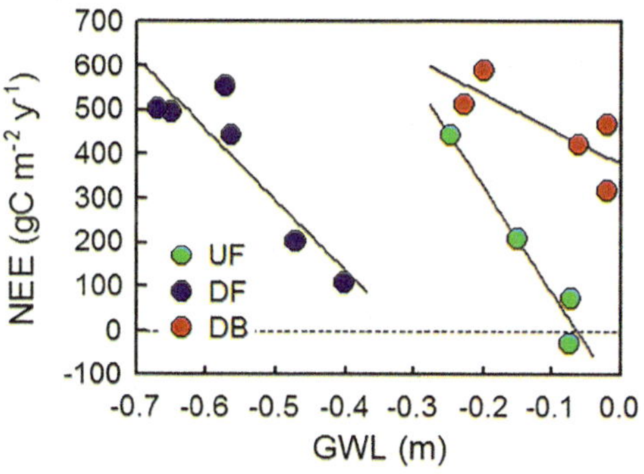

Fig. 1.20 Correlation between groundwater level (GWL) and net ecosystem exchange (NEE) in three ecosystems: UF (undrained forest), DF (drained forest) and DB (drained and burned field). Obtained from Osaki et al. (2023a)

grassland drained and burnt) (Fig. 1.20). Additionally, Osaki's team monitored GWLs and surface soil moisture (Osaki et al. 2021a, 2023a), and they found a linear correlation between the GWL and surface soil moisture, from which they have been developing a GWL mapping model applied to surface soil moisture from a satellite sensing data set (see more in Sect. 1.6 "iMRV System on Eco-evaluation").

1.4.3 Triharmony/Trilemma of Nutrient Assets

Peatland soils and water are characterized as having low nutrient concentrations. Additionally, even if nutrients are rich, a high GWL restricts the oxygen supply to roots, and roots cannot absorb nutrients because roots require oxygen to generate energy for active absorption metabolism. Thus, these ecosystems are most limited by nutrient supply among global ecosystems.

Osaki's group (Osaki et al. 2023b) described numerous reasons for low nutrients in peatland in more detail as follows. First, in terms of geography, peatland geomorphology controls the source of water. Peatland regions form peatland domes, areas of relatively higher topography (approximately 10 m in elevation in the center, decreasing to river level along a radius of approximately 10 km) created by thick layers of peat soils. As natural water courses (rivers) flow around these structures, they do not serve as a water source for the peat soils within the dome (Fig. 1.21). The second reason is that rainwater, i.e., precipitation, which is naturally nutrient poor, is the primary water source for peat soils. The relatively acidic pH of precipitation (~5.6) also contributes to nutrient leaching (Osaki et al. 2021a) (Fig. 1.22). The third reason is the physical trait of oxygen; the oxygen concentration in peat water is severely deficient due to extremely low oxygen solubilization in water, resulting in reduced nutrient absorption by plants (Figs. 1.6 and 1.7). Finally, regarding the chemical properties of organic compounds, carbon-rich peat fibers, mainly lignin, desorb nutrients at the low pH of peat soil (<4), which exacerbates this low cation (positive ion) absorption capacity, even if nutrients are applied as fertilizers, resulting in limited exchange efficacy. Thus, nutrient stocks in water and soil are

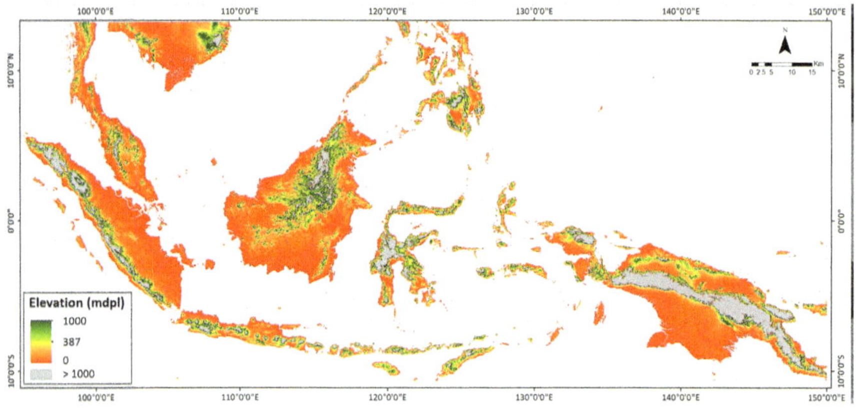

Fig. 1.21 Elevation model of the continent in Southeast Asia. Obtained from Osaki et al. (2023b)

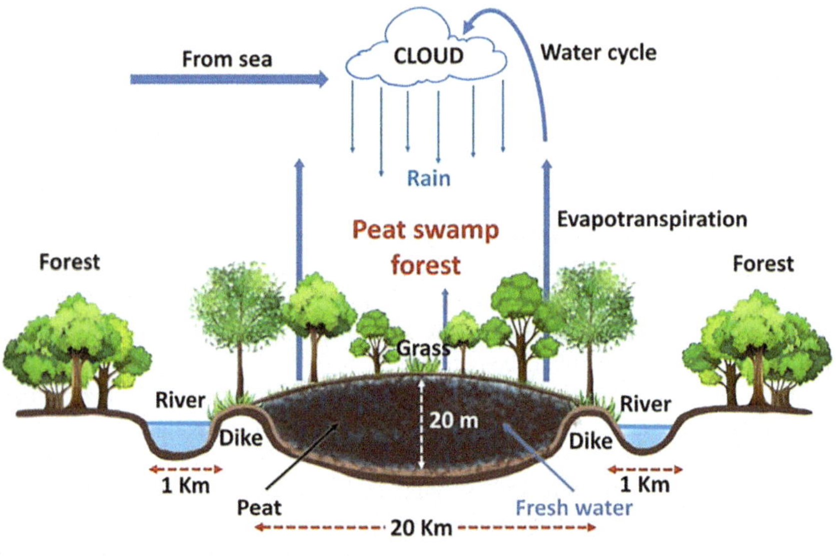

Fig. 1.22 Peat dome structure and water cycle (no nutrient charge system). Obtained from Osaki et al. (2023b)

very poor in tropical peatlands, especially around the Maritime Continents (MC) in Southeast Asia (Osaki et al. 2021a; Osaki 2022).

Nevertheless, biomass in native tropical peatlands is abundant and contains adequate nutrients. Therefore, we define tropical peatlands as paradoxical ecosystems (see Sect. 1.2 "Tropical Peatland Paradoxes"). From this paradox related to nutrients, we found that nutrient stocks in water and soil are very poor, but this nutrient shortage is compensated by nutrient cycling (called nutrient flow) on the peatland surface. Peatland plants have developed adaptation strategies to maximize

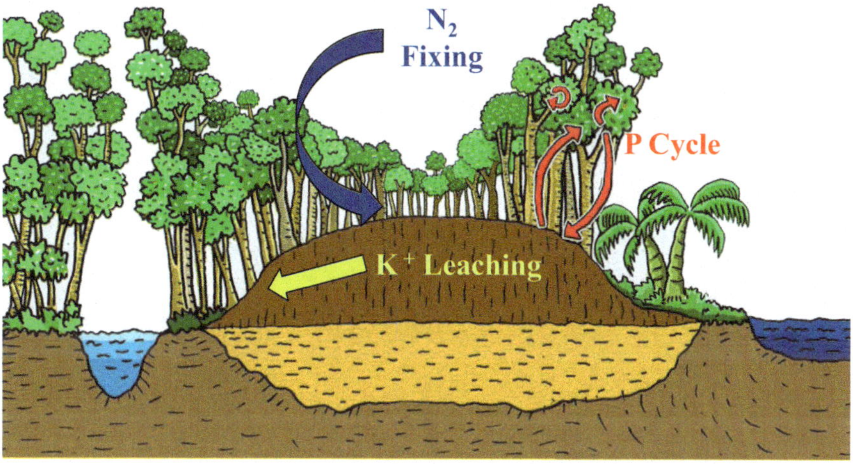

Fig. 1.23 N, P and K cycling in the peat dome. Obtained from Osaki et al. (2023b)

nutrient utilization. In particular, they have evolved to recycle nutrients from litter by symbiosis with microorganisms to decompose organic matter on the peatland surface, to fix nitrogen (N_2) in association with free-living N_2-fixing bacteria in the rhizosphere (the thin layer of soil immediately adjacent to the roots), and to take up oxygen via the development of aerial and mound roots (Osaki et al. 2021a, 2022c; Osaki 2022).

Osaki's team reported nutrient and oxygen acquisition mechanisms as follows (Osaki et al. 2021a, 2022c; Osaki 2022). Oxygen, nitrogen (N), phosphorus (P), and potassium (K) are key nutrients for plant growth (Fig. 1.23). The very low oxygen availability in peatlands is a direct reflection of the high GWL (because the solubility of oxygen in water is very low). Pristine peatland forests have adapted to low-oxygen conditions in two main ways. First, they have evolved aerial roots, which enable plants to absorb oxygen from the air. Second, the formation of mound roots in the litter mound allows both oxygen absorption from the air and nutrient absorption from the forest litter that accumulates on the mound.

Isotope data from peatland forests in Thailand and Indonesia show that tropical plants fix nitrogen from the air effectively because plants prefer the lighter ^{14}N isotope over its heavier ^{15}N counterpart when N_2 gas from air is fixed by microorganisms. Nitrogen in the atmosphere is predominantly ^{14}N; moreover, during the decomposition of litter, ^{14}N is released before ^{15}N, leaving a heavier ^{15}N in soil. As a result of the change in ^{15}N (delta ^{15}N) in leaves, in tropical peatlands, the main nitrogen source is identified as air through aerial roots in association with symbiotic N_2-fixing free bacteria (Osaki et al. 2021a).

Melastoma (*Melastoma malabathicum*) is a unique plant because of its Al accumulation and tolerance to strongly acidic soils (Watanabe et al. 1997, 1998, 2005a, 2006; Osaki et al. 2021a). Melastoma plants exude high amounts of organic

Fig. 1.24 Root exudates (mucigel) from Melastoma plants. Modified from Osaki et al. (2021a)

acids and mucilage, which contribute to Al accumulation and Al^{3+} toxicity tolerance (Watanabe et al. 2002; Watanabe et al. 2005a,b, 2008) and to symbiosis with microorganisms, especially N_2-fixing bacteria. In N-free hydroculture, endophytic nitrogen-fixing bacteria were scarce inside the leaves, stems, and roots of Melastoma plants; however, many free-living bacteria were found in the mucilage in the rhizosphere (Watanabe et al. 2008) (Fig. 1.24). Endophytic nitrogen-fixing bacteria have been reported in many papers; however, Osaki's team never found apparent nitrogen fixation by endophytic nitrogen-fixing bacteria under complete N-free hydroculture, even by legume plants. Among hundreds of candidate plants (mainly temperate plants), only Melastoma plants could grow with free-living N_2-fixing bacteria in complete N-free hydroculture (Fig. 1.25). As a result, aerial roots of Melastoma plants are essential for N_2 fixation in the rhizosphere (organic acid exudation and mucilage exudation for symbiosis with N_2-fixing bacteria).

In disturbed peatlands, the rhizosphere system and symbiotic system are completely destroyed, and it is almost impossible to naturally rehabilitate peatlands (especially peatland surfaces). Thus, we recommend that Melastoma plants be planted as companion plants to recover microflora.

Another critical plant nutrient is phosphorus, with deficiencies impacting metabolic disorders. In tropical peatlands, phosphorus is mainly concentrated in aboveground biomass and forest litter. Plants obtain phosphorus from the litter decomposition of other plant species, with the rate of decomposition and level of

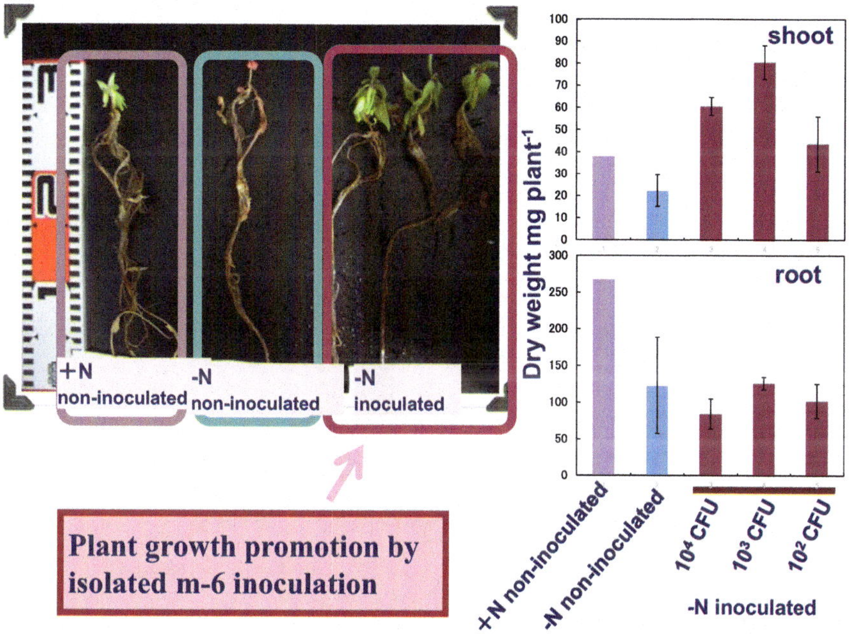

Fig. 1.25 Symbiotic association between Melastoma plants and N_2-fixing bacteria

available phosphorus depending on local conditions and plant species. Forest fires seriously degrade phosphorus stocks. The main mechanisms for absorption are mycorrhizal symbiosis (phosphorus uptake by symbiotic fungi that colonize the roots) and the production of acid phosphatases and organic acids. Acid phosphatase decomposes organic phosphorus accumulated on the peatland surface or litter mound, while organic acids and acid phosphatases solubilize insoluble phosphorus compounds (Fig. 1.26).

Potassium is the third major plant nutrient. As the cation (mainly K^+) adsorption capacity of peat soil (organic soil) is extremely low under low pH conditions (less than 4), most potassium, even that from commercial fertilizers, is leached, and only a small amount is absorbed by plants (Figs. 1.3 and 1.23). Serious potassium deficiencies have been observed in oil palm plants. The severity of such deficiencies is exacerbated because potassium is critical for healthy root development; poor root development in turn leads to even lower potassium absorption, creating a negative feedback cycle. As research into nutrient uptake in tropical peatlands has largely focused on nitrogen and phosphorus, the potassium cycle is less well understood in peatlands (Osaki et al. 2021a).

Osaki's team, who studied nutrient cycling in peatlands, subsequently developed an innovative nutrient management system for agriculture and silviculture called AeroHydro Culture, involving cultivation using high GWLs through a stock-based WM system, with nutrient and oxygen supplementation occurring via natural compost and aerial-like roots induction (see Sect. 1.5.1 "AeroHydro Culture") (Osaki

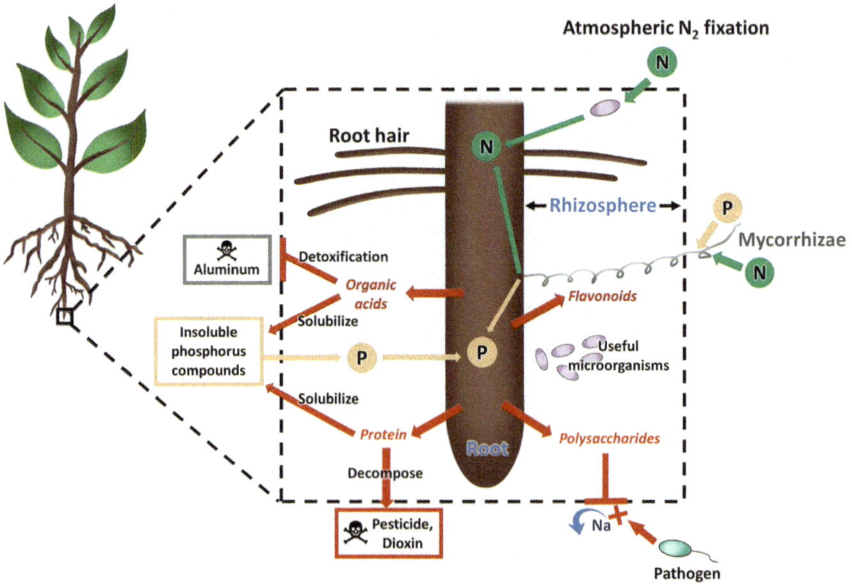

Fig. 1.26 Rhizosphere function. As the surface area of roots is not large enough to absorb nutrients, roots must form symbiotic systems with microorganisms, which is known as rhizosphere function. Obtained from Osaki et al. (2022c)

et al. 2021a, 2022b,c; Osaki 2022). This approach is in contrast to conventional methods that focus on drainage-based WM systems. As a proponent of eco-management and NbSs for environmental problems, Osaki's team is on a mission to innovate agricultural and silviculture practices and eco-management in the peatlands and forests around the MCs in Southeast Asia.

1.5 Land-Surface Management

Tropical peatlands are globally important reservoirs of water and carbon that supply balanced amounts of three elements: water, oxygen, and nutrients. When these three elements, water, oxygen, and nutrients, are supplied in a good balance, carbon stock (sequestration) accumulates in high amounts, called the triharmony of carbon assets (see Sect. 1.4.2). In contrast, once these three element (water, oxygen, nutrients) balances are disturbed, the trilemma of carbon assets appears to accelerate minus carbon flow (emissions) (see Sect. 1.4.2).

Osaki's team found that the triharmony/trilemma of water, carbon, and nutrients assets occurs mainly on peat surfaces, not in peat soil. The key factor is root growth on peatland surfaces, on which roots form "aerial roots" and "mound roots" to absorb oxygen and nutrients (through nutrient recycling by microorganisms and N_2-fixing bacteria), and roots cannot grow on land surfaces under dry conditions.

Osaki's team concluded that land-surface management (LSM) is the first priority to reserve, rehabilitate, restorate, and regenerate tropical peatland ecosystems and should be followed up by stock-based WM systems.

The general principle behind LSM technology is how to control and manage the three basic elements (water, oxygen, and nutrients) on the land surface, even in different ecotypes such as wetlands and drylands. In low-quality soil (e.g., peat, acid sulfate, sandy, saline, and arid soils) and degraded agricultural lands (which are abandoned, degraded, and/or contaminated), plant roots employ generally an "avoidance strategy." The "avoidance strategy" is the extension of roots along the land surface rather than a tolerance strategy that involves growing into soil. This "avoidance strategy" has largely worked together to advance the symbiotic system, nutrient cycling, and oxygen absorption in the rhizosphere (the thin area around the roots) on the land surface; however, other key points of this "avoidance strategy" should be maintaining appropriate moisture conditions on the land surface, controlling high GWLs, or mulching the land surface with litters and/or porous fibrous materials (grass mulching, etc.).

Furthermore, Osaki's team insisted on the constitutive principle of LSM technology, which is related to the material exchange process. When material exchange occurs between objects and living organisms at a global scale, the material surface is a key merkmar of this process, with a larger surface area increasing the potential for material exchange capacity, indicating that a fractal structure is, in fact, superior. Osaki's team (Osaki et al. 2023b) considered the leaf surface as a single unit of the terrestrial land surface; on that basis, plant roots represent approximately 35 area units; however, as leaves are exponentially more efficient at CO_2 assimilation than roots are at nutrient uptake, roots and the soil sphere are major limiting factors for Earth's metabolism. Moreover, given the very thin layer of soil (approximately 30–50 cm) compared with the Earth's size, water and nutrient carrying capacities are very small; however, metabolic activity is very high for regulating the Earth's metabolism, indicating that metabolic activity in the soil sphere is very sensitive and fragile. Protecting the physical, chemical, and biological health of these soil spheres is critical for maintaining rhizosphere function and ensuring that this fragile mechanism within the global cycle of water, nutrients, and carbon does not become damaged beyond repair (Fig. 1.27).

Thus, the "soil sphere" is a key center of Earth's metabolism; however, as the land surface layer is similar to earth skin, LSM technology is most important for Earth's metabolism and ecosystem health, similar to earth skin care.

1.5.1 "AeroHydro Culture" in Peatland

Osaki's team reported details on "AeroHydro Culture" technology in an article (Osaki et al. 2023b) and book (Wetadewi 2021) as follows (also see Chap. 5). With knowledge of nutrient and oxygen cycling in pristine tropical peatland ecosystems, Osaki's team developed a new cultivation system that was both productive

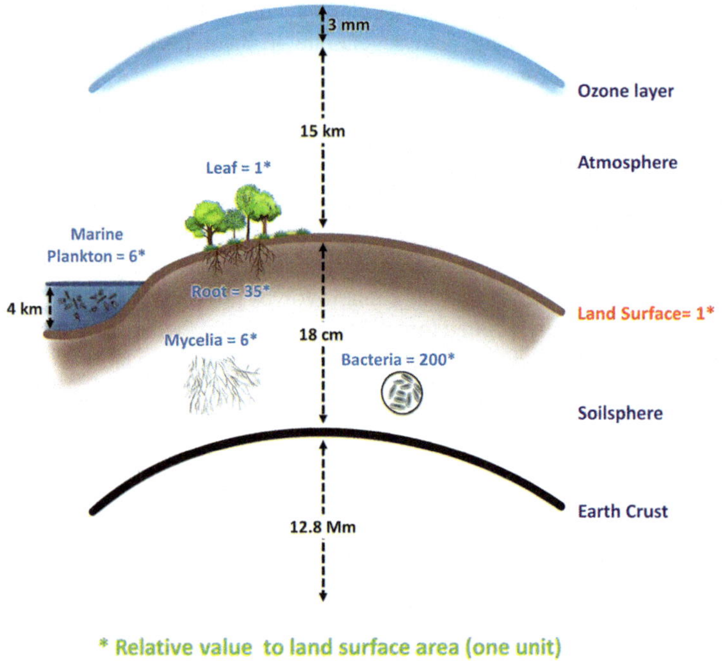

Fig. 1.27 Land surface unit of materials and living organisms when the terrestrial land surface is 1 unit. The leaf surface area is 1 unit and thus similar to the terrestrial land surface. However, as the root surface area is 35 units, root function is limiting to plant growth. The soil sphere is extremely thin (average thickness 30–50 cm), so it has a low water- and nutrient-holding capacity. Obtained from Osaki et al. (2022c)

and protective. Conventional water management in tropical peatlands (and elsewhere) is based on water drainage (drainage-based WM), which is inappropriate for peatland management. Unfortunately, where drainage-based WM was applied to peatland areas, large-scale severe disasters have followed: (1) peat fires (hot fires), peat degradation by microorganisms has occurred (cold fires), (2) high levels of GHGs (CH_4 and N_2O) are emitted with chemical fertilizer application, (3) nutrient loss causes environmental pollution, and (4) plant growth has declined. Consequently, Osaki's team disapproves of drainage-based WM in tropical peatland in the case of the unprecedented worst trilemma management of tropical peatland ecosystems. In particular, chemical fertilizers are both environmentally and economically damaging (nutrient pollution and CH_4 and N_2O emissions), especially under drainage-based WM, which causes more damage than stock-based WM (see Sect. 1.4 "Eco-evaluation Based on 'Triharmony/Trilemma'").

To mitigate this tragic disaster caused by intensive peatland development, Osaki's team proposed an innovative culture system, defined as "AeroHydro Culture," in tropical peatlands. By mimicking plant strategies in the high GWL environment of a pristine tropical peatland, Osaki's team developed a creative way to establish

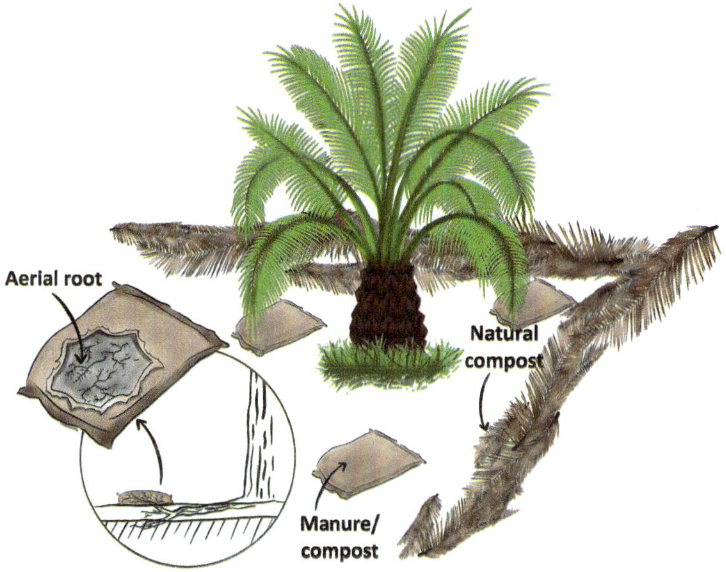

Fig. 1.28 Example of AeroHydro Culture for oil palm at a high groundwater level in a peatland. Grass compost (with biochar, mycorrhizae, zeolite, and plant growth-promoting compounds) bags and natural compost (oil palm fronds) were placed on the land surface around the oil palm plants. Modified from Osaki et al. (2022c)

"AeroHydro Culture" in a high GWL environment, which is guaranteed by stock-based WM and no application of chemical fertilizers.

AeroHydro culture technology is simple, inexpensive, and effective. Nutrients and oxygen are added from the land surface setting via bags containing natural materials (e.g., natural compost formed of leaves, grasses, and weeds; biochar; microbes and fungi) (Wetadewi et al. 2021; Osaki et al. 2022a,c,d) (Fig. 1.28) (also see Chap. 5. An Innovated Restoration Technology of Tropical Peatland; AeroHydro Culture (AHC)).

The application of this AeroHydro Culture technology to experimental plots in Indonesia has been very successful. Details are as follows (Wetadewi et al. 2021; Osaki et al. 2023b). Within just 2 months, the team observed the growth of aerial and mound roots, similar to those observed in pristine tropical peatlands. Different types of aerial roots were even observed around the trunks of oil palm, similar to the trunk-aerial root in sago palm (which has high N_2-fixing ability). Leaves and fronds were greener and more numerous. For the species *Shorea balangeran*, the average plant height increased from 107 to 191 cm in just 6 months. The harvested yield of oil palms increased from 16 t/ha/year to 21 t/ha/year in a one-year period.

AeroHydro Culture has abundant benefits because of the typical triharmony model. The degradation of tropical peatlands by desiccation (via both the use of drainage systems [drainage-based WM] and reduced rainfall owing to IOD abnormality) has also resulted in (1) increased CO_2 emissions from these areas via both microbial decomposition and forest/peat fires and (2) increased CH_4 and N_2O emissions from the application of a large amount of chemical fertilizers, especially in oil palm plantations. AeroHydro Culture with a high-water environment (stock-based WM) has the added benefit of mitigating CO_2 emissions from agricultural land, even reversing this process via significant carbon sequestration with a greater amount of biomass. The adoption of this triharmony model will assist in achieving not only a carbon-neutral society but also an innovative carbon-negative society (Osaki et al. 2022b).

1.5.2 HydroCycle Culture in Dryland and Semiarid Areas

In peatlands and wetlands, Osaki's team developed an innovative AeroHydro Culture system based on LSM, of which the key concept is mimicking plant (trees) strategies in the high GWL environment of a pristine tropical peatland. However, after drainage-based WM was applied, peatland was degraded, and it was difficult to apply HydroCycle Culture methods. After peat degradation, some of the worst problem subsoils appear; one is acid sulfate soil (extremely low pH), and the other is sandy podzolic soil (extremely poor nutrient content) (Fig. 1.6). Once these soils appear, they are impossible to reverse. Conventionally, we try to rehabilitate this barren land by applying soil improvement technology, such as pH adjustment by Ca application or the application of a large amount of nutrients. However, from observations in barren land ecosystems, we found that trees can grow even under these conditions. Trees and grass roots extend on the land surface under a thin organic layer; however, trees and grass roots never intrude into the problem soil, from which we understand that plants choose a more "avoidance strategy" than a "tolerance strategy," from which we also understand that LSM is essential in various plant ecosystems.

This new LSM technology is defined as HydroCycle Culture, which can be applied to dry or semiarid land as well as problem soils (acid sulfate soil and sandy podzolic soil) (Fig. 1.29) (Osaki et al. 2022d). In the model study for this LSM technology approach, a porous mulch (e.g., a cotton-based mulch) is laid over the soil (Fig. 1.30). When porous mulch containing water is irradiated by the sun, the liquid water inside the cotton pores vaporizes by solar energy and removes latent heat. The mulch surface becomes hot, while the bottom of the mulch cools because of latent heat removal, with condensation creating water droplets at the bottom of the mulch (Fig. 1.31). Thus, the porous mulch on the land surface can also extract moisture using solar radiation by capturing dew, which becomes trapped below the mulch where it can be utilized by surface roots. Experiments have shown that this approach can increase the weight of biomass by up to 70%. However, as it is still not sufficient to produce a large amount of biomass, the HydroCycle Culture system supports primary land protection and increases the water-holding capacity of soil.

Another case of the laten heat effect is porous metal-organic material, developed by the Massachusetts Institute of Technology. Porous $[Zr_6O_4(OH)_4(fumarate)_6]$ can

Fig. 1.29 Effect of cotton sheet thickness on plant growth (upper) and the slope-greening construction (Greencosmo construction) method of mulching with a cotton sheet (SoilCott Registered Trademark) (Bottom). Obtained from Osaki et al. (2022d)

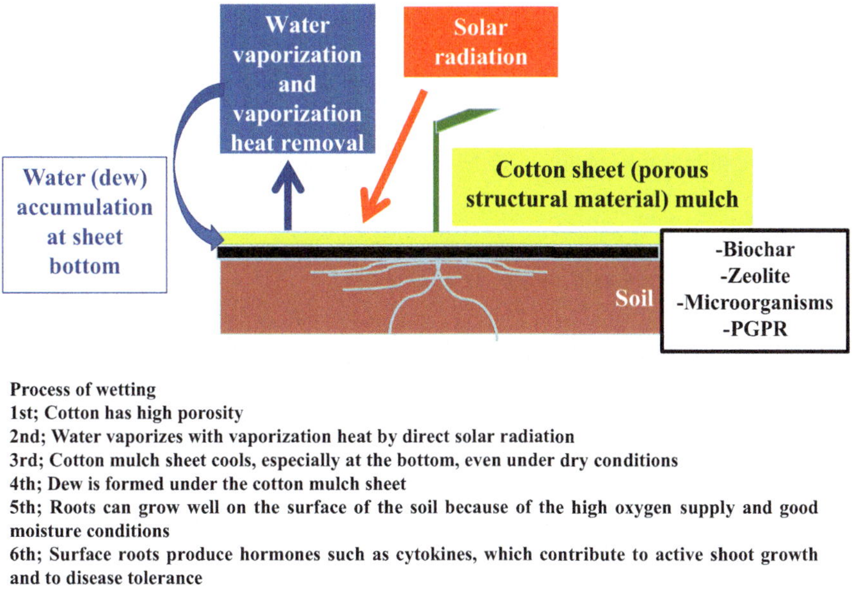

Fig. 1.30 Principle of water cycle culture. Mulching with high-porosity fiber, such as cotton, and applying biochar, compost/organic matter, and microorganisms under cotton mulch sheets

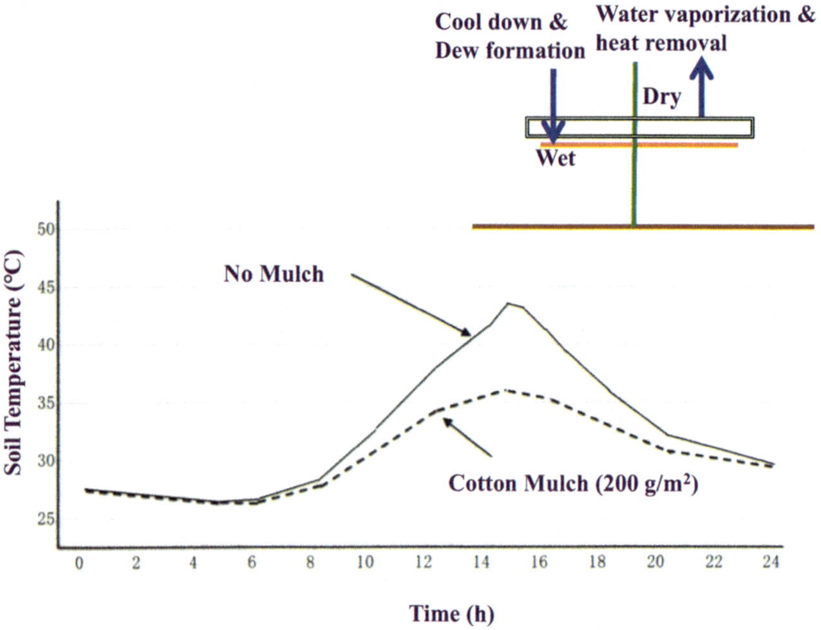

Fig. 1.31 Soil temperature difference between cotton with and without mulch

capture 2.8 L of water/day using only solar radiation as its energy input, even when the relative humidity is as low as 20% (Kim et al. 2017).

1.5.3 Z-Tillage Culture in Arable Land

Zero-tillage was proposed by Masanobu Fukuoka (Fukuoka 1975) for arable land. Tillage involves preparation for germination, enrichment of the soil, and restriction of weed growth. By combining this zero-tillage culture with the use of natural mulch (e.g., spreading grasses on the soil surface), high soil moisture is maintained, and then, the soil is preserved for generations of plants (Osaki et al. 2022d). Osaki's group extended a new concept of LSM to arable land, which includes zero-tillage with grass mulching culture (called Z-tillage Culture).

1.6 Informatics-Based Equator/Earth Observation System for Eco-evaluation

The iMRV system is an integrated monitoring, reporting, and verifying system. Pristine peatlands offer a stable and important reservoir for water and carbon; however, this is under threat from drainage-based WM and anthropogenic climate change. The triharmony management of peatlands depends on the continued balance among carbon, water, and nutrients. Unfortunately, intensive agriculture and poor agricultural practices have destroyed this balance over large areas for many years. Osaki's team argues that preserving surviving areas and restoring already damaged areas should be a priority in developing climate change mitigation strategies. To this end, they propose an iMRV system to track dynamic changes in water and carbon cycles, biomass and biodiversity, forest degradation (including deforestation), and the impacts of climate change and human interference. The iMRV system can operate at a high resolution ($<100\ m^2$) and in real time using a range of satellite sensors operating in different sections of the electromagnetic spectrum. The outputs offer the ability for three-dimensional (3D) analysis and are ground-truthed using drone/UAV (unmanned aerial vehicle) technology and other satellite products. While this approach can be applied globally, the team has focused much of its attention on tropical peatland areas for numerous reasons. Primary among them, tropical peatlands have been degraded at an alarming rate owing to commercial agriculture (primarily oil palm), slash-and-burn practices, and illegal logging. However, warm and humid climate conditions offer the potential for relatively rapid regeneration, assuming that appropriate remediation approaches are taken (e.g., the implementation of the **stock-based WM system**). As such, a major new initiative to complement the iMRV system is the informatics based on Equator Observation System (iEOS), which uses several different spectral products, 3D photogrammetry, and a GHG sensor aboard the Equator orbiter. The satellite collects data from areas along the equator several times a day, allowing for the creation of mosaic images that avoid clouds. This approach mitigates one of the main challenges of equatorial remote sensing: high levels of cloud cover. The iEOS offers significantly improved abilities to measure variables such as the GWL, soil moisture, SOC, carbon flux, CH_4 emissions, plant nutrient phenology, and plant diversity (Tsuji et al. 2021; Osaki 2022).

A critical element of any new scientific discipline is the development of robust systems for evaluation. For field science, there are two main approaches. The first approach is monitoring plant phenology (i.e., seasonal and environmental changes in plants and impacts of soil conditions on plant leaves; nutrients, water, and carbon status; microorganism activity; and soil structure) using informatics based on iEOS; that is, remote sensing tools such as satellites and airborne measurements are used to support ground-based data collection (Tsuji et al. 2021). For example, peat subsidence can be measured with ground-based leveling surveys, airborne LiDAR, and satellite-based interferometric synthetic aperture radar (InSAR) (Fig. 1.32).

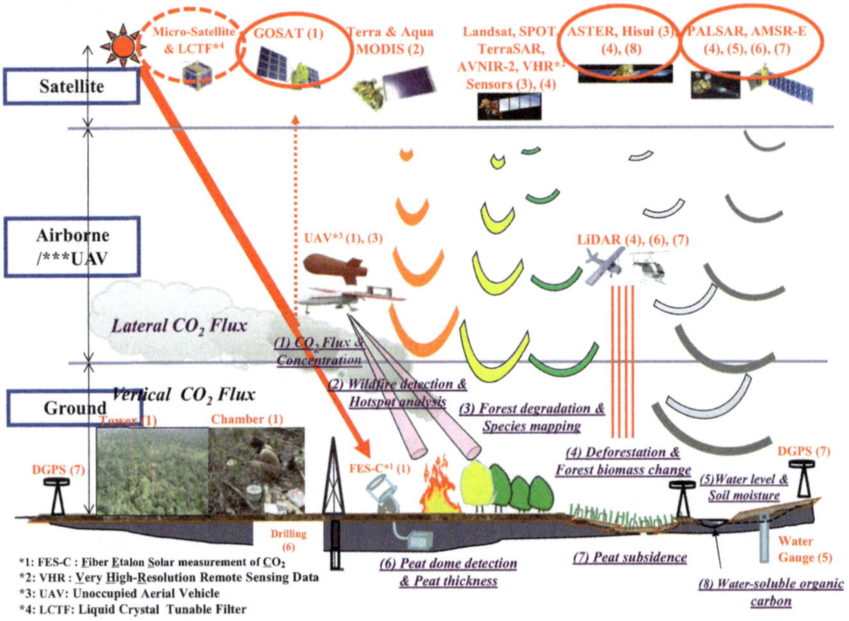

Fig. 1.32 Integrated monitoring, reporting, and verifying (iMRV) system, based on Hirose et al. (2016). Obtained from Osaki et al. (2022c)

1.6.1 Carbon Emission Mapping

Generally, the estimation of carbon emissions in peatlands is based on the carbon stock multiplied by an emission factor. Because calculations based on emission factors with carbon stocks obtained from satellite imagery such as Landsat are inexpensive and fast and cover a wide area, we developed a method for estimating carbon emissions that also has the same advantages. The method uses a satellite launched by NASA called the SMAP satellite. Calculations using this method produce a good estimation if an eddy covariance tower is installed in the area. This measurement shows a strong correlation between the GWL and NEE, and the GWL and peat subsidence with carbon emissions (Hirano et al. 2012; Mezbahuddin et al. 2014). A decrease in the GWL of approximately 0.1 m/year is correlated with carbon emissions of 89 gC/m^2/year (Hirano et al. 2012). The GWL is the first component of GHG emission because GHG emissions are regulated by redox potential: low redox potential (reduction) under high GWL and high redox potential (oxidation) under low GWL. Therefore, peatland carbon emissions occur in the dry season when the peatlands dry up. The drought level is expressed using soil moisture parameters. A strong relationship exists between drought and decreasing GWLs. Therefore, soil moisture information (map) from satellite data is useful to estimate GWL map. Once we can get GWL map, we can estimate the NEE map using

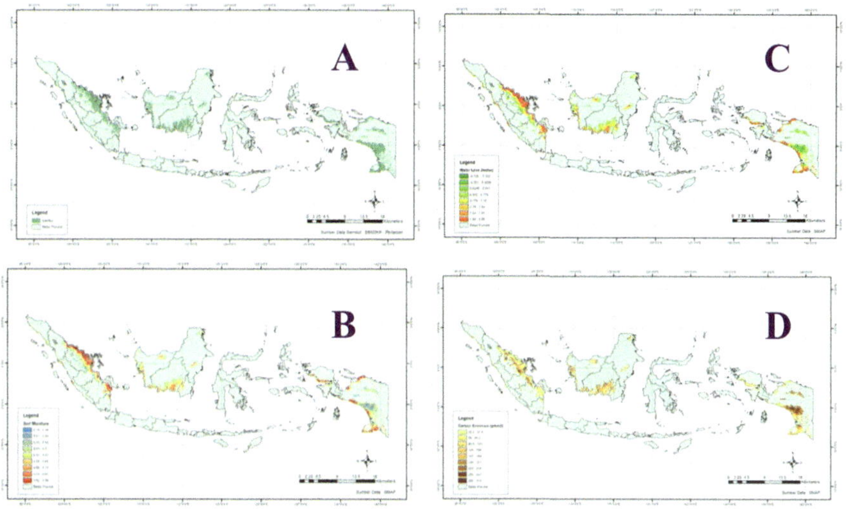

Fig. 1.33 NEE (CO_2 emissions) map of a peatland on the Indonesian Maritime Continent. (**a**): Peatland area; (**b**): soil moisture content, February 2018; (**c**): groundwater level measured in February 2018; and D: NEE (CO_2 emission), February 2018. Obtained from Osaki et al. (2023a)

NEE-GWL coefficiency value (Fig. 1.33). Hamada et al. (2016) and Tsuji et al. (2021) show detailed method.

In other method, peat subsidence monitoring is a good indicator of peatland degradation, which is correlated with CO_2 emissions from peatland. Peatland subsidence can be measured using ground-based leveling surveys, airborne LiDAR (light detection and ranging), and satellite-based interferometric synthetic aperture radar (InSAR).

1.6.2 "Omics" Analysis

At the other end of the scale is "omics" analysis (genomics, transcriptomics, proteomics, and ionomics) of the soil sphere, rhizosphere, and land surface, which is the use of advanced analytical techniques (e.g., gene sequencing, nuclear magnetic resonance, and liquid chromatography–mass spectrometry) to quantify the chemistry, structure, and genetics of soil components (Fig. 1.34). "Omics" analysis is very important to understand healthy/unhealthy land conditions; however, as "omics" analysis is applied in soil, it is difficult to bind remote sensing data from airspace. Plants respond to environmental factors, including soil vital conditions, which appear in plant form or leaf colors. This is defined here as "phenology," including seasonal and environmental changes in plants and symptoms of plant leaves affected by soil conditions, nutrients, water and carbon status, microorganism activity, and

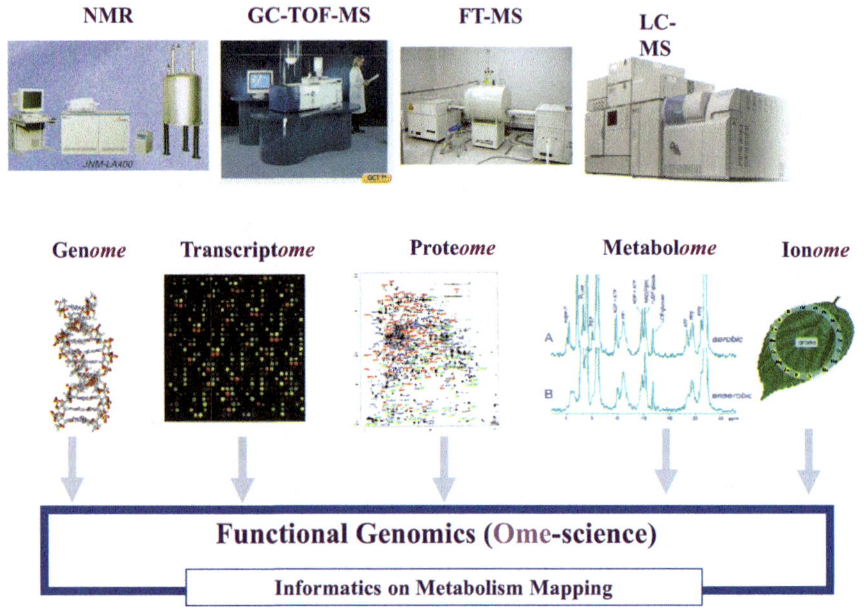

Fig. 1.34 "Omics" analysis includes genomics, transcriptomics, proteomics, and ionomics for soil sphere metabolism analysis. Obtained from Osaki et al. (2022c)

soil structure. Thus, we define "physiology" as metabolic processes that occur inside plants, microbes and soil spheres/rhizospheres, and "phenology" as plant responses caused by metabolic processes affected by environmental conditions.

Hyperspectral sensing technology in "informatics based on Equator observation systems (iEOS)" can detect the "phenology" of forests, contributing to the analysis of biodiversity, forest degradation, and healthy/unhealthy land.

1.7 Fieldlogy

"Fieldlogy" is proposed as a new transdisciplinary field of science based on the "triharmony/trilemma model" concept and "land surface management (LSM)" technology. The "fieldlogy" concept was first proposed by Osaki's team (Osaki et al. 2022c). Osaki (2022) argued that the Earth's system must be maintained and regenerated by the transdisciplinary forest function driven by NbS (nature-based solution) rather than ITbS (industrial technology-based solution). To this end, Osaki's team developed a new scientific discipline, known as "fieldlogy," focused on the pivotal transdisciplinary function of forests. Fieldlogy includes teleconnections among the land/atmosphere/ocean and rhizosphere/soilsphere dynamics through forest physiology/phenology (Osaki et al. 2022c). As "fieldlogy" is a new concept proposed recently, here, we introduce this "fieldlogy" concept, mainly citing Osaki's article (Osaki et al. 2022c) as follows.

1.7.1 Fieldlogy as Transdisciplinary Field Science

The "traditional" sciences are single disciplines, each distinct and freestanding (e.g., "physics," "chemistry," "biology"). Dual disciplines incorporate aspects of two subdisciplines (e.g., biochemistry), while mulch disciplines incorporate aspects of more than two subdisciplines (e.g., biogeochemistry). Transdisciplinary science fuses many other disciplines, accounting for feedback systems, self-organization, mutual interactions, cause-and-effect sequences, and multiplicity (Osaki et al. 2022c). Within fieldlogy, we have developed the concept of the trilemma state, in which Earth's environment, ecology, and resources overlap, forming subcategories that include ecological destruction, environmental pollution, and resource depletion; at the center is the human dimension (Osaki et al. 2022b). We argue that the solutions of the Global North, mainly ITbS, need to be replaced by those utilized by the Global South (Global Central recommended by Osaki's team), true NbS (tNbS), minimizing industrial renewable energy, especially that generated by solar panels, without biomass energy, to address environmental and climate change. Osaki's team called this tNbS approach "triharmony" (Osaki et al. 2022b). Forests mainly provide a wealth of functions that are critical to human survival, including controlling the flows of energy, carbon, water, and nutrients. Nevertheless, forest value is rarely quantified beyond the potential forest logging value. No attention is given to the value provided by carbon sequestration, biomass production (solar energy storage), atmospheric cooling, water storage in soil (run-off protection and flood protection), soil erosion protection, and a range of other functions (Osaki et al. 2022a). To this end, Osaki's team developed a method for quantifying the natural capital provided by forests in terms of the monetary value (Hojo et al. 2021; Osaki and Kasuga 2022).

1.7.2 Key Eco-Zones for Fieldlogy

While forests are found across the surface of the plant, fieldlogy is strongly focused on certain eco-zones. Most obviously, forest biomass is concentrated in the tropics, where climate conditions are primed for plant growth (e.g., abundant water, warm temperatures, fertile soils, and high soil carbon). Moreover, the tropics contain 76% of Earth's major river deltas (e.g., Nile Delta, Ganges Delta), which are the recipients of nutrient-rich sediments from eroding mountain ranges and volcanic deposits (Fig. 1.35) (Osaki et al. 2022c). Geographical conditions are also conducive to forest growth potential in the fertile crescents of Asia, Africa, and old Mesopotamia (Fig. 1.36) (Osaki et al. 2022c). Straddling the tropics and areas to the north, these

Fig. 1.35 The world's major river deltas. Obtained from Osaki et al. (2022c)

crescents are areas of nutrient-rich soils derived from river transport and volcanic ash and rocks.

Rainfall was recognized as one of the most important factors for classifying climate and its global quantitative description by Köppen (1918), who divided the tropical climate (A; classified by temperature in his first global climate map in 1884) into rainforest (Af) and savanna (Aw), according to vegetation (Beck et al. 2018) (Fig. 1.37). This separation is also reasonable because, essentially, temperature and rainfall represent the sensible (radiative) and latent heating processes, respectively, which maintain climate. The Earth's climate system includes latent heat input from the ground to the atmosphere through the cloud-precipitation cycle, which is also a part of the water cycle, connecting the land and oceans through the atmosphere (Osaki et al. 2021b).

Of particular note are the MCs, which fall within both the Asian Fertile Crescent and the tropics (Osaki et al. 2021a, 2022c). The MCs include the islands, peninsulas, and shallow seas of Southeast Asia, including Indonesia, the Malay Peninsula, Borneo Island, Thailand, Vietnam, and the Philippines (Fig. 1.38). This area is particularly noteworthy for the richest biodiversity in terms of peatland forests, mangroves, coral, sea grass, and fish, which support abundant biomass and bioproduction (Osaki et al. 2022c). Within the tropical zone, MCs are the most sensitive and highly uncertain element. The MCs are described primarily by Ramage (1968) and consist of the warm ocean region (called the tropical warm pool (TWP) or Indo-Pacific warm pool). As the MC area includes the TWP, extremely high

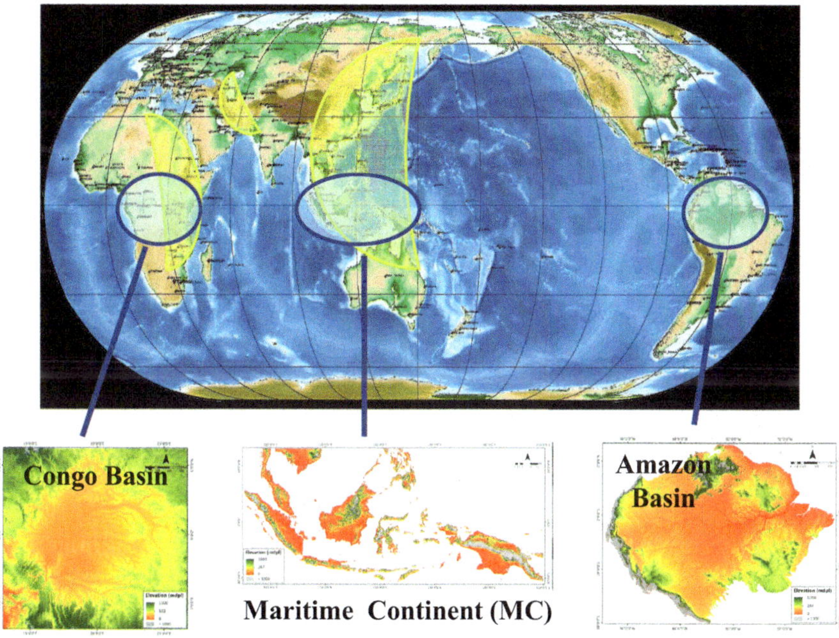

Fig. 1.36 Fertile Crescent and elevation model of wet–/peatland in Maritime Continent (MC) of Southeast Asia, Congo Basin and Amazon Basin. African Rift Valley crescent (left), Mesopotamian Crescent (center) and Asian Crescent (right). Modified from Osaki et al. (2022c)

Fig. 1.37 Tropical rainforest (Af) map based on the Köppen–Geiger climate classification. Obtained from Osaki et al. (2021b) (modified from Beck et al. 2018)

evaporation from the sea is observed, and this moisture is transported into coastal and near-coastal areas, where heavy precipitation occurs. As a result, plants grow well in these areas, and high evapotranspiration rates occur, contributing to the circulation of water in coastal and inland areas. Thus, a large water reservoir is established in the coastal and near-coastal zones of the MCs. In ecosystems with large water reservoirs, plant growth is vigorous, and litter decomposition is restricted

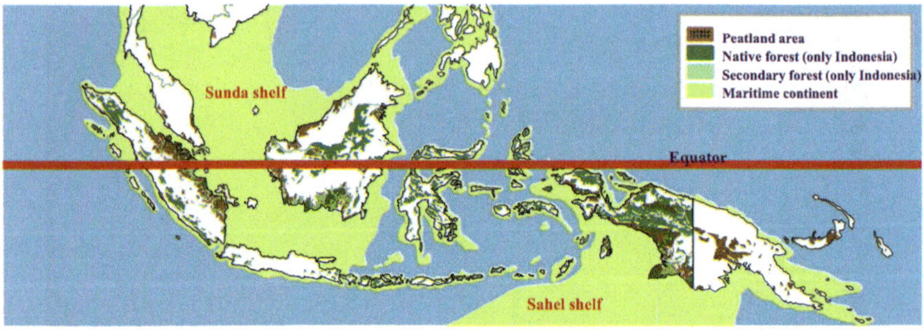

Fig. 1.38 Map of the Maritime Continents (MCs) and peatland distribution. Obtained from Osaki et al. (2022c)

by low microorganism activity owing to low oxygen concentrations (due to the low solubility of oxygen in water), creating ecosystems with high carbon reservoirs called tropical peatlands.

Osaki's team (Osaki et al. 2021b) described water cycle mechanisms from sea to inland. The mechanisms driving the water cycle were confirmed using heavy oxygen isotopes (oxygen-18), and it was found that half of the rainfall over the Amazon came from the transpiration of the forest itself (Salati et al. 1979). Additionally, van der Ent et al. (2010) computed that, on average, 40% of terrestrial precipitation originates from land evaporation and that 57% of all terrestrial evaporation returns as precipitation over land. Moisture evaporated from the Eurasian continent is responsible for 80% of China's water resources. In South America, the Río de la Plata basin depends on evaporation from the Amazon forest for 70% of its water resources. The main source of rainfall in the Congo Basin is moisture evaporated over East Africa, particularly the Great Lakes region. The Congo basin, in turn, is a major source of moisture for rainfall in the Sahel. Furthermore, it has been demonstrated that due to local orography, local moisture recycling is a key process near the Andes and the Tibetan Plateau.

The "biotic pump" theory, proposed by Makarieva and Gorshkov (2010), pertains to the importance of forests in the water cycle in determining the levels of rainfall. The "biotic pump" theory suggests that intense condensation, associated with high evaporation from natural forest cover, maintains regions of low atmospheric pressure on land, causing moist air to flow from ocean to land, which compensates for river runoff. The "biotic pump" governs continental moisture supplies through horizontal pressure gradients. High evaporation rates and extensive natural forests guarantee stable and high-throughput hydrological cycles, protecting continents against devastating floods, droughts, hurricanes, and tornadoes; therefore, sustaining natural forests is a sound strategy for promoting water security and stabilizing the climate. Makarieva and Gorshkov (2010) described how the water vapor produced by trees drives wind patterns: (1) winds cross the continent, taking moist air from Europe, through Siberia, and into Mongolia and China; (2) winds deliver rains that keep the very large rivers of eastern Siberia flowing; and (3) winds deliver the precipitation

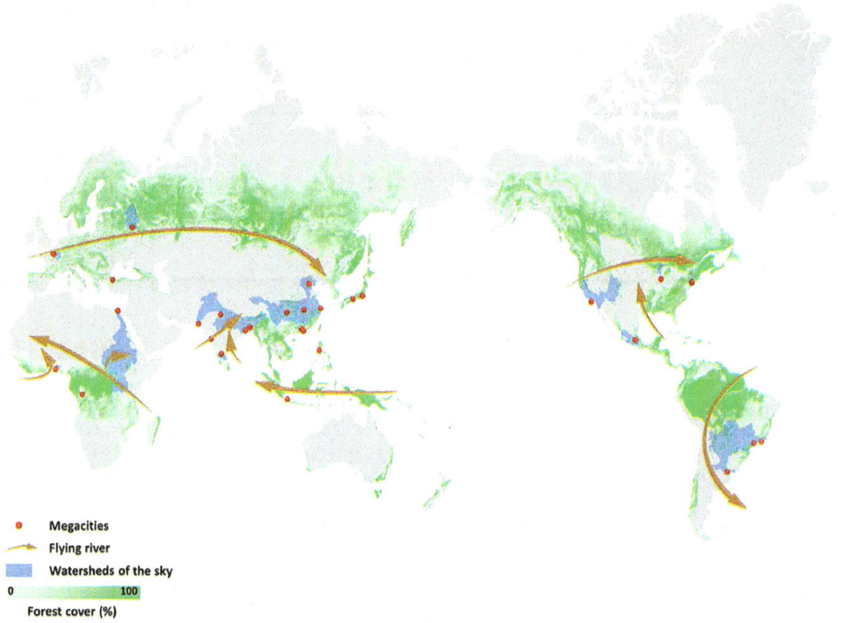

Fig. 1.39 Flying rivers and watersheds of the sky based on the biotic pump theory: forests generate not only rain but also wind, and the prevailing winds, which collect water vapor, are produced by forests and deliver rain to distant water basins. Modified from Pearce (2020) and Keys et al. (2018). (1) The Amazon's flying river provides 70% of the rain in southeastern South America. (2) Some of China's rain (80%) comes from the west, thanks to a trans-Siberian flying river. (3) A total of 19 out of 29 megacities depend on terrestrial evaporation (mainly from forests) for more than one third of their water supply (i.e., watersheds of the sky)

that waters China's northern plain, the breadbasket of the most populous nation on earth (Fig. 1.39). On a global scale, the Coriolis force is important for sea currency and wind currency, driving them clockwise in the Northern Hemisphere and counterclockwise in the Southern Hemisphere. It is therefore hypothesized that the flying river on the Indonesian Islands should be attracted to the Coriolis force of both hemispheres.

1.7.3 Fieldlogy Stewardship of Natural Capital

According to Osaki's team (Osaki et al. 2022c), last but certainly not least, fieldlogy places significant emphasis on the utilization and stewardship of natural capital, the assets provided by the natural world. Among the most important of these assets are cultivable land, minerals (nutrients), water, and biomass such as forests (carbon). Such assets are ultimately the basis of all wealth (including oil, gas, solar energy, carbon, water, land, etc.); however, while most economic systems have exploited

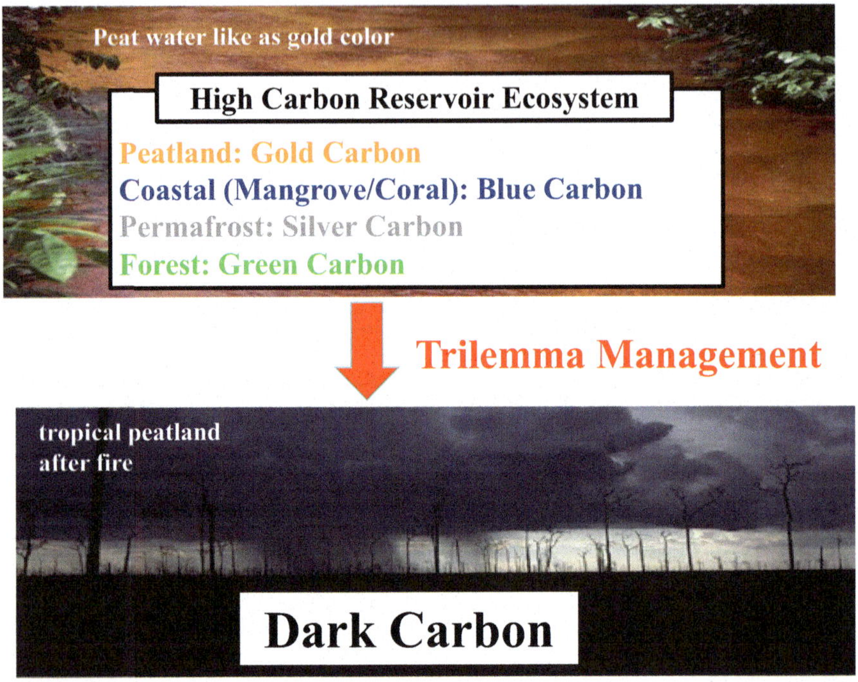

Fig. 1.40 High Carbon Reservoir Ecosystem clarified with colors. Presented paper in UNFCCC workshop on "Technical and scientific aspects of ecosystems with high-carbon reservoirs not covered by other agenda items under the Convention", 24 to 25 October 2013, Bonn, Germany

these assets for growth and development, Osaki's team advocates an approach that can achieve long-term sustainability and Earth's regeneration. One facet of this approach is the differentiation and quantification of carbon stores. The most well-known carbon reservoir (clarified here with colors) is fossil fuels (black carbon); however, other sources and stores include blue carbon (stored in seas and oceans), gold carbon (stored in peat), silver carbon (stored in permafrost), and green carbon (stored in plants and forests) (Fig. 1.40) (Osaki et al. 2021a). Among those on land, trees have the highest potential for carbon removal from the atmosphere. Therefore, the conservation, rehabilitation, and restoration of forests promote green carbon and offer the potential for a carbon-negative society. On the contrary, if we fail to conserve and manage the colored carbon ecosystem, we encounter the Trilemma ecosystem, called "Dark Carbon" society (carbon-positive society through ecosystem corruption).

Osaki and colleagues have proposed the "land surface management (LSM)" approach to cultivation (Osaki et al. 2021a, 2022d). In developing this approach, they studied natural systems in which plants thrive despite challenging conditions (e.g., peat/wetland, dryland, saline or acidic soils, extremely poor nutrient soil). Based on their findings, they developed new approaches that mimic natural processes to allow innovative cultivation.

Common threads among all of these approaches are the harvesting of solar energy to drive cycles of water and nutrients, the use of organic materials (carbon), and the protection of biodiversity. Under the prevailing global economic model, value chains trace the route of a product from its natural source to market (e.g., forest to wood product) without assigning an economic value to the reservoir of the natural resource. Moreover, the end consumer has no knowledge of, or link to, the ecosystems from which the product ultimately stemmed (Osaki et al. 2022b). At its heart, fieldlogy asks us as a society to reconstruct these value chains. Placing emphasis on the value of natural resources, we have an opportunity to foster a future that can achieve long-term sustainability and Earth's regeneration.

1.7.4 Four-Quadrant Model on Fieldlogy

Osaki's team splits the different approaches used to address environmental change into four quarters, or quadrants, formed by two intersecting straight lines (or axes): the envelopment-development and the global-regional (Fig. 1.41) (Osaki et al. 2022b). Quadrants I and II cover the global and regional application of development mechanisms, respectively. Osaki's team highlights TechnoGarden in Germany as an example of quadrant II. TechnoGarden is a modified three-farm system where crops for feed, crops and pasture for animal grazing, and biomass from plants and forests are supported by renewable energy facilities (e.g., biomass boilers and energy, solar panels, and wind power). Quadrants III and IV cover the global and regional application of "envelopment mechanisms." The Gaia concept is an example of quadrant III, and the Forest CoCycle concept (where materials, energy cycles, and

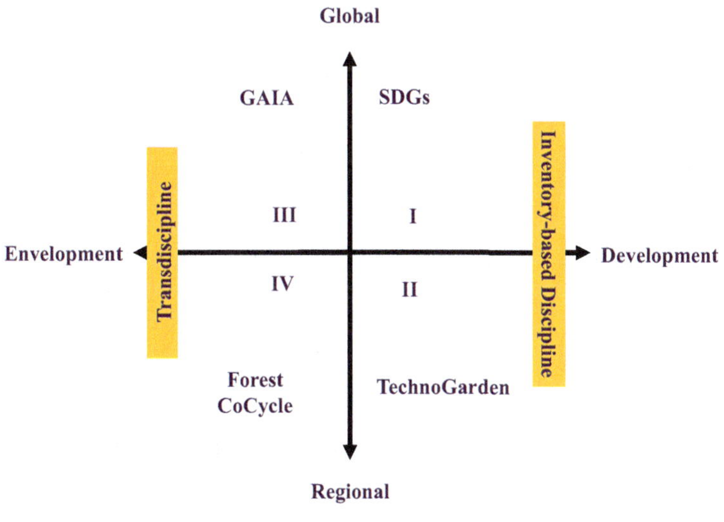

Fig. 1.41 Four-quadrant model. Obtained from Osaki et al. (2022b)

cyclic forest management are based on coexistence, cooperation, cohabitation, and cyclicity) is an example of quadrant IV. Among these four quadrant (quarter) analyses, tropical peatlands are an important key ecosystem in quadrant IV.

1.8 Conclusion

As tropical peatlands are complicated and integrated ecosystems, specific concepts and management systems are required for these peatlands. Inventory-based disciplines, such as single, dual, and mulch disciplines, are necessary at the primary stage of study and provide various information on tropical peatlands. However, tropical peatlands are composed of a wholly water (oxygen)–carbon–nutrient ecosystem, and it is impossible to separate this ecosystem into elements (inventory). Instead of inventory information depth, we propose a transdisciplinary approach to tropical peatland study, which is similar to separating "informatics" (transdisciplinary science) from "information" (inventory-based disciplines).

If we focus on only a single element (inventory-based disciplines) such as water (oxygen), carbon, or nutrients in tropical peatlands, we now understand that tropical peatlands have various "paradoxes" and, on the contrary, encourage us to understand them as wholly water (oxygen)–carbon–nutrients ecosystems (transdisciplinary ecosystems), for which we introduce a new model called the triharmony/trilemma concept. If we apply the ITbS and NbS systems to understand and manage tropical peatlands, they promote falling into the trilemma ecosystem and creating the triharmony ecosystem, respectively.

Tropical peatlands are characterized as high water–carbon storage ecosystems. However, water–carbon balance itself is one kind of paradox because high water conditions are normally associated with poor nutrient conditions (low oxygen and nutrient leaching by the desorption mechanisms of organic matter at low pH). Then, the carbon assimilation of plants (mainly trees) decreases, causing low productivity of carbon compound material because carbon accumulation is too low. However, a large amount of carbon accumulates in tropical peatlands. Another case of the water–carbon paradox is arable land, which normally has moderate (or lower) amounts of water, and high carbon assimilation and accumulation (stock) in soil are expected; however, the carbon stock is limited in arable land. Therefore, we need two criteria (indexes) to describe carbon sequestration (accumulation, storage) in soil, as shown in Fig. 1.42. One criterion is the water axis, and the other is the nutrient axis in the global carbon sequestration model in Fig. 1.42. In this model, water includes various functions in ecosystems: oxygen (causing different Eh values) and nutrients (causing adsorption and desorption). We already know that the water-nutrient model is a paradox when we discuss carbon sequestration on land. From this water–nutrient paradigm model, we find a very important sphere called the land surface sphere (or rhizosphere on the land surface), not only the soil sphere. The land surface sphere (or rhizosphere) is the interface of plant–microbe interactions, nutrient cycling, and

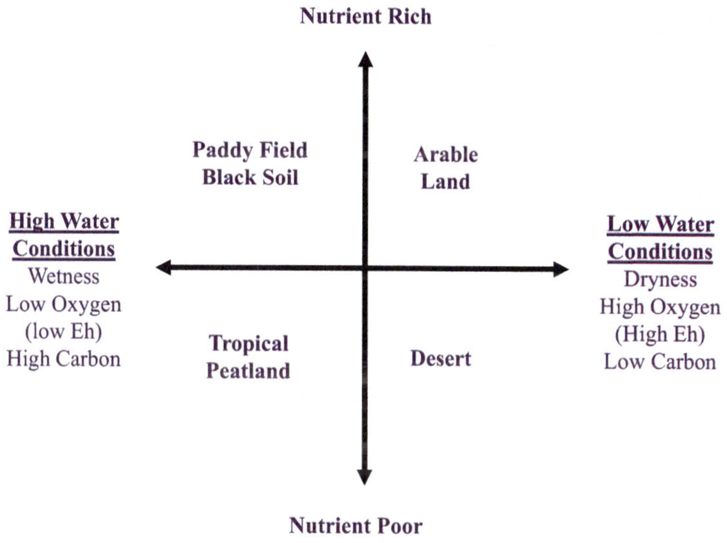

Fig. 1.42 Global paradoxical ecosystem categories

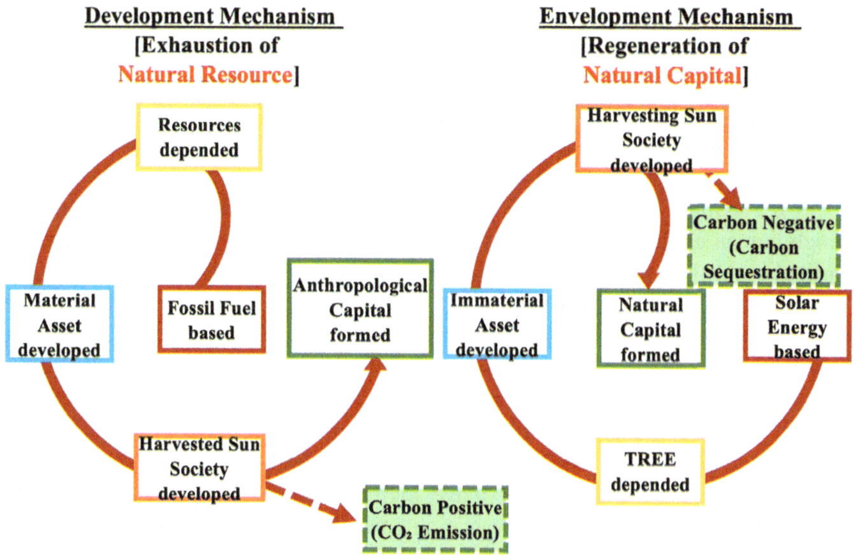

Fig. 1.43 Development mechanism (left) and envelopment mechanism (right) of sun harvesting societies. Obtained from Osaki et al. (2022b)

gas exchange. Thus, we identify "land surface" as the "skin" of the land (soil sphere) or earth system.

According to Osaki et al. (2022c), a critical aspect of triharmony is the conversion of solar energy to forms that can sustain life, called a "harvesting sun" society. In contrast, a "harvested sun" society is driven by development mechanisms through which the human exploitation and consumption of natural resources (fossil fuels in particular) are carbon positive and lead to increased climate change (Fig. 1.43). However, today, many initiatives are required to focus on carbon-neutral goals, whereby natural resources are conserved and humans can mitigate or adapt to climate change.

Finally, "fieldlogy" is proposed as a new transdisciplinary field science based on the "triharmony/trilemma model" concept and "land surface management (LSM)" technology (see Sect. 1.7 "Fieldlogy," and Osaki et al. 2022c). The "fieldlogy" concept was first proposed by Osaki's team (Osaki et al. 2022c). Osaki (2022) argued that the Earth's system must be maintained and regenerated by the transdisciplinary forest function based on NbS rather than ITbS. To this end, Osaki's team (Osaki et al. 2022c) developed a new scientific discipline, known as "Fieldlogy," focused on the pivotal transdisciplinary function of forests. "Fieldlogy" also includes teleconnections among the land/atmosphere/ocean and rhizosphere/soilsphere dynamics through forest physiology/phenology. "Fieldlogy" in the tropics covers not only a single ecosystem but also a mulch ecosystem, peatlands, wetlands, mangroves, and coral, called the Maritime Continents in Southeast Asia, or Big Satoyama surrounding the Himalayan Mountain-Tibetan Plato, involving the Monsson, large rivers and big river mouths, and flying river, which cover the water dynamics in Asia.

Finally, the term "envelopment mechanisms" was proposed by Osaki's team (Osaki et al. 2022b) as a new concept of "Earth regeneration," especially in the equator/tropical (including subtropics) zone, guaranteed by iMRV (informatics) and LSM (NbS technology). "Envelopment mechanisms" are basically operated by "sun energy" as much as possible (called the "harvesting sun society") (Osaki et al. 2022b). A "harvested sun" society is driven by "development mechanisms" in which human exploitation and the consumption of natural resources (fossil fuels in particular) result in positive carbon and lead to enhanced climate change (Fig. 1.43). Instead, today, many initiatives focus on carbon neutral goals, whereby natural resources are conserved and humans can mitigate or adapt to climate change. However, Osaki's team suggests that our ultimate goal should be to become a 'harvesting sun' society, which follows "envelopment mechanisms" characterized by the cyclic use of natural resources, a reduction in climate change impact, and a carbon-negative society. Such "envelopment mechanisms" work well mainly in tropical zones, which continue to have high natural capital potential even after environmental degradation or devastation (primarily owing to high temperatures and abundant moisture). In contrast, temperate and boreal zones, where natural capital is lost to development activities, cannot be recovered due to their relatively low temperatures. It is for this reason that technological (or development) mechanisms (ITbS) were first applied in these regions and have since been slowly adopted globally, despite the more suitable NbS options afforded to the largely tropical Global South (called Global Central here) (Osaki et al. 2022b).

Acknowledgments The ideas for an innovative "tropical peatlands eco-evaluation" were mainly obtained from projects such as (1) the Japan Society for Promotion of Science (JSPS) Core University Program on "Environmental Conservation and Land Use Management of Wetland Ecosystem in Southeast Asia" between Hokkaido University, Japan, and the Research Center for Biology (LIPI), Indonesia, from April 1997 to March 2007; (2) the Science and Technology Research Partnership for Sustainable Development (SATREPS) Project on "Wildfire and Carbon Management in Peat-Forest in Indonesia" between Hokkaido University, Japan, and the National Standardization Agency (BSN), Indonesia, founded by the Japan Science and Technology Agency (JST) and Japan International Cooperation Agency (JICA) from September 2009 to March 2014; (3) the IJ-REDD project founded by JICA in 2014; and (4) the JICA-JPS-BRG Program of the Japan International Cooperation Agency (JICA, Japan), the Japan Peatland Society (JPS, Japan) and the Peatland Restoration Agency (BRG, Indonesia) from October 2017 to March 2018. The reports, proceedings, and guidebooks for these projects are available on the website of the Japan Peatland Society (JPS) as follows:

https://jps.sakura.ne.jp/jspsproc/jspsproc.html.

Through the above projects, we published two books on **"Tropical Peatland"** (Springer), which contribute to this chapter, and this third book, "Tropical Peatland Eco-evaluation," as follows:

1. "Tropical Peatland Ecosystems" (2015) eds. by M. Osaki and N. Tsuji, Nature Springer Singapore Pte Ltd., ISBN-10: 443155680X, ISBN-13: 978-4431556800
2. "Tropical Peatland Eco-management" (2021) eds. by Osaki M, Tsuji N, Foead N, Rieley J, Nature Springer Singapore Pte Ltd., ISBN-10: 9813346531, ISBN-13: 978-9813346536.

The key concepts of natural capital/assets, transdisciplinary science, triharmony/trilemma, and envelopment/development mechanisms were consolidated based on the analysis and evaluation of the following articles of research outreach and research features:

1. Osaki M (2022) Natural-capital-based societies in the tropics: Harnessing forest function to combat climate change. Research Outreach, January 21. https://researchoutreach.org/articles/natural-capital-based-societies-tropics-harnessing-forest-function-combat-climate-change/. https://doi.org/10.32907/RO-128-2187568700
2. Osaki M, Kasuga T (2022) Nature as an asset: Natural capital in 'FutureCity' Shimokawa. Research Outreach, April 5. https://researchoutreach.org/articles/nature-asset-natural-capital-futurecity-shimokawa/. https://doi.org/10.32907/RO-129-2508888606
3. Osaki M, Takahashi H, Kato T, Tsuji N, Putri NA (2022) Transdisciplinary forest science and the rise of nature-based Solutions, Research Features, September 23. https://researchfeatures.com/transdisciplinary-forest-science-nature-based-solutions-nbs/. https://doi.org/10.26904/RF-143-3252558210
4. Osaki M, Tsuji, N, Kato T, Putri NA (2022) Transdisciplinary human-nature science towards triharmony, Research Features, September 23. https://researchfeatures.com/transdisciplinary-human-nature-science-towards-triharmony/. https://doi.org/10.26904/RF-143-3252534021
5. Osaki, M, Kawakita N, Kato T, Tsuji N, Sisva S (2022) Transdisciplinary field science based on Land-Surface Management (LSM), Research Outreach, October 17. https://researchoutreach.org/articles/transdisciplinary-field-science- based-land-surface-management-lsm/. https://doi.org/10.32907/RO-132-3403781504
6. Osaki M, Kato T, Tsuji N, Putri NA, Sisva S (2022) Fieldlogy Science: A new transdisciplinary science for Earth's regeneration, Research Outreach, December 5. https://researchoutreach.org/articles/fieldlogy-science-new-transdisciplinary-science-earths-regeneration/. https://doi.org/10.32907/RO-133-3649942836
7. Kato T, Asep YA, Silsigia S, Tsuji N, Osaki M (2023) Triharmony/Trilemma of Water Assets in tropical peatland, Research Outreach, February 10. https://researchoutreach.org/articles/

triharmony-trilemma-water-assets-tropical-peatland/. https://doi.org/10.32907/RO-134-3957184811
8. Osaki M, Kato T, Takahashi H, Sulaiman A, Tsuji N (2023) Triharmony/Trilemma of Carbon Assets in tropical peatland, Research Outreach, February 10. https://researchoutreach.org/articles/triharmony-trilemma-carbon-assets-tropical-peatland/. https://doi.org/10.32907/RO-134-3991159188
9. Osaki M, Kato T, Turjaman M, Asep YA, Tsuji N (2023) Triharmony/Trilemma of Nutrients Assets in tropical peatland, Research Outreach, February 10. https://researchoutreach.org/articles/triharmony-trilemma-nutrients-assets-tropical-peatland/. https://doi.org/10.32907/RO-134-3989821456

References

AJASS-Association of Japanese Agricultural Scientific Societies (1975) Rice in Asia. University of Tokyo Press, Tokyo, pp 54–65

Anderson JAR (1964) The structure and development of the peat swamps of Sarawak and Brunei. J Trop Geogr 18:7–16

Beck HE, Zimmermann NE, McVicar TR, Vergopolan N, Berg A, Wood E F (2018) Present and future Köppen-Geiger climate classification maps at 1-km resolution. Nature Scientific Data. Köppen–Geiger climate classification map for Tropical, rainforest (Af), 5. https://doi.org/10.1038/sdata.2018.214

Couwenberg J (2009) Methane emissions from peat soils (organic soils, histosols) -Facts, MRV-ability, emission factors-, Wetlands International. http://www.imcg.net/media/download_gallery/climate/couwenberg_2009b.pdf

Ehara H, Kakuda K, Miyazaki A, Naito H, Nakamura S, Nitta Y, Okazaki M, Sasaki Y, Toyota K, Watanabe A, Watanabe M, Yamamoto Y, Goto Y, Kimura SD (2021) Sago palm in peatland (chapter 17). In: Osaki M, Tsuji N, Foead N, Rieley J (eds) Tropical peatland eco-management. Nature Springer Singapore Pte Ltd.. ISBN-10:9813346531, ISBN-13:978 9813346536

Fukuoka M (1975) The natural way of farming: The theory and practice of green philosophy. www.rivendellvillage.org/Natural-Way-Of-Farming-Masanobu-Fukuoka-Green-Philosophy.pdf

Furukawa H, Supiandi S (1986) Agricultural landscape in the lower Batang Hari, Sumatra part one: stratigraphy and geomorphology of coastal swampy lands. Kyoto Univ Res Inf Reposit 23:3–37. (In Japanese)

Hamada Y, Tsuji N, Kojima Y, Qirom M, Sulaiman A, Firmanto, Jagau Y, Irawan D, Naito R, Sari E. (2016) Guidebook for estimating carbon emissions from tropical peatlands in Indonesia. Indonesia-Japan Project for Development of REDD+ Implementation Mechanism. https://www.jica.go.jp/project/english/indonesia/015/materials/c8h0vm0000c2fxdl-att/materials_01.pdf

Haraguchi A, Yabe K (2002) Vertical and horizontal distribution of redox potential of soil in wetland forests in Lahei, Central Kalimantan, Indonesia. Tropics 11:91–100

Haraguchi A, Shimada S, Takahashi H (2000) Distribution of peat and its chemical properties around Lahei in the catchments of the Mangkutup River, Central Kalimantan. Tropics 10:265–727

Hartoto DI, Iwakuma T (2002) Implementation of meta-population concept for conservation of biodiversity through development of inland fishery reserves in Jambi Province. In annual report on "environmental conservation and land use Management of Wetland Ecosystem in Southeast Asia", during April 2001-march 2002, Core University program, March 2002

Hirano T, Segah H, Kusin K, Limin S, Takahashi H, Osaki M (2012) Effects of disturbances on the carbon balance of tropical peat swamp forests. Glob Chang Biol 18:3410–3422. https://doi.org/10.1111/j.1365-2486.2012.02793

Hirose K, Osaki M, Takeda T, Kashimura O, Ohki T, Segah H, Gao Y, Evri M (2016) Contribution of hyperspectral applications to tropical peatland ecosystem monitoring. In: Osaki M, Tsuji N (eds) Tropical peatland ecosystems. Springer, pp 421–431

Hojo A, Tsuji N, Kasuga T, Osaki M (2021) Natural capital evaluation in the FutureCity of Shimokawa, northern Japan, based on forest economics. Environ Monit Assess 193:793. https://doi.org/10.1007/s10661-021-09434-y

Jayadi S, Hata H (2009) Rehabilitation of fired peat swamp forest using livestock. In Proceedings of International Workshop on Wild Fire and Carbon Management in Peat-Forest in Indonesia, March 5th and 6th, 2009, Jakarta, Indonesia

Kato T, Osaki M, Tsuji N, Silsigia S (2021) Principles of eco-management in a large-scale ecosystem of tropical peatland (Chapter 2). In: Osaki M, Tsuji N, Foead N, Rieley J (eds) Tropical peatland eco-management. Nature Springer Singapore Pte Ltd.. ISBN-10:9813346531, ISBN-13:978 9813346536

Kato T, Asep YA, Sisva S, Tsuji N, Osaki M (2023) Triharmony/trilemma of water assets in tropical peatland, research outreach, February 10, https://researchoutreach.org/articles/triharmony-trilemma-water-assets-tropical-peatland/. https://doi.org/10.32907/RO-134-3957184811

Keys PW, Wang-Erlandsson L, Gordon LJ (2018) Megacity precipitation sheds reveal tele-connected water security challenges. PLoS One 13(3):e0194311. https://doi.org/10.1371/journal.pone.0194311

Kim H, Yang S, Rao SR, Narayanan S, Kapustin EA, Furukawa H, Umans AS, Yaghi OM, Wang EN (2017) Water harvesting from air with metalorganic frameworks powered by natural sunlight. Science 356(6336):430–434. https://doi.org/10.1126/science.aam8743

Köppen W (1918) Klassifikation der Klimate nach Temperatur, Niederschlag und Jahresablauf (classification of climates according to temperature, precipitation and seasonal cycle). Petermanns Geogr Mitt 64:193–203. http://koeppen-geiger.vu-wien.ac.at/pdf/Koppen_1918.pdf

Lai DYF (2009) Methane dynamics in northern peatlands: a review. Pedosphere 19:409–421

Lina SB, Okazaki M, Kimura SD, Yano Y, Yonebayashi K, Igura M, Quevedo MA, Loreto AB (2008) Nitrogen uptake efficiency by sago palm (Metroxylon sagu) as estimated by isotopic method. Sago Palm 16(1):21–25

Makarieva A, Gorshkov V (2010) The biotic pump: condensation, atmospheric dynamics and climate. Int J Water 5(4):365–385. https://doi.org/10.1504/IJW.2010.038729

Mezbahuddin M, Grant R, Hirano T (2014) Modelling effects of seasonal variation in water table depth on net ecosystem CO2 exchange of a tropical peatland. Biogeosciences 11:577–599

Miyamoto K, Kohyama TS, Rahajoe JS, Mirmanto E, Simbolon H (2016) Forest structure and productivity of tropical heath and peatland forests. In: Tropical peatland ecosystems. Springer, Tokyo, pp 151–166

Osaki M (2014) Concept of Satoyama system on resilience and vulnerability for bioproduction ecosystem in Asia. Glob Environ Res 18:113–123

Osaki M (2022) Natural-capital-based societies in the tropics: harnessing forest function to combat climate change. Research outreach, January 21, https://researchoutreach.org/articles/natural-capital-based-societies-tropics-harnessing-forest-function-combat-climate-change/. https://doi.org/10.32907/RO-128-2187568700

Osaki M, Kasuga T (2022) Nature as an asset: natural capital in 'FutureCity' Shimokawa. Research outreach, April 5, https://researchoutreach.org/articles/nature-asset-natural-capital-futurecity-shimokawa/. https://doi.org/10.32907/RO-129-2508888606

Osaki M, Inoue I, Haraguchi A, Yabe K, Limin SH, Jaya S, Dohong S, Jaya A, Umbing RA, Sabiham S, Sudadi U, Suwarno, Kalsim DK (1999) A brief information of the visited sites. In Annual Report for April 1998–March 1999, Core University Program on Environmental Conservation and Land Use Management of Wetland Ecosystem in Southeast Asia, p 48–75

Osaki M, Kato T, Kohyama T, Takahashi H, Haraguchi A, Yabe K, Tsuji N, Shiodera S, Rahajoe JS, Atikah TD, Oide A, Matsui K, Wetadewi RI, Silsigia S (2021a) Basic information about tropical peatland ecosystems (chapter 1). In: Osaki M, Tsuji N, Foead N, Rieley J (eds) Tropical peatland eco-management. Nature Springer Singapore Pte Ltd., pp 3–62, ISBN-10: 9813346531, ISBN-13:978 9813346536

Osaki M, Tsuji N, Kato T, Yamanaka MD, Sulaiman A, Silsigia S, Wetadewi RI (2021b) Natural capital-based societies in the tropics (chapter 6). In: Osaki M, Tsuji N, Foead N, Rieley J (eds) Tropical peatland eco-management. Nature Springer Singapore Pte Ltd., pp 3–62. ISBN-10: 9813346531, ISBN-13:978 9813346536

Osaki M, Takahashi H, Kato T, Tsuji N, Putri NA (2022a) Transdisciplinary forest science and the rise of nature-based solutions, research features, September 23, https://researchfeatures.com/transdisciplinary-forest-science-nature-based-solutions-nbs/. https://doi.org/10.26904/RF-143-3252558210

Osaki M, Tsuji N, Kato T, Putri NA (2022b) Transdisciplinary human-nature science towards triharmony, research features, September 23, https://researchfeatures.com/transdisciplinary-human-nature-science-towards-triharmony/. https://doi.org/10.26904/RF-143-3252534021

Osaki M, Kato T, Tsuji N, Putri NA, Sisva S (2022c) Fieldlogy science: a new transdisciplinary science for Earth's regeneration, research outreach, December 5, https://researchoutreach.org/articles/fieldlogy-science-new-transdisciplinary-science-earths-regeneration/. https://doi.org/10.32907/RO-133-3649942836

Osaki M, Kawakita N, Kato T, Tsuji N, Sisva S (2022d) Transdisciplinary field science based on land-surface management (LSM), research outreach, October 17, https://researchoutreach.org/articles/transdisciplinary-field-science- based-land-surface-management-lsm/. https://doi.org/10.32907/RO-132-3403781504

Osaki M, Kato T, Takahashi H, Sulaiman A, Tsuji N (2023a) Triharmony/trilemma of carbon assets in tropical peatland, research outreach, February 10, https://researchoutreach.org/articles/triharmony-trilemma-carbon-assets-tropical-peatland/. https://doi.org/10.32907/RO-134-3991159188

Osaki M, Kato T, Turjaman M, Asep YA, Tsuji N (2023b) Triharmony/trilemma of nutrients assets in tropical peatland, research outreach, February 10, https://researchoutreach.org/articles/triharmony-trilemma-nutrients-assets-tropical-peatland/. https://doi.org/10.32907/RO-134-3989821456

Pearce F (2020) A controversial Russian theory claims forests don't just make rain—they make wind. Science 368:1302–1305. https://doi.org/10.1126/science.abd3856

Ramage CS (1968) Role of a tropical "maritime continent" in the atmospheric circulation. Mon Weather Rev 96:365–370. https://doi.org/10.1175/1520-0493(1968)096<0365:ROATMC>2.0.CO;2

Salati E, Dall'Olio A, Matsui E, Gat JR (1979) Recycling of water in the Amazon Basin: an isotopic study. Water Resour Res 15(5):1001–1000. https://doi.org/10.1029/WR015i005p01250

Sulaiman A, Osaki M, Takahashi H, Yamanaka YD, Susanto RD, Shimada S, Kimura K, Hirano T, Wetadewi RI, Sisva S, Kato T, Kozan O, Kubo H, Awaluddin A, Tsuji N (2023) Peatland groundwater level in the Indonesian maritime continent as an alert for El Niño and moderate positive Indian Ocean dipole events. Sci Rep 13:939. https://doi.org/10.1038/s41598-023-27393-x

Tampubolon AP, Turjaman M, Osaki M (2021) Sago palm practice as natural AeroHydro culture (Chapter 12). In: Osaki M, Tsuji N, Foead N, Rieley J (eds) Tropical peatland eco-management. Nature Springer Singapore Pte Ltd.. ISBN-10:9813346531, ISBN-13:978 9813346536

Tsuji N, Kato T, Osaki M, Sulaiman A, Ajie GS, Kimura K, Shigenaga Y, Hirose K, Silsigia S (2021) Evaluation of eco-management of tropical peatlands (Chapter 5). In: Osaki M, Tsuji N, Foead N, Rieley J (eds) Tropical peatland eco-management. Nature Springer Singapore Pte Ltd., pp 163–196, ISBN-10:9813346531 ISBN-13:978 9813346536

van der Ent RJ, Savenije HHG, Schaefli B, Steele-Dunne SC (2010) Origin and fate of atmospheric moisture over continents. Water Resour Res 46(9). https://doi.org/10.1029/2010WR009127

Watanabe T, Osaki M, Tadano T (1997) Aluminum-induced growth stimulation in relation to calcium, magnesium, and silicate nutrition in *Melastoma malabathicum* L. Soil Sci Plant Nutr 43:827–837

Watanabe T, Osaki M, Yoshihara T, Tadano T (1998) Distribution and chemical species of aluminum in the Al accumulator plant, *Melastoma malabathicum* L. Plant Soil 201:165–173
Watanabe T, Jansen S, Osaki M (2005a) The beneficial effect of aluminum and the role of citrate in Al accumulation in *Melastoma malabathricum*. New Phytol 165:773–780
Watanabe T, Misawa S, Osaki M (2005b) Aluminum accumulation in the roots of *Melastoma malabthicum*, an aluminum-accumulating plant. Can J Bot 83:1518–1522
Watanabe T, Jansen S, Osaki M (2006) Al–Fe interactions and growth enhancement in *Melastoma malabathricum* and *Miscanthus sinensis* dominating acid sulphate soils. Plant Cell Environ 29: 2124–2132
Watanabe T, Misawa S, Hiradate S, Osaki M (2008) Characterization of root mucilage from *Melastoma malabathricum*, with emphasis on its roles in aluminum accumulation. New Phytol 178:581–589
Watanabe T, Osaki M (2002) Mechanisms of adaptation to high aluminum condition in native plant speciesgrowing in acid soils: a review. Commun. Soil Sci. Plant Anal. 33:1247–1260. https://doi.org/10.1081/CSS-120003885.
Wetadewi RI, Osaki M, Turjaman M, Antonius S, Goenadi DH, Nursyamsi D, Maswar, Surayah L, Kato T (2021) Principles of AeroHydro culture (chapter 7). In: I Osaki M, Tsuji N, Foead N, Rieley J (eds) Tropical peatland eco-management. Nature Springer Singapore Pte Ltd., pp 249–283, ISBN-10:9813346531, ISBN-13:978 9813346536

Part II
Eco-Evaluation of Assets in Tropical Peatland

Chapter 2
Integrated Eco-Evaluation Practices of Industrial Forest Plantation in Peatland of West Kalimantan, Indonesia

Tsuyoshi Kato, Niken Andika Putri, Sisva Silsigia, Asep Andi Yusup, Sy. Abdul Jabar, Natalia Desi Piana Soeka, and Mitsuru Osaki

Abstract This chapter focuses on peatland management and conservation on a business basis. Peatlands play an important role in Earth's ecosystems, as peatlands are ecosystems that can preserve Earth's biodiversity, improve water quality, and mitigate climate change. However, if these areas are not properly managed, they can cause serious environmental problems. Poorly managed peatlands can release large amounts of greenhouse gases and increase the risk of peat fires. Therefore, understanding and properly managing peatlands and preventing fires is critical. This chapter discusses the impacts of peatland management on ecosystems and the best practices for managing them. It also details how to assess and monitor peatland conditions. Forest plantation projects on peatlands are important to produce wood sustainably, restore their ecosystems, and enhance their capacity to absorb carbon dioxide. However, such activities should be properly managed and monitored, paying particular attention to the following:

Soil conservation: Peatlands have unique soil conditions, indicating that forest plantation activities should consider soil characteristics. For example, peat soils are acidic and have a high moisture content, so selecting plants adapted to these conditions is essential.

Monitoring: Forest plantation projects require regular monitoring. Thus, it is essential to monitor the growth of newly planted plants and adjust management measures as necessary. The groundwater level must be constantly monitored to ensure that planted trees grow properly and to prevent peat fires. Moreover, the managed area's greenhouse gas emissions and biodiversity should also be monitored.

T. Kato (✉) · N. A. Putri
Sumitomo Forestry Co., Ltd., Tokyo, Japan
e-mail: KATO_tsuyoshi5@star.sfc.co.jp

S. Silsigia · A. A. Yusup · S. A. Jabar · N. D. P. Soeka
PT. Wana Subur Lestari/PT. Mayangkara Tanaman Industri, Jakarta, Indonesia

M. Osaki
Hokkaido University, Sapporo, Japan

M. Osaki et al. (eds.), *Tropical Peatland Eco-evaluation*,
https://doi.org/10.1007/978-981-99-6790-2_2

Long-term perspective: Forest plantation projects on peatlands should aim for short-term results and long-term ecological health. This requires a sustainable management plan and regular monitoring.

These elements will serve as basic guidelines to ensure that forest plantation projects on peatlands contribute to timber production, ecosystem conservation, and climate change action.

Keywords Peat fire · Drainage-based water management (drainage-based WM) · Stock-based water management (stock-based WM) · Groundwater level (GWL) · Biodiversity

2.1 Introduction

When conducting tree or palm plantation activities on peatlands, continuous monitoring is crucial to prevent peatland degradation and ensure sustainable and stable harvests. Once the peatland is drained or dried, a moderate oxygen supply causes the deposited organic matter to be decomposed by microorganisms (Könönen et al. 2018; Arai et al. 2014). This process, called oxidation, produces carbon dioxide (CO_2) as microorganisms feed on organic matter and gain energy. Although the groundwater level in peatlands has been intentionally manipulated, lowering the groundwater level significantly impacts the initiation and progression of peat fires (Putra et al. 2018; Silviana et al. 2020). As groundwater levels drop, peatlands dry out, and microbial decomposition is accelerated. This dry peat is highly susceptible to fire and can spread quickly when ignited. Peat fires are hazardous and can spread to the surface and underground since the underground peat layer is dry and highly flammable. Once a fire begins, it can spread deep underground and be difficult to extinguish. In addition, fires can last for long periods and release carbon dioxide (Graham et al. 2022; Qin et al. 2022). When groundwater levels are high, peatlands are wet, and the fire risk is significantly reduced. The high moisture content prevents peat from drying out and slows fire progression (Prat-Guitart et al. 2016; Palamba et al. 2020). Proper groundwater level management is, therefore, essential for peat fire prevention. If the groundwater level in a peatland is too high, it can adversely affect the growth and survival of plantation trees, as tree roots need oxygen. Thus, if the groundwater level is too high, oxygen is eliminated from the soil, potentially causing the roots to become oxygen deprived. This can inhibit tree growth and ultimately lead to tree death. Additionally, when the groundwater level is too high, the root system can become flooded, and the tree may face problems such as root rot. This can threaten tree health and growth. Trees with groundwater levels that are too high may experience slow growth and, consequently, reduced tree productivity. Since each tree species has different adaptations to different groundwater levels, proper tree selection and management are essential (Tata et al. 2022).

Tree growth monitoring is essential in forest management. Measurements of tree height and trunk diameter, as well as observations of tree canopy health, are conducted to monitor tree growth and health. This provides the determination of tree growth conditions, harvest time, and early detection of pests and diseases (Dau et al. 2015). Soil is essential to forestry management, and its health must be monitored. Soil monitoring involves taking samples and measuring the soil's physical and chemical properties. This allows us to determine soil fertility, nutrient status, and acidity changes and develop measures to enhance soil management. Forest managers deal with these issues on a daily basis and have spent a lot of time and money on them. However, in peatlands, they must now consider not only the growth of plantation trees but also the groundwater level, the reduction of CO_2 emissions, the prevention of peat fires, biodiversity, and water issues.

At the Convention on Biological Diversity (CBD)/COP15 held in Montreal, Canada, in December 2022, the "2050 Vision" was presented as the global framework for the post-2020 period, and the "2030 Mission" was presented as a way to achieve it.

The approved "2030 Mission" is as follows:

To take urgent action to halt and reverse biodiversity loss to put nature on a path to recovery for the benefit of people and planet by conserving and sustainably using biodiversity and ensuring the fair and equitable sharing of benefits from the use of genetic resources, while providing the necessary means of implementation (CBD 2022).

To achieve the "2030 Mission," the "2030 Targets," consisting of 23 action guidelines, were defined. Of the 23 targets, 14–23 were to provide solutions for countries, companies, and consumers to realize the "2030 Mission." In particular, "Target 15" refers directly to business activities. Risks, dependencies, and impacts on biodiversity must be regularly monitored, assessed, and transparently disclosed.

Today, it is becoming almost mandatory for companies to report on their CO_2 emissions and biodiversity. This is because corporate environmental responsibility and corporate activities often involve environmental impacts. This includes CO_2 emissions from production processes, raw materials used, and their impact on biodiversity. Reporting is a means for companies to demonstrate their awareness of their environmental responsibility and their willingness to understand and manage their impacts. Second, transparency and credibility are also important. A company's reporting of its environmental impacts is essential to increasing its transparency and credibility (Wong et al. 2021). Customers, investors, and stakeholders want to understand how a company addresses environmental issues. Scientific reporting methods are needed to help companies identify their CO_2 emissions and impacts on biodiversity and to report them fairly and transparently. For example, the Greenhouse Gas (GHG) Protocol is the international standard for companies to calculate and report their greenhouse gas emissions. Companies report separately on direct emissions (Scope 1), indirect emissions (Scope 2, e.g., electricity purchases), and emissions in the supply chain (Scope 3). On the other hand, biodiversity indicators are tools to measure a company's impact on biodiversity (Katic et al. 2023), considering factors such as consumption of ecosystem services, land use change, and drainage and waste management. Indicators such as the biodiversity impact index were included.

Peatlands are vital ecosystems and contribute to biodiversity conservation. Peatlands are home to unique plant and animal species, forming unique ecosystems (Johnson et al. 2005; Dennis and Aldhous 2004; Ng et al. 1994). The conservation and protection of biodiversity must be considered in peatland management. Currently, the above reporting is not required for peatland management. Nevertheless, peatland management significantly impacts CO_2 emissions and biodiversity, so scientific monitoring will undoubtedly be required in the future. Appropriate monitoring will also enable the construction of economically and environmentally successful peatland management businesses.

This chapter introduces knowledge and data obtained from more than 10 years of forest plantation projects on tropical peatlands in West Kalimantan, Indonesia and the peatland management methods based on these findings. Although the project has a total concession area of 115,000 ha, only approximately 20% of the area has been used to plant commercial tree species, and the rest is conserved. We hope that this economically and environmentally successful case study will serve as a business reference.

In closing, it should be firmly stated that the techniques and observation methods described in this chapter are intended only for peatlands severely ecologically impacted by anthropogenic activities or peat fires. There is no intention to encourage new development of undisturbed peat ecosystems.

2.2 Peatland Eco-Management

PT. Wana Subur Lestari (WSL) and PT. Mayangkara Tanaman Industri (MTI) are industrial forest plantation companies in West Kalimantan, Indonesia (Fig. 2.1). Both companies are subsidiaries of a Japanese company, Sumitomo Forestry Co., Ltd. This company covers approximately 115,000 ha, and more than half of the area is covered by peat. WSL-MTI began its activities in 2009 and, from the beginning, has practiced sustainable peatland management, which has been supported by observation and monitoring activities based on scientific methods.

At the beginning of the operation, WSL-MTI carried out various field surveys to collect detailed data at the landscape level. Field surveys collected information and data on topography, peat distribution and depth, and peat chemical and physical properties. A detailed topographic map, peat depth distribution, and information on peat properties are fundamental data in designing the large-scale eco-management of tropical peatlands. One of the principles of eco-management is to maintain a constant groundwater level (GWL) throughout the year through a proper water management system. WSL-MTI has kept the GWL at the legislatively mandated 40 cm throughout the year by implementing a stock-based water management system, which will be described later. The high GWL maintains the high moisture content of the peat soil, generating an environment that is less prone to peat fires. As a result, in the 2015 and 2019 El Niño events, no fire was recorded in the planted areas of these two companies.

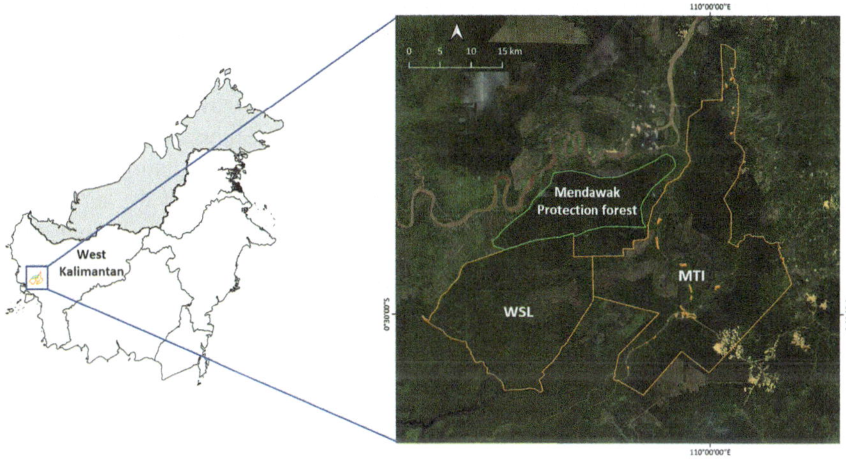

Fig. 2.1 Location of the PT. Wana Subur Lestari (WSL), PT. Mayangkara Tanaman Industri (MTI), and Mendawak Protection forest

Peatland monitoring is also conducted outside the concession areas in the adjacent protected forest, namely, the Mendawak Protection Forest. The protected forest is approximately 18,700 ha in area and is located on the northern side of the two forest plantations and the southern side of palm oil companies facing the Kapuas River (Fig. 2.1). The protected forest was legally allocated as a watershed in 2014 (Decree of the Minister of Forestry 2014) and is managed by the local government. With permission, WSL-MTI has conducted several surveys, including topographic, peat depth and peat characteristics, as well as flora and fauna studies in that protected forest.

The Mendawak protection forest is covered by peat soil and is considered one of the giant peat domes in the coastal peatland in West Kalimantan. The peat depths can reach up to 13 m with an elevation of 13.5 m, making this peatland essential for water reservoirs whose function is to supply water to the entire peatland area in the dry season. However, illegal logging activities have been frequently confirmed inside protected forest areas, leading to forest degradation in some areas. Thus, monitoring this area is crucial to prevent further degradation that threatens its existence and functionality.

2.2.1 *Indonesian Laws and Regulations in Peatland Management*

Indonesia, the world's third-largest tropical peatlands, has been trying to manage its peatlands for the past 30 years. In fact, since 2014, there have been 11 laws and regulations regarding peatland management. Initially, most laws and regulations focused on fire prevention in peatlands. Nevertheless, after the enormous fires across Indonesia in 2015, decrees on enforcing a moratorium on peatland development and a reassessment of available peatland areas were issued in 2016 (ex., Government Regulation Number 57 of 2016 concerning Amendment to Government Regulation Number 71 of 2014 concerning Peatland Ecosystem Protection and Management). Moreover, it was made mandatory for forest and oil palm plantation companies operating on peatlands to keep the groundwater level 40 cm below the surface and regularly monitor it at several locations. Notably, these regulations comprised a penalty clause that included the possibility of operations being suspended. Even if the government imposes mandatory standards and monitoring obligations through policy, private companies operating forest and oil palm plantations on peatlands face challenges that must be addressed. This is because peat fires caused by improper groundwater level management and unstable groundwater levels result in reduced productivity.

2.2.2 *Water Management*

In their previous book, Tropical Peatland Eco-Management, the authors introduced conventional, faulty groundwater level management and new stock-based water

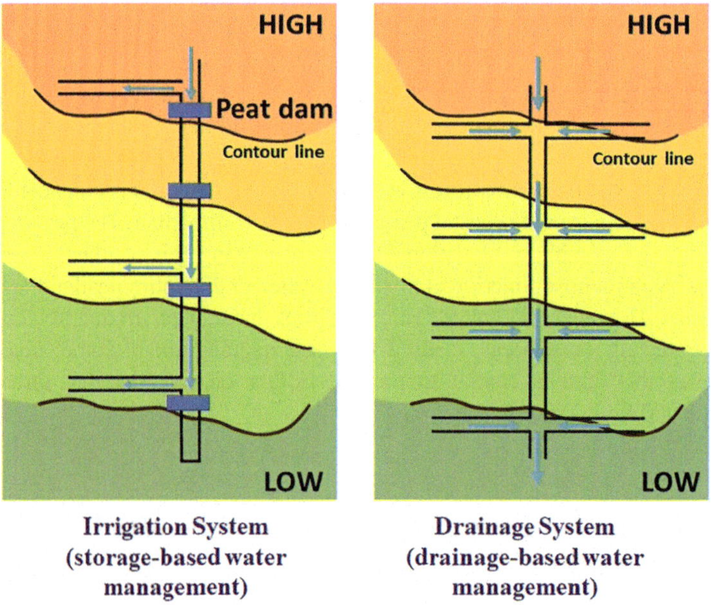

Fig. 2.2 Comparison of the new water management system (storage-based water management, left) with the conventional water management system (drainage-based water management, right) (Kato et al. 2021a)

Fig. 2.3 The canal network in one palm oil and forest plantation company is seen from satellite imagery. Their water management system clearly ignores the topography of the site

management (Fig. 2.2) (Kato et al. 2021a). The first is called "drainage-based water management" (drainage-based WM), which aims to drain water from the tropical peatland ecosystem as quickly as possible. Meanwhile, "stock-based water management" (stock-based WM) seeks to keep water in the tropical peatland ecosystem for as long as possible. Many businesses have adopted drainage-based WM because it is a simple system that does not require detailed and specific ground data, and the canal layout can be designed to fit the investment budget. The drainage-based WM can drain water by collecting it in a straight wide canal (primary canal) built from the top to the bottom of the peat dome's slope.

On the other hand, stock-based WM requires detailed information and ground data, such as topography, peat depth, peat characteristics, land use, and biodiversity, to design the canal network and water management infrastructures. The key to this system is to maintain a peat dome's overall water balance (or budget), which is accountable for the peat dome's inputs, outputs, and storage (Kato et al. 2021a). WSL-MTI has successfully implemented this system in their concession areas, referring to previous examples and intensive field surveys on peat and vegetation.

As previously stated, many companies have adopted drainage-based WM. Unfortunately, most companies continue to use the conventional system despite knowing the negative impacts of the drainage-based WM system. This can be seen clearly using satellite imagery. For example, Fig. 2.3 shows how one oil palm and forest plantation company operating in peatland completely ignores field characteristics such as topography in their water management system.

Under natural conditions, tropical peatlands serve as freshwater reservoirs, moderate water levels, buffer against saltwater intrusion, reduce storm flow, and maintain river flows even during the dry season; however, excessive drainage and deforestation in tropical peatlands have disrupted their natural functions, causing the naturally fire-resilient peatland to be susceptible to fire (Wösten et al. 2008). Groundwater level decline is considered one of the main drivers associated with drought, fire hazards, and peat subsidence, releasing significant amounts of carbon into the atmosphere. On the other hand, drainage of the water-saturated peat is required for both development and plantation forestry on peatlands. As a result, determining critical groundwater levels in tropical peatlands is essential not only for reducing drought severity and peat fires, which cause severe air pollution, but also for increasing tree productivity.

Extensive research has been conducted to determine critical groundwater level depths associated with drought and fire hazards. Previous research has found that groundwater levels cannot exceed 40 cm below the surface (Usup et al. 2004; Wösten et al. 2008), as deeper groundwater levels result in increased peat oxidation and fire susceptibility. Additionally, it was found that the moisture content of peat soil increased with depth, with the moisture content of surface peat being approximately 100% and increasing to approximately 120% at a depth of 10 cm. Even though the moisture content was nearly constant down to a depth of 40 cm, it increased sharply to approximately 220% at depths less than 40 cm from the surface (Usup et al. 2004). When the groundwater level dropped below 40 cm, the soil moisture content significantly decreased, resulting in reduced capillary rise and an increased risk of desiccation. Desiccation can cause peat soil to crack and shrink, reducing its water-holding capacity and causing the peat to become drier and more prone to ignition. The deep peat soil layer contains more debris than the surface layers, producing coarse-textured peat material. The high wood content and low bulk density of deeper peat layers enable oxygen to be supplied to the deeper peat layers when smoldering combustion occurs, increasing the occurrence of deep peat fires, which are difficult to extinguish (Usup et al. 2004).

Another study by Wösten et al. (2008) found that during a dry El Niño year in 1997, areas with identified deep groundwater levels coincided with areas on fire as detected by radar images. The surface peat is only moderately to lightly humified (hemic to fabric), with many large pores that lose water quickly when the groundwater level is slightly lowered.

Despite having over a third of the world's tropical peatlands, Indonesia is one of the top GHG emitters from peatlands (UNEP 2022). As a result, the Government of Indonesia has issued several regulations to address the underlying causes of peatland fires. The government issued Government Regulation No. 57/2016, Article 23 paragraph 3, which states that the peatlands' groundwater level must be no deeper than 40 cm below the ground surface, among others (MoEF 2016). The decision came from the previous studies mentioned above. Furthermore, based on Government Regulation Number 15/2017, the person in charge of the business and/or activity shall conduct GWL measurement and report it to the MoEF (MoEF 2017).

2.2.3 *Peat Fire Monitoring*

Peat fire outbreaks are often caused by dry seasons and strong droughts, especially those associated with El Niño (Worden et al. 2013; Field et al. 2016; Hein et al. 2022; Usup and Hayasaka 2023). In 2015, as mentioned above, with a strong El Niño, severe peat fires spread throughout Indonesia. As a result, based on a presidential instruction, those operating in peatlands were required to implement "canal blocking." The importance of canal blocking was widely promoted in the media (see Fig. 2.4). One company even constructed thousands of peat dams to implement canal blocking. As a result, something unexpected happened.

At the time, the idea was to halt water flow so that the peatlands would not dry out, and it was thought that canal blocking, or the construction of peat dams, was the ultimate solution. However, the peat fires did not cease. This is because, as many peat researchers know, tropical peatlands have a gradual but dome-like shape, with an elevation increase of 1 m for every 1 km distance. Therefore, as the water flowing from high to low was stopped by canal blocking, dryness increased in the lower elevation areas at the base, although the water was secured near the center of the dome. Since the local people could not live in the center of the peatland, settlements were formed along the river. Canal blocking significantly dried up the riverside areas, and increased settlements created more opportunities for using fire. In areas where water is stored by canal blocking, fires will not start unless there is a fire source. In general, the fire sources were in settlements that had been further dried out by canal blocking, where peat fires spread, and the more the canal was blocked, the greater the fires spread, which was tragic.

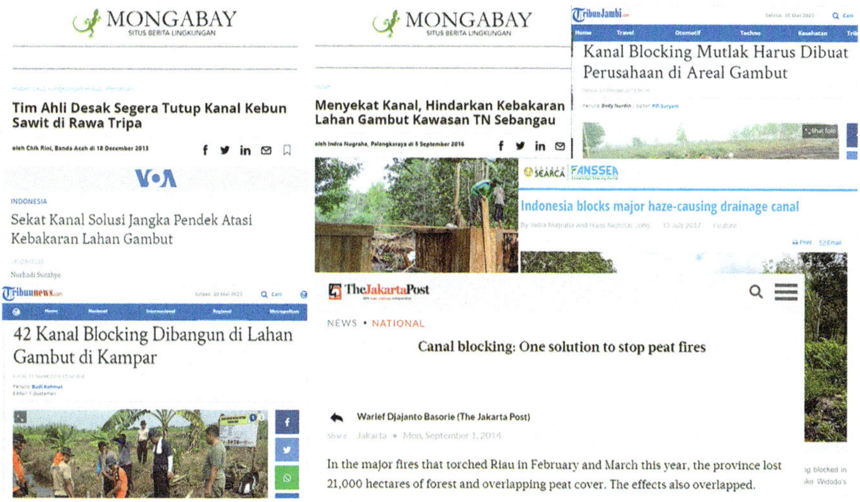

Fig. 2.4 The importance of canal networks promoted by the media

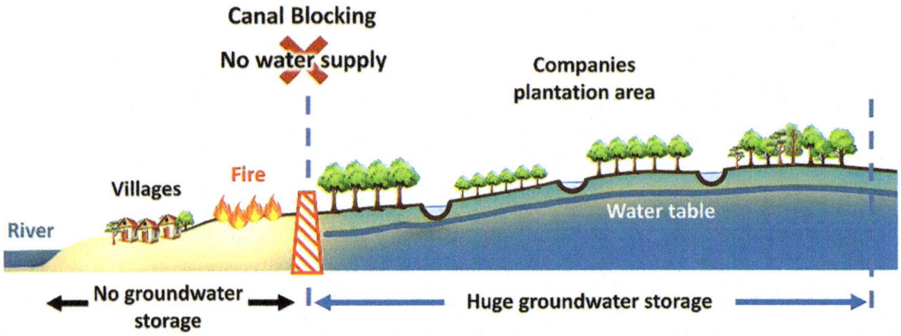

Fig. 2.5 Illustration of the problem caused by canal blocking in plantation areas

The massive fires of 2015 naturally ended with the start of the rainy season. However, the peat fires of 2015 have not been fully verified. A peat decree (Government Regulation Number 57 of 2016) and a stipulation were issued to maintain the groundwater level 40 cm below the surface. Moreover, companies must protect their plantations from fire. Thus, peat dams were constructed to hold plenty of groundwater, so fires would be less likely to start. This was effective within the concession. However, this affected the surrounding communities. As the water flow was completely halted in the concession area at higher elevations, there was a water shortage in the villages at lower elevations. As a result, peat fires started to occur during the dry season, even in years without an El Niño. For that reason, private companies are strongly urged to consider groundwater movement not only within the concession but also at the landscape level. Figure 2.5 illustrates the problems occurring in one plantation project.

Peatland use must be planned for one peat ecosystem, which is one of the main reasons why WSL-MTI plans to have the same constant groundwater level throughout the year, regardless of where it is dug. Stock-based WM should not be completed only in the project area but also within the entire ecosystem.

2.2.4 Peat Fire Risk Indicators

The fire danger rating (FDR) is commonly used in Indonesia for fire risk. This is introduced by the Canadian Forest Fire Danger Rating System (CFFDRS). Furthermore, the National Danger Rating System (NFDRS) of the United States is a well-known method for evaluating forest fire risk. Nevertheless, its methods are based on different observation and calculation methods than those used in Canada. Thus, they may lead to different results. The CFFDRS was introduced across Canada in the 1970s and is used in many parts of the world because of its simplicity and the required data and generated output (Stocks et al. 1989). FDR is an essential tool for identifying fire risks in the field. Although designed for a given region or environment, different regions or environments may generate different results.

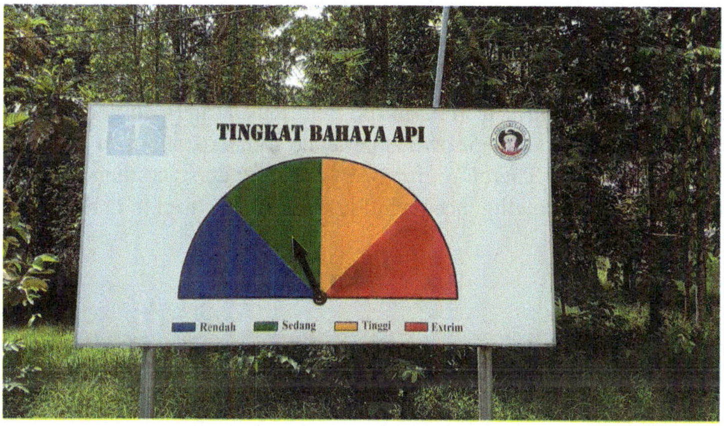

Fig. 2.6 The visualization of the FDR board in the WSL-MTI areas

The fire weather index (FWI) is an index for evaluating weather conditions for forest fires and is the main component in determining the FDR. The FWI is an indicator of fire danger that considers weather elements and fuel conditions. The fuel condition here refers to the state of what burns during a fire. The FWI consists of three elements: Fine Fuel Moisture Code (FFMC), Duff Moisture Code (DMC), and Drought Code (DC). FFMC is a moisture index for fine fuels (dry leaves, grass, branches, etc.) and is calculated by considering temperature, relative humidity, precipitation, and the previous day's FFMC value. DMC is an indicator of the moisture content of organic matter on the ground surface and is calculated by considering the previous day's DMC value, precipitation, and temperature. The humus layer on the ground surface is formed from organic matter such as dead leaves, branches, and plant debris; the drier it is, the higher the fire risk. DC is an indicator of the humidity in the ground. The higher the DC is, the drier the underground fuels are, and the higher the fire risk is. In Indonesia, the method of calibrating FWI originated in Canada and has been shown as FDR (Parwati et al. 2011). As an example, see Fig. 2.6 for the visualization of FDR in the WSL-MTI areas.

While its FDR has undoubtedly been a valuable indicator in the field over the past decade, the time and cost of acquiring the data necessary for FWI is another problem. Satellite-based methods are also used for wide-area evaluation, but local evaluation still requires FWI to use local data. The actual FDR is determined by adjusting the target to be burned by the fire and for each of the codes shown above. Therefore, one challenge is for local governments, companies, farmers, etc., to determine the FDR independently. There is a specific example of the difficulties in making FDR decisions. The surveyor's observations often determine the drought code (DC), but the methodology is subjective rather than objective. As a result, they have often been over- or underestimated by surveyors.

If the fire risk is underestimated, even if unintentional, it carries a significant fire risk at the field level. In FDR, the on-site monitoring and frequency of patrols greatly differ between MEDIUM and HIGH/EXTREME. Farmers often burn peatlands even though the risk is moderate with an FDR of Medium. Farmers who have lived on the land for a long time know when to burn effectively by their experience and intuitiveness. We have learned in the field that farmers do not rely on official weather data, but rather on their own experience and perceptions to determine when a fire is most effective. Therefore, for everyone to understand their experience and intuitiveness, artificial intelligence techniques should be used to determine fire risks in the future.

2.3 Ground Truth Monitoring

The WSL-MTI has conducted several monitoring activities since the operation began. The accumulated data used are not only to design water management but also to study the impact of ongoing management on peat conditions, biodiversity, tree growth, CO_2 emissions, and the effect of climate change on the peatland ecosystem.

2.3.1 Tree Growth

Forest plant inventories are carried out when plants are 1, 6, 12, 24, 36, and 48 months old or up to a year before harvesting. Inventory carried out for plants 1–48 months old is called Plantation Monitoring Assessment (PMA). Moreover, inventories for plants a year before the harvesting period are named the Pre-Harvesting Inventory (PHI). Inventory activities aim to acquire the plant's growth data, such as diameter, height, standing stock, and plant survival rate.

Generally, the sampling intensity used in PMA and PHI activities is 1% of the compartment area. However, if the plant diversity in a compartment is relatively high, the sampling intensity used is greater than 1%, depending on the diversity level. The most common sampling plot design applied to plantation forests is circular, with a radius of 11.28 m and an area of 0.04 ha. The number of sample plots (N) is calculated using the following formula:

$$N = \frac{\mathrm{SI} \times \text{Compartment area}}{\text{Plot area}}$$

N: Number of sample plots.

SI: Sampling intensity (1%).

Plot area: 0.04 ha.

The number of sample plots in a compartment is always rounded up. For example, if the total number of sample plots based on the above formula results in 6.2, the total actual number of sample plots taken would be 7. The plots in the compartment

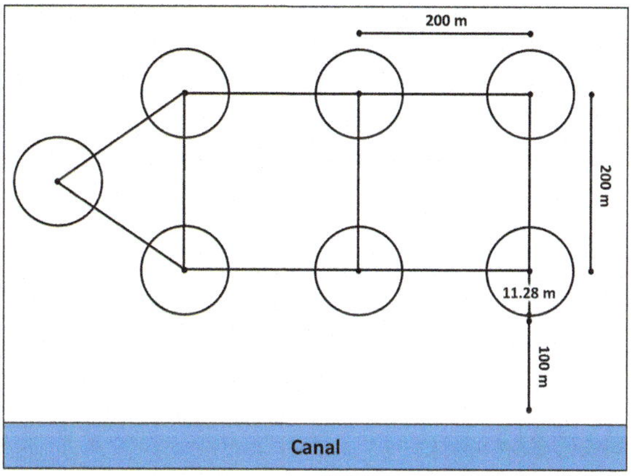

Fig. 2.7 Plot design of PMA samples in the compartment

are approximately 100 m from the canal's side, with a 200-meter distance interval between plots (systematic sampling). With a sampling intensity of 1%, 1 sample plot covering 0.04 ha represents every 4 ha plant condition in the compartment. Figure 2.7 displays a PMA plot design.

In PMA 1 (PMA in 1-month-old plants), the data collected focus on standing stock and survival rate (diameter and height are not measured due to being too small). Standing stock, survival rate, and tree height are measured in PMA 6. Ideally, if the survival rate is 100%, one hectare of the area will have a standing stock of 1,333 plants with a planting distance of 2.5 × 3 m. From PMA 12–48 months, diameter at breast height (DBH) is measured in addition to standing stock, survival rate, and tree height. The tree volume and mean annual increment (MAI) are calculated starting at 36 months old to understand a compartment's potential for production.

DBH measurements in the field are technically performed using a diameter tape or a phi-band. Diameter data are taken perpendicularly at 1.3 m from the tree's base. The plant height is estimated using a simple measuring stick and a haga hypsometer. Each measuring parameter, such as tree height, diameter, stocking, survival rate, volume, and MAI, has a standard/benchmark that can be used to identify a compartment's high and low potential. All trees in the sample plot must be documented for subsequent examination. If the data for each parameter are not corresponsive against the benchmark, improvements are made.

2.3.2 *Groundwater Level*

Since the massive forest and peat fires in 2015, the Government of Indonesia has taken significant steps to prevent peat fires. In addition to Government Regulation No. 57/2016, which regulates peat land use based on peat depth and limits groundwater level to 40 cm below the surface, the Indonesian Government has been putting up various efforts to control forest and peat fires. For instance, the Indonesian

Government established procedures for measuring groundwater levels in the peatland ecosystem through Government Regulation Number 15/2017, consisting of 14 articles (MoEF 2017). Based on the provisions of Article 2, groundwater level measurements on the peatland ecosystem shall be determined at the control points, called organization points, carried out by the Director General. The GWL measurement points shall be set minimally at 15% of the total number of main crops or production blocks.

Under the Indonesian Government's regulation, WSL-MTI has implemented several intensive monitoring systems, including manual and automatic systems, to maintain the GWL and enhance the water management system. The location of the GWL measurement was chosen based on specific field data and information used for developing water zones in production areas, in addition to the total number of production blocks.

For manual GWL monitoring, WSL-MTI utilizes Dipwell, a simple yet inexpensive and highly accurate method. This method requires the installation of permanent pipes into the substrate layer at each target point. The GWL is then estimated by measuring the dry length of the wooden stick inserted into the opening portion of the pipe, excluding the length of the pipe exposed to the ground. The manual measurement is conducted twice a month, and the result is then sent to the government for verification. Meanwhile, automatic monitoring is conducted using data loggers, which record data hourly. The daily average data from the automatic measurements are also submitted to the government for verification. Based on government regulations, the GWL measurement points need to be from at least 142 locations scattered in WSL-MTI areas, given that the total production blocks in the WSL-MTI are 945 units. However, approximately 3–5 replications or even more are also set in each location, resulting in 821 registered GWL measurement points. The distribution of the GWL measurement instruments is represented in Fig. 2.8.

In addition to the aforementioned systems, WSL-MTI has implemented a real-time monitoring system called SESAME II (Sensory Data Transmission Services by Midori Engineering). The system comprises a parent unit and multiple child stations, with sensors to measure GWL and soil moisture content, temperature, and rainfall. Those datasets are acquired at 10-min intervals. The parent unit is placed in a 25-m-high tower, while the child stations are distributed evenly over the target area. The datasets are transmitted to the parent unit via wireless communication, and the parent unit further sends the datasets to an internet service using satellite-based communication. For the detailed concept of the real-time monitoring system using SESAME II, see Kato et al. (2021b).

WSL and MTI each have ten units of child stations for observation (see Fig. 2.9). This system provides monitoring of spatial and temporal changes in the GWL and its behavior in response to rainfall changes. For example, Figs. 2.10 and 2.11 show the relationship between GWL and rainfall from June 2018 to July 2022 in the upper (9.5 m in elevation) and lower (5.9 m in elevation) slopes of the peat dome, respectively.

The upper slope of the peat dome is allocated as a hydrobuffer zone located between the conservation and plantation areas. The GWL in this area is set to

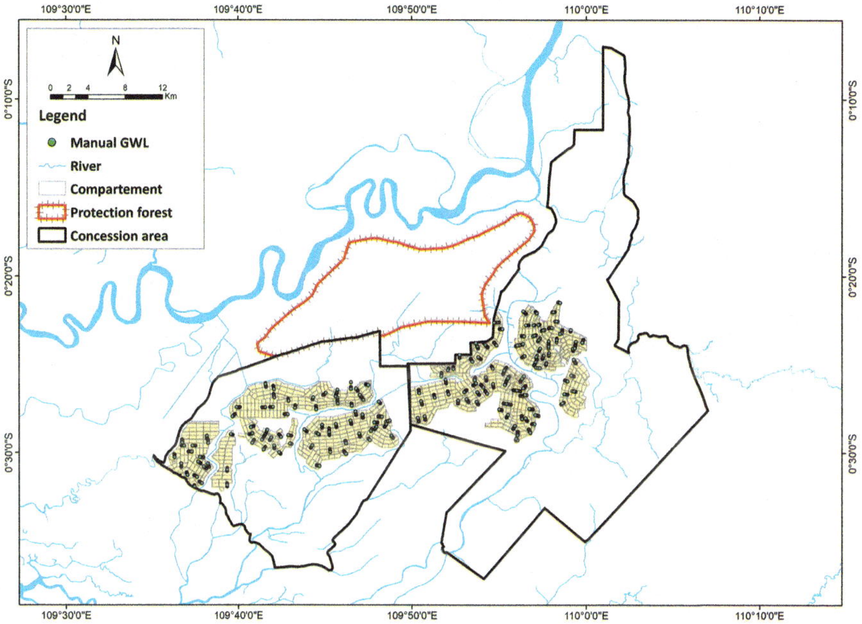

Fig. 2.8 The distribution of the manual groundwater level measurement points across WSL-MTI areas

0–20 cm because it functions as a water reservoir, supplying water to the lower areas required to maintain a high GWL. Meanwhile, the lower slope is a plantation area located at the foot of the peat dome. As shown in Figs. 2.10 and 2.11, the GWL in the plantation area is more stable than the GWL in the hydrobuffer area, despite a significant drop (no more than 40 cm) during the 2019 El Niño event. The high fluctuation in the upstream region is directly associated with precipitation; when precipitation is high, water is preserved; and when precipitation is low during the dry season, water is released to the downstream region. This explains why this type of water management is referred to as "stock-based water management."

2.3.3 CO_2 Emissions

Tropical peatlands have significantly contributed to atmospheric CO_2 levels and have become a major source of carbon emissions (Murdiyarso et al. 2010). Most carbon emissions come from drainage peatlands and peat fires. It is estimated that CO_2 emissions from drained peatland decomposition are between 355 and 855 Mt./year (Hooijer et al. 2010). The impact of peat fires goes far beyond that; for example, according to the Global Fire Emission Database (GFED), the 2015 fires in Indonesia's peat swamp forests alone emitted almost 1.750 million metric tons of

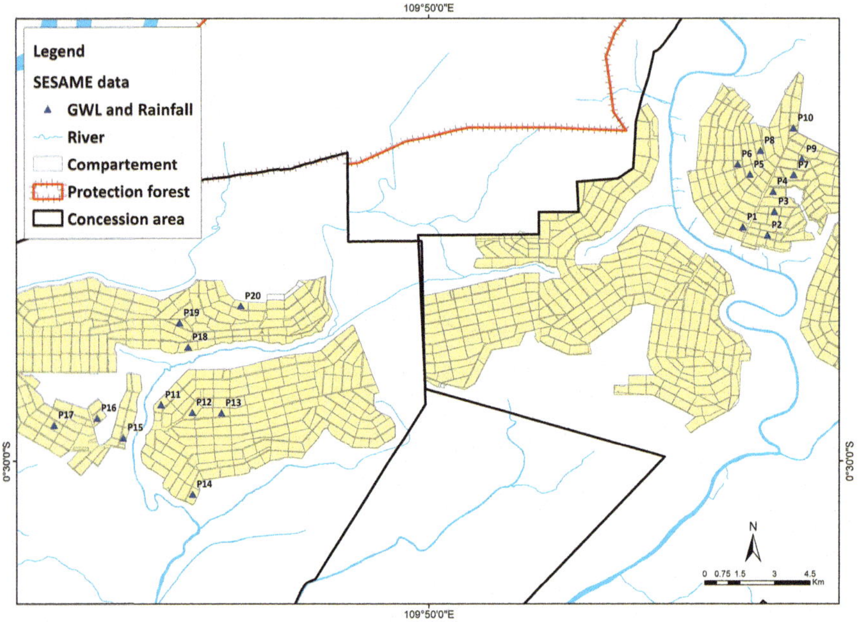

Fig. 2.9 The distribution of automatic groundwater level measurement points using SESAME II across WSL-MTI areas

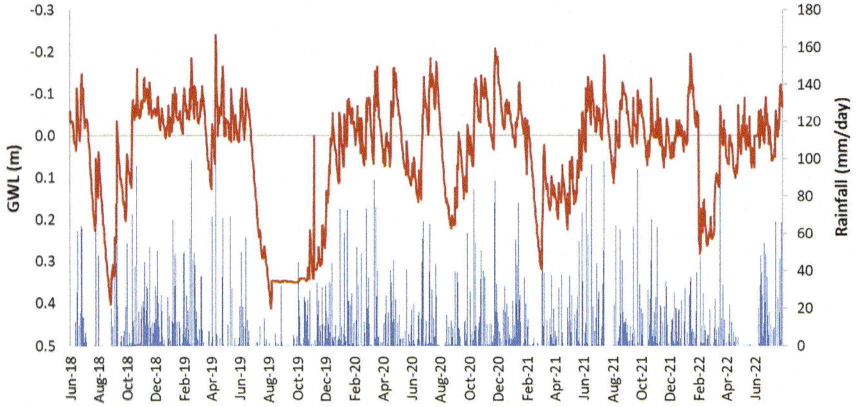

Fig. 2.10 The groundwater level and rainfall fluctuations in the upper slope of the peat dome. This figure is modified from Fig. 3.30 in Kato et al. (2021b) with additional data

carbon dioxide equivalent (mtCO_2e) (World Bank 2016). In this regard, monitoring carbon emissions in tropical peatlands is critical in addition to restoration and sustainable management efforts. Furthermore, under the 2015 Paris Agreement to limit global warming, each country must measure and report progress toward their

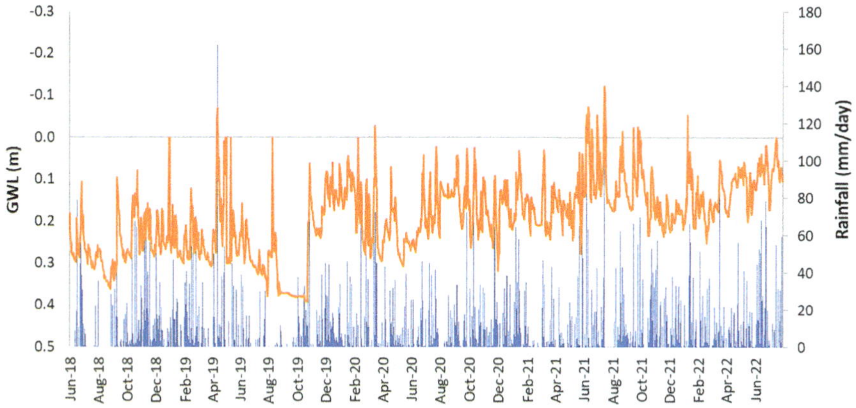

Fig. 2.11 The groundwater level and rainfall fluctuations in the lower slope of the peat dome. This figure is modified from Fig. 3.31 in Kato et al. (2021b) with additional data

pledged reductions in emissions under a transparent framework. To secure the transparency of the inventory, a system is necessary to compare and evaluate the inventory in independent ways. Carbon emissions can be estimated through three approaches, including satellite-, UAV (unmanned aerial vehicle)-, and ground-based measurements.

Unlike satellite- and UAV-based measurements, ground-based measurements of GHG are typically localized to specific sites or networks of measurement stations, offering a more detailed view of an ecosystem's GHG concentrations. Ground-based measurements are crucial in validating satellite and UAV data and providing site-specific information to enhance climate change projection models. Among the available ground-based GHG measurements, the eddy covariance flux tower is the most widely known approach for measuring ecosystem-scale fluxes of various gases (Fortuniak et al. 2021; Guevara-Escobar et al. 2021; Krasnova et al. 2022). This technique has been used in carbon cycling studies in various ecosystems to estimate CO_2 emissions based on the gas exchange between the Earth's surface and the atmosphere. It involves measuring the turbulent gas fluctuations in the atmosphere by using a fast-response CO_2 analyzer mounted on a tower and a sonic anemometer to collect data on wind speed, temperature, pressure, and other parameters at a high frequency. The net ecosystem exchange (NEE) of CO_2 can be calculated by subtracting ecosystem respiration from gross primary productivity. The former can be estimated using nighttime data when photosynthesis is absent, while the latter can be estimated using daytime data when respiration is negligible. Following some corrections, the NEE is scaled up to the entire ecosystem using footprint analysis and spatial interpolation.

Although ground-based measurement is more accurate than other methods (Guo et al. 2012), it has some spatial and vertical coverage limitations. Satellite-based measurements offer global coverage, including remote or inaccessible locations, and provide frequent measurements, often daily or near daily. Ground-based measurements, on the other hand, provide more localized emission information and are more commonly focused on near-surface measurements. When all approaches are combined, it is possible to comprehensively understand GHG emissions and their spatial-temporal variations.

2.4 Remote Sensing

During the last decade, remote sensing methods have been increasingly employed in peatland studies due to the enhancement of new remote sensing data sets and the rapid development of open-source software (Czapiewski and Szumińska 2022). Remote sensing studies in peatlands have demonstrated the potential for providing information about peatland distribution or extent (Vernimmen et al. 2020; Dargie et al. 2017; Anda et al. 2021), fire disturbance impacts (Miettinen et al. 2017; Huijnen et al. 2016), vegetation structure (Lähteenoja and Page 2011; Brown et al. 2018), water conditions (Rahman et al. 2017; Gumbricht et al. 2017), and GHG emissions (Lees et al. 2018; Kalacska et al. 2018; Dąbrowska-Zielińska et al. 2022). The latter is discussed in this section.

2.4.1 Satellite Sensing

For the past decades, space-based earth observation (EO) has provided information about the state of the surface and atmosphere, which has been useful for various applications. One example is using satellite-based EO for climate change studies, which provides data on climate variables such as temperature, precipitation, sea surface temperature, sea ice extent, and atmospheric composition, including GHG concentrations. Satellites use various active and demonstration devices to measure GHG emissions in the Earth's atmosphere. Satellite-based measurements can provide a distinct opportunity due to their global coverage and continuous monitoring capabilities, providing information about atmospheric GHG concentrations over time (Hardwick and Graven 2016). Satellite-based measurements help to fill information gaps on GHG emissions that other approaches find difficult to fill. The field of satellite-based EO is rapidly evolving due to technological advancements, data processing capacity, and new collaborations. Many space agencies are connecting EO with cloud computing platforms, resulting in a big data system that enables complex computation of emission data and forecast modeling (Nakalembe et al. 2021). These large data processing capabilities, as well as the combination of different types of satellite and ground data, provide accurate and timely GHG estimations and increased transparency and accountability on contributions to climate change mitigation. Some products, such as GOSAT, OCO-2, and Sentinel-5P, are publicly available.

GOSAT (Greenhouse gases Observing SATellite, also known as "IBUKI") is the world's first satellite to observe the concentration of carbon dioxide (CO_2) and methane (CH_4), the major greenhouse gases, from space. It is a joint project of the Japan Aerospace Exploration Agency (JAXA), the Ministry of the Environment of Japan (MOE), and the National Institute for International Studies (NIES). It was successfully launched on January 23, 2009, and provides smooth observations. On October 29, 2018, GOSAT-2, also known as "IBUKI-2," was launched as the successor mission of the previous GOSAT. GOSAT-2 is equipped with cutting-edge sensors and instruments to collect detailed GHG measurements. It features the thermal and near-infrared sensor for carbon observation-Fourier transform spectrometer-2 (TANSO-FTS-2), which measures the concentrations of CO_2 and CH_4, with high precision by observing the sunlight reflected off the Earth's surface. GOSAT and GOSAT-2 have revealed the global distribution of carbon dioxide and methane, their absorption and emission, and their seasonal and interannual variations by analyzing the data observed by GOSAT. This information is used to contribute to climate change science and climate change policy.

In addition to GOSAT, OCO-2 (Orbiting Carbon Observatory-2) is an American environmental science satellite launched on July 2, 2014, and has been operating since. It is the second most successful high-precision ($<0.3\%$) CO_2 observation satellite after GOSAT. Both satellites are affected by clouds and aerosols, but OCO-2 is designed with a spatial analysis capability of 3 m^2, smaller than GOSAT's 10 m^2, to better avoid cloud effects. OCO-3, a follow-up mission to OCO-2, was launched in 2018 onboard the International Space Station (ISS), making it the first space-based instrument to measure CO_2 from a near-polar orbit. OCO-2 and OCO-3 measure CO_2 concentrations in the Earth's atmosphere using a high spectrometer that operates in the near-infrared spectral range. OCO-2 can determine atmospheric CO_2 concentrations by measuring the sunlight reflected by Earth's surface and atmosphere at specific wavelengths absorbed by CO_2. OCO-3 data products include global maps of atmospheric CO_2 concentrations and are useful to characterize sources and sinks on regional scales and quantify their variability over the seasonal cycle.

Another notable satellite mission for GHG measurement is Copernicus, previously known as GMES (Global Monitoring for Environment and Security). Copernicus is the European Earth observation program that aims to benefit European citizens by observing the Earth and its environment. It is managed by the European Commission (EC) and implemented by the Member States, the European Space Agency (ESA), the European Meteorological Satellite Organization (EUMETSAT), the European Centre for Medium-term Forecasting (ECMWF), the EU and Mercator Ocean (a private nonprofit organization for ocean analysis and forecasting). The observations are mainly carried out with the ESA-developed Sentinel series; the data are open and free. The Copernicus program combines satellite data with ground-based, airborne, and seaborne measurement systems to produce timely, quality information, services, and knowledge. This allows for a more comprehensive understanding of environmental changes and helps service

Table 2.1 Comparison of GOSAT-2, OCO-3, and Sentinel-5P

Mission	Target GHGs	Spectral bands	Spectral resolution	Nadir footprint size
GOSAT-2	CO_2 and CH_4	0.75–0.77 μm 1.59–1.67 μm 2.04–2.08 μm 2.33–2.38 μm 5.50–14.3 μm	0.2 cm^{-1} 0.2 cm^{-1} 0.2 cm^{-1} 0.2 cm^{-1} 0.2 cm^{-1}	2.5 km
OCO-3	CO_2	0.76–0.77 μm 0.77–0.79 μm 0.80–0.82 μm	4.7 nm 7.5 nm 5.2 nm	1.3×2.3 km^2
Sentinel-5P	CO_2, CH_4, O_3	0.27–0.32 μm 0.32–0.38 μm 0.40–0.74 μm 0.75–0.77 μm 1.59–1.68 μm 2.10–2.25 μm 2.31–2.38 μm	0.6 nm 0.6 nm 0.4 nm 0.4 nm 0.5 nm 0.5 nm 0.5 nm	7×3.5 km^2

providers, public authorities, and other international organizations enhance the quality of life of citizens.

As mentioned, the Copernicus program's heart is the Sentinel satellites' constellation. Sentinel satellites provide data on various parameters, including land cover, vegetation, air quality, sea surface temperature, ocean color, ice cover, and GHG emissions. On October 13, 2017, a mission, Sentinel-5P (TROPOMI, TROPOspheric Monitoring Instrument), was launched. TROPOMI is an imaging spectrometer that measures the composition of the Earth's atmosphere from space by measuring the intensity of sunlight reflected by the Earth's atmosphere and surface at different wavelengths. It operates in the ultraviolet (UV), visible (VIS), near-infrared (NIR), and shortwave infrared (SWIR) spectral ranges, allowing it to detect and quantify a wide range of atmospheric constituents, such as carbon dioxide (CO_2), methane (CH_4), and ozone (O_3). One of the main objectives of Sentinel-5P is to study the sources and sinks of GHGs covering the entire Earth's surface at high spatial (ranging from 3.5×7 km to 7×7 km) and high temporal (1 day) resolutions. The comparison of GOSAT-2, OCO-3, and Sentinel-5P is shown in Table 2.1.

2.4.2 *UAV Sensing*

UAVs are increasingly utilized for various environmental monitoring applications, including the measurement of GHG emissions. UAV-based measurements have several advantages, such as flexibility, accessibility, and capturing high-resolution data over specific areas of interest. UAV-based measurements can also be used to validate and calibrate ground- or satellite-based GHG measurements. The accuracy and reliability of the different measurement methods can be assessed by comparing UAV-derived data with ground- or satellite-based measurements to ensure consistency and improve overall measurement capabilities.

One example of using UAV-based carbon measurements is estimating the volume of peat lost due to subsidence. UAVs equipped with photogrammetry sensors and a set of the Global Navigation Satellite System (GNSS) can provide highly accurate models and the quantification of peat subsidence rates. The carbon emissions associated with peat subsidence can be calculated by multiplying the volume of peat loss by the carbon density and other related parameters.

UAV-based measurements, however, face some challenges, including flight endurance, payload capacity, and the requirement for precise sensor calibration. Nonetheless, as UAV technology advances, it provides promising opportunities to further our understanding of GHG emissions and assist with mitigation efforts.

2.5 Biodiversity Assessment

Biodiversity assessment and monitoring are important for understanding species' presence and flourishing, habitat health, and ecosystem functions and services. Methods include (i) species surveys and monitoring, (ii) biological indicators, (iii) habitat assessment, (iv) ecosystem service assessment, and (v) use of biodiversity information systems. Species surveys and monitoring are methods used to regularly assess the number, distribution, and habitat quality of a particular species or group of species. This is often done using field surveys or remote sensing techniques. Biological indicators are tools used to assess the health of a particular ecosystem, using a specific species or species group as an indicator. For example, the presence or number of certain aquatic insects is often used to indicate water quality. Habitat assessment is a method for evaluating habitat quality and quantity for a species, usually using tools such as geographic information systems (GIS) and satellite image analysis. Ecosystem service evaluation assesses the quantity and value of services that a particular ecosystem provides (e.g., carbon fixation, water quality regulation, recreation). Biodiversity information systems are databases and platforms for aggregating, analyzing, and sharing biodiversity data. For example, the Global Biodiversity Information Facility (GBIF) is an extensive network providing worldwide biodiversity data. These methods are needed to understand the current state of biodiversity, track changes, evaluate the effectiveness of conservation measures, and help companies understand the impact of their business activities on biodiversity and promote sustainable practices.

2.5.1 Biodiversity in Peatlands

Tropical peatlands are well known for providing numerous environmental services, including supporting regional and global biodiversity (Ribeiro et al. 2021; Tan et al. 2023; Yu et al. 2010; Dommain et al. 2010; Chu et al. 2020). It is home to a wide range of plants and wildlife, including endemic, rare, and endangered species, and it is one of the most important habitats for the last large populations of orangutans (Morrogh-Bernard et al. 2003). However, large areas of tropical peatland have degraded over the last two decades due to both natural and anthropogenic disturbances, such as excessive logging, extensive drainage, and agricultural conversion (Hooijer et al. 2006; Miettinen and Liew 2010; Takakai et al. 2006). The threats are aggravated by demographic, economic, and social changes and climate change, leading to peatland drying and subsequent fire (Wijedasa et al. 2017; Page et al. 2002). Not only will the disturbances release significant amounts of carbon, economic losses, and health impacts, but they will also result in unbalanced natural systems and functions due to changes in tropical peat forest structure and species composition. A study by Volkova et al. (2023) found that it takes two to three decades after fires for burnt peat forests to reach levels of richness and diversity similar to those of relatively undisturbed reference peat forests.

Most of the forests in the WSL-MTI had already degraded due to various human impacts before the companies began their activities, resulting in the transformation of the forests' natural states from pole forests, mixed swamp forests, and riparian forests into degraded secondary forests, shrubs, and bushes. Even worse, the southern parts of the sites, which are close to the sea, have been completely degraded and replaced by nipa palms. To avoid further changes and forest losses, WSL-MTI designated approximately 50% of the total area of 115,000 ha as conservation areas. The area is managed without drainage, and frequent intensive monitoring is conducted.

WSL-MTI has conducted continuous field surveys across the concession areas and the adjacent protection forest, including flora and fauna surveys, by developing permanent plots and fixed-point observations using camera traps. Moreover, we regularly monitor orangutans and Proboscis monkeys, as they are categorized as critically endangered and endangered species, respectively, based on the IUCN Red List of Threatened Species (IUCN 2023).

2.5.2 *Flora*

Peat swamp forests in Indonesia have several distinct forest types. Pole forest is a vegetation type that characterizes peat swamp forests in tropical areas of Southeast Asia and Indonesia. The word "Pole" describes the typical vegetation type of Pole Forest, i.e., a dense growth of relatively young, slender trees. These forests are generally found in deep, moist peat soils. Plant species diversity is relatively low, with some dominant species, and is relatively low in height at approximately 10 m, with branches that are concentrated in the upper parts of peat domes. Surrounding the Pole forests are mixed forests with tall trees, medium-height trees, and shrubs of various species. These forests are found in areas with more stable peat soils and enhanced nutritional conditions. A variety of tree species are found, and they grow relatively tall. These forest types are generally shaped by factors such as peat soil depth, water table, nutrient status, and the balance between peat formation and

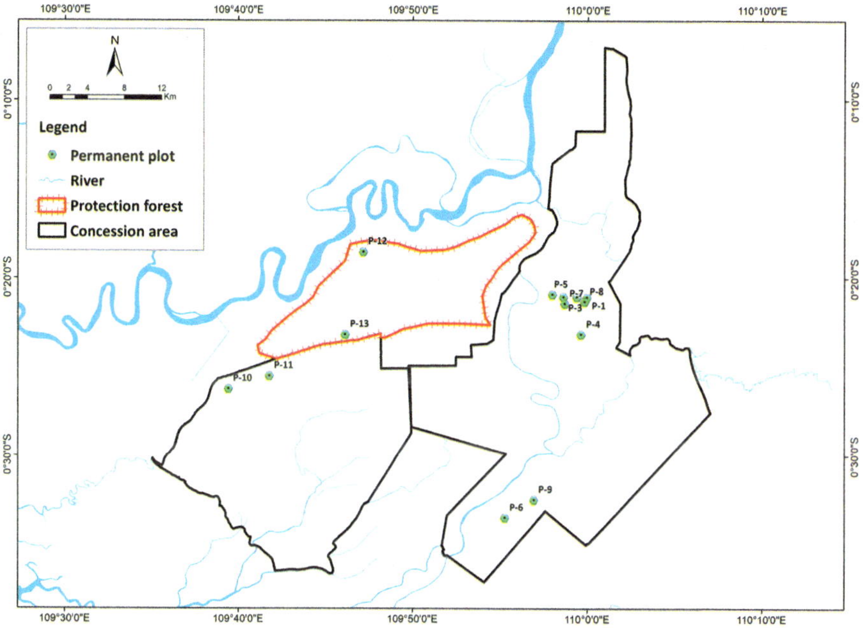

Fig. 2.12 Distribution of the permanent plot in WSL-MTI areas and Mendawak protection forest

decomposition. Trees also play an important role in the forest ecosystem, and their loss can disrupt the ecosystem, including the loss of wildlife habitat and changes in the food chain. After logging, soils are more exposed to erosion and drought. In addition, the destruction of root systems can reduce soil water retention and stability.

Anthropogenic logging and fires severely impact peat swamp forests. After they occur, the condition of the remaining forest changes. For example, logging leads to a decrease in plant species and especially affects large tree species, which changes the structure and species diversity of the forest. The loss of large trees allows light to reach the forest understory, thereby facilitating the invasion of herbaceous plants and ferns. On the other hand, fire is extremely destructive, destroying most vegetation in the area. In large fires, entire forests can be lost. Fires burn the soil surface, converting organic matter to ash, which may temporarily increase the nutrient content of the soil, but the nutrients may deplete. Swidden agriculture on peatlands may last only approximately a year. After a fire, especially a large fire, the forest may take a long time to recover. In addition, the open lands caused by fire may provide opportunities for invasive species, but repeated fires can result in lands where herbaceous plants and ferns thrive, causing difficulties in restoring native forests.

The vegetation survey was conducted inside a one-hectare plot (100 m × 100 m) to collect ground reference data, including species composition and individual tree measurements. Eleven and two permanent plots were established within WSL-MTI's conservation areas and Mendawak protection forest, respectively (see Fig. 2.12).

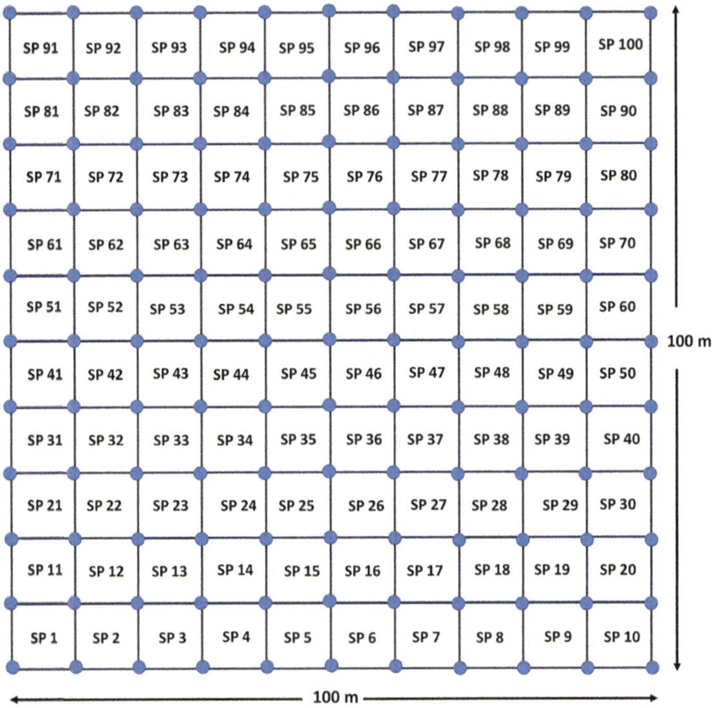

Fig. 2.13 Design of permanent sample plot of 100 m × 100 m

Each permanent plot was divided into 10 m × 10 m subplots to study species and forest structures. All trees with a breast height (DBH) diameter of ≥5 cm were identified and given a specific reference number for the subplot in each permanent plot. Tree species were identified directly in the fields, and some individuals were sampled for identification in the laboratory. The design of the permanent sample plot and distribution of the permanent plot can be seen in Fig. 2.13.

Based on the surveys from eleven permanent plots, WSL-MTI recorded 315 species from 66 families across the WSL-MTI conservation area. The dominant species found were *Diospyros evena*, *Syzygium oligomyrum*, *Syzigium densinervium*, and *Calophyllum lanigerum*. Among the species observed, there are eleven, four, and two species categorized as vulnerable, endangered, and critically endangered species based on the IUCN Red List of Threatened Species (IUCN 2023). The vulnerable species include *Alangium havilandii*, *Cantleya corniculata*, *Ctenolophon parvifolius*, *Dyera polyphylla*, *Melanochyla caesia*, *Mezzettia parviflora*, *Mezzettia umbellata*, *Monocarpia euneura*, *Psydrax dicoccos*, *Tetramerista glabra*, and *Vatica manggachapoi*. The endangered species include *Cephalomappa paludicola*, *Hopea griffithii*, *Shorea pachyphylla*, and *Vatica nitens*. Meanwhile, the critically endangered species include *Gonystylus bancanus* and *Shorea platycarpa*.

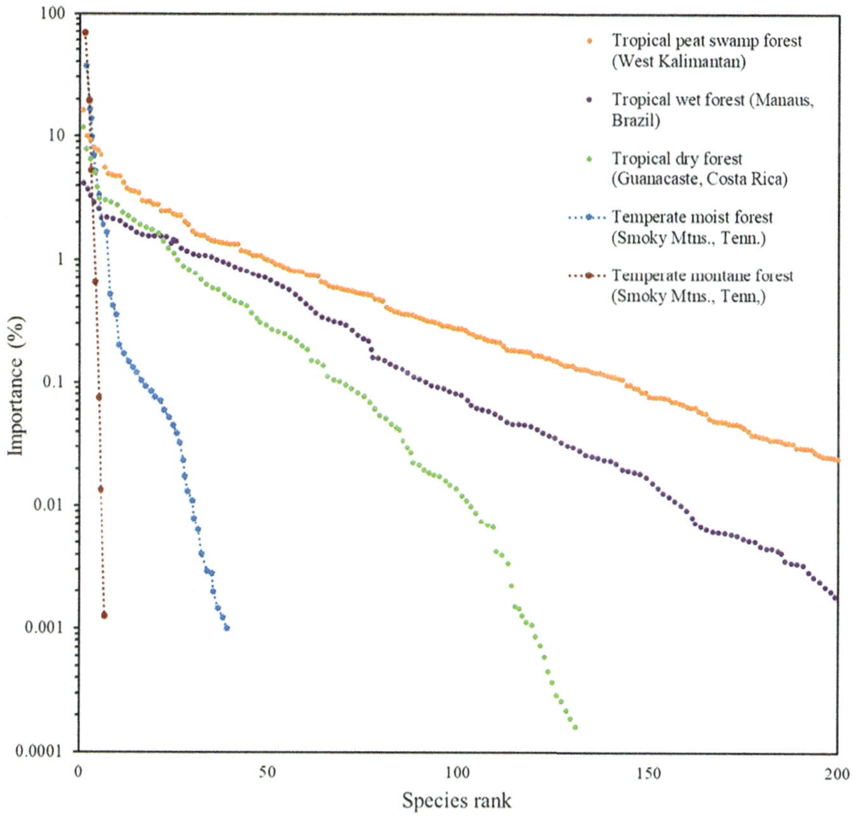

Fig. 2.14 Comparison of the dominance-diversity curves for the conservation area of MTI and other forest ecosystems. This figure is modified from Fig. 8 in Hubbell (1979) with additional data

Figure 2.14 shows a graph by Hubbell (1979) showing the dominance-diversity curves of tropical and temperate forests overlaid with the results of nine permanent plots in conservation forests in the MTI concession. For peat swamp forests, the community's most abundant species (rank-1 species) make up 10% of the total. Then, the percentage of species present decreases exponentially as species rank increases (i.e., as species become less common). The "slow rightward" slope indicates that the proportion of individual species declines with rank, but the decline is not rapid. In other words, the ecosystem may maintain rich biodiversity, and some species may be less dominant than others. The comparison between the importance value of rank-1 species was slightly higher than the 4.7 percent of tropical Amazonian forests and significantly lower when compared with temperate forests. The rank-1 species in the peat swamp forest was *Diospyros evena*, a species of the family Ebenaceae. The conservation forests of WSL-MTI may be more biodiverse than those of the Amazon because the slope continues to be gradual.

2.5.3 Fauna

Tropical peatlands not only support a wide variety of flora but are also home to numerous faunas, including several endangered and rare species. Although several studies on the faunal diversity of tropical peatlands have been conducted (Morrogh-Bernard et al. 2003; Posa et al. 2011), there is still much that is unknown, and the biodiversity has long been underestimated, with little attention to its conservation. Several species are probably going extinct before they are found and identified, given the threat that both natural and anthropogenic disturbances pose to tropical peatlands. Furthermore, without understanding the species composition that forms or inhabits tropical peatlands, there will be no successful restoration projects. From this, WSL-MTI has been conducting a series of fauna surveys to better understand species distribution, to record the (long-term) impacts of ongoing peatland management and the recovery of biodiversity, and to comprehend how this affects the provision of ecosystem services. The surveys include both direct and indirect observations, such as expert field monitoring of Orangutans and Proboscis monkeys, monitoring using camera traps and UAVs, and aquatic fauna inventories. Please see Chap. 4 of the previous book of Tropical Peatland Eco-Management for detailed information on peatland animal species found in the WSL-MTI area (Kato et al. 2021c).

2.6 Conclusion

"Integrity" is now strictly required in forest valuation and forest-derived sink carbon credits (forest carbon credit). To achieve this, quantitative assessment and monitoring are essential to accurately calculate the amount of CO_2 a forest absorbs and track this amount over time. This requires robust field surveys and remote sensing techniques to measure parameters such as forest area, tree height and diameter, and species type and distribution. Then, sustainability and permanence require that forest management generate carbon credits that have long-lasting effects. When forests are logged or lost due to natural disasters or disease, carbon is released into the atmosphere, and the value of the credits is lost. Therefore, it is essential to implement sustainable forest management. As discussed in this chapter, transparency and verification are essential to project design, implementation, and reporting to ensure the integrity of forest projects and carbon credits. It is also necessary to ensure that credits are calculated and reported fairly and accurately through a third-party verification process. Social and environmental considerations are also important. Forest projects should assess impacts on local communities and biodiversity and take mitigation measures as necessary. Forest stakeholders and local communities should be involved, consent should be obtained, and biodiversity should be conserved.

Through these actions, we hope to ensure that forest management and appropriate peatland management will genuinely contribute to climate change mitigation, as well as to biodiversity and water issues.

References

Anda M, Ritung S, Suryani E et al (2021) Revisiting tropical peatlands in Indonesia: semi-detailed mapping, extent and depth distribution assessment. Geoderma 402:115235. https://doi.org/10.1016/j.geoderma.2021.115235

Arai H, Hadi A, Darung U et al (2014) Land use change affects microbial biomass and fluxes of carbon dioxide and nitrous oxide in tropical peatlands. Soil Sci Plant Nutr 60:423–434. https://doi.org/10.1080/00380768.2014.903576

Brown C, Boyd DS, Sjögersten S et al (2018) Tropical peatland vegetation structure and biomass: optimal exploitation of airborne laser scanning. Remote Sens 10:1–21. https://doi.org/10.3390/rs10050671

CBD (2022) COP Biodiversity Convention. 2050 Vision and 2030 Mission. https://www.cbd.int/gbf/vision/

Chu D, Grasby SE, Song H et al (2020) Ecological disturbance in tropical peatlands prior to marine permian-triassic mass extinction. Geology 48:288–292. https://doi.org/10.1130/G46631.1

Czapiewski S, Szumińska D (2022) An overview of remote sensing data applications in peatland research based on works from the period 2010–2021. Land 11(1):24. https://doi.org/10.3390/land11010024

Dąbrowska-Zielińska K, Misiura K, Malińska A et al (2022) Spatiotemporal estimation of gross primary production for terrestrial wetlands using satellite and field data. Remote Sens Appl Soc Environ 27:100786. https://doi.org/10.1016/j.rsase.2022.100786

Dargie GC, Lewis SL, Lawson IT et al (2017) Age, extent and carbon storage of the Central Congo Basin peatland complex. Nature 542:86–90. https://doi.org/10.1038/nature21048

Dau JH, Mati A, Dawaki SA (2015) Role of forest inventory in sustainable forest management: a review. Int J For Hortic 1:33–40

Decree of the Minister of Forestry (2014) No SK 4192 of 2014

Dennis C, Aldhous P (2004) A tragedy with many players. Nature 430:396–398. https://doi.org/10.1038/430396a

Dommain R, Couwenberg J, Joosten H (2010) Hydrological self-regulation of domed peatlands in Southeast Asia and consequences for conservation and restoration. Mires Peat 6:1–17

Field RD, van der Werf GR, Fanin T, Fetzer EJ, Fuller R, Jethva H, Levy R, Livesey NJ, Luo M, Torres O, Worden HM (2016) Indonesian fire activity and smoke pollution in 2015 show persistent nonlinear sensitivity to El Niño-induced drought. Proc Natl Acad Sci 113(33): 9204–9209

Fortuniak K, Pawlak W, Siedlecki M et al (2021) Temperate mire fluctuations from carbon sink to carbon source following changes in water table. Sci Total Environ 756:144071. https://doi.org/10.1016/j.scitotenv.2020.144071

Graham LLB, Applegate GB, Thomas A et al (2022) A field study of tropical peat fire behaviour and associated carbon emissions. Fire 5:62. https://doi.org/10.3390/fire5030062

Guevara-Escobar A, González-Sosa E, Cervantes-Jimenez M et al (2021) Machine learning estimates of eddy covariance carbon flux in a scrub in the Mexican highland. Biogeosciences 18: 367–392. https://doi.org/10.5194/bg-18-367-2021

Gumbricht T, Roman-Cuesta RM, Verchot L et al (2017) An expert system model for mapping tropical wetlands and peatlands reveals South America as the largest contributor. Glob Chang Biol 23:3581–3599. https://doi.org/10.1111/gcb.13689

Guo M, Wang X, Li J et al (2012) Assessment of global carbon dioxide concentration using MODIS and GOSAT data. Sensors (Switzerland) 12:16368–16389. https://doi.org/10.3390/s121216368

Hardwick S, Graven H (2016) Satellite observations to support monitoring of greenhouse gas emissions. Grantham Brief Pap

Hein L, Spadaro JV, Ostro B et al (2022) The health impacts of Indonesian peatland fires. Environ Heal A Glob Access Sci Source 21:1–16. https://doi.org/10.1186/s12940-022-00872-w

Hooijer A, Silvius M, Wosten H, Page S (2006) Peat-CO_2: assessment of CO_2 emissions from drained peatlands in South East Asia. Wetlands International, Delft Hydraul, p 36

Hooijer A, Page S, Canadell JG et al (2010) Current and future CO 2 emissions from drained peatlands in Southeast Asia. Biogeosciences 7:1505–1514. https://doi.org/10.5194/bg-7-1505-2010

Hubbell SP (1979) Tree dispersion, abundance, and diversity in a tropical dry forest. Science 203: 1299–1309

Huijnen V, Wooster MJ, Kaiser JW et al (2016) Fire carbon emissions over maritime Southeast Asia in 2015 largest since 1997. Sci Rep 6:26886. https://doi.org/10.1038/srep26886

IUCN (2023) IUCN Red List of Threatened Species. https://www.iucnredlist.org/

Johnson AE, Knott CD, Pamungkas B et al (2005) A survey of the orangutan (Pongo pygmaeus wurmbii) population in and around Gunung Palung National Park, West Kalimantan, Indonesia based on nest counts. Biol Conserv 121:495–507. https://doi.org/10.1016/j.biocon.2004.06.002

Kalacska M, Arroyo-Mora JP, Soffer RJ et al (2018) Estimating peatland water table depth and net ecosystem exchange: a comparison between satellite and airborne imagery. Remote Sens 10: 687. https://doi.org/10.3390/rs10050687

Katic PG, Cerretelli S, Haggar J et al (2023) Mainstreaming biodiversity in business decisions: taking stock of tools and gaps. Biol Conserv 277:109831. https://doi.org/10.1016/j.biocon.2022.109831

Kato T, Osaki M, Tsuji N, Silsigia S (2021a) Principles of eco-management in a large-scale ecosystem of tropical peatland. In: Osaki M, Tsuji N, Foead N, Rieley J (eds) Tropical peatland eco-management. Springer, Singapore

Kato T, Silsigia S, Yusup AA, Osaki M (2021b) Large-scale practice on tropical peatland eco-management. In: Osaki M, Tsuji N, Foead N, Rieley J (eds) Tropical peatland eco-management. Springer, Singapore

Kato T, Silsigia S, Yusup AA, Osaki M (2021c) Coexistence of humans and nature in tropical peatlands. In: Osaki M, Tsuji N, Foead N, Rieley J (eds) Tropical peatland eco-management. Springer, Singapore

Könönen M, Jauhiainen J, Straková P et al (2018) Deforested and drained tropical peatland sites show poorer peat substrate quality and lower microbial biomass and activity than unmanaged swamp forest. Soil Biol Biochem 123:229–241. https://doi.org/10.1016/j.soilbio.2018.04.028

Krasnova A, Krasnov D, Peter H, et al (2022) Eddy-covariance carbon fluxes of a heterogeneous forest: one tower-two heights. EGUSphere, 1–34

Lähteenoja O, Page S (2011) High diversity of tropical peatland ecosystem types in the Pastaza-Marañón basin, Peruvian Amazonia. J Geophys Res Biogeosciences 116. https://doi.org/10.1029/2010JG001508

Lees KJ, Quaife T, Artz RRE et al (2018) Potential for using remote sensing to estimate carbon fluxes across northern peatlands – a review. Sci Total Environ 615:857–874. https://doi.org/10.1016/j.scitotenv.2017.09.103

Miettinen J, Liew SC (2010) Degradation and development of peatlands in peninsular Malaysia and in the islands of Sumatra and Borneo since 1990. L Degrad Dev 21:285–296. https://doi.org/10.1002/ldr.976

Miettinen J, Shi C, Liew SC (2017) Fire distribution in peninsular Malaysia, Sumatra and Borneo in 2015 with special emphasis on peatland fires. Environ Manag 60:747–757. https://doi.org/10.1007/s00267-017-0911-7

MoEF (2016) Regulation of the Government of the Republic of Indonesia - Ministry of Environment and Forestry (MoEF) Number 57 of 2016 Concerning Amendment to Government

Regulation Number 71 of 2014 Concerning Peatland Ecosystem Protection and Management. 1–34

MoEF (2017) Regulation of the Government of the Republic of Indonesia - Ministry of Environment and Forestry (MoEF) Number 15 of 2017 Concerning Procedures for Measurement of Groundwater at The Point of Peat Ecosystem. 1–11

Morrogh-Bernard H, Husson S, Page SE, Rieley JO (2003) Population status of the Bornean orangutan (Pongo pygmaeus) in the Sebangau peat swamp forest, Central Kalimantan, Indonesia. Biol Conserv 110:141–152. https://doi.org/10.1016/S0006-3207(02)00186-6

Murdiyarso D, Hergoualc'h K, Verchot LV (2010) Opportunities for reducing greenhouse gas emissions in tropical peatlands. Proc Natl Acad Sci U S A 107:19655–19660. https://doi.org/10.1073/pnas.0911966107

Nakalembe C, Becker-Reshef I, Bonifacio R et al (2021) A review of satellite-based global agricultural monitoring systems available for Africa. Glob Food Sec 29:100543. https://doi.org/10.1016/j.gfs.2021.100543

Ng PKL, Tay JB, Lim KKP (1994) Diversity and conservation of Blackwater fishes in peninsular Malaysia, particularly in the North Selangor peat swamp forest. Hydrobiologia 285:203–218

Page SE, Siegert F, Rieley JO et al (2002) The amount of carbon released from peat and forest fires in Indonesia during 1997. Nature 420:61–65. https://doi.org/10.1038/nature01131

Palamba P, Ramadhan ML, Perdana LR (2020) The proceedings of 11th Asia-Oceania Symposium on Fire Science and Technology. Proc 11th Asia-Oceania Symp Fire Sci Technol. https://doi.org/10.1007/978-981-32-9139-3

Parwati MP, Kusumaning Ayu DS, M. Rokhis Khomarudin (2011) Dinamika Kondisi Potensi Bahaya Kebakaran Hutan/Lahan di Sumatera dan Kalimantan Tahun 2009–2011 Berdasarkan Fire Weather Index (FWI) dari Data Satelit Penginderaan Jauh. rosiding Semin. Nas. Pemanfaat. Data Penginderaan Jauh Untuk Mendukung Pembang. Nas. 83–90

Posa MRC, Wijedasa LS, Corlett RT (2011) Biodiversity and conservation of tropical peat swamp forests. Bioscience 61:49–57. https://doi.org/10.1525/bio.2011.61.1.10

Prat-Guitart N, Rein G, Hadden RM et al (2016) Effects of spatial heterogeneity in moisture content on the horizontal spread of peat fires. Sci Total Environ 572:1422–1430. https://doi.org/10.1016/j.scitotenv.2016.02.145

Putra EI, Cochrane MA, Vetrita Y et al (2018) Determining critical groundwater level to prevent degraded peatland from severe peat fire. IOP Conf Ser Earth Environ Sci 149:012027. https://doi.org/10.1088/1755-1315/149/1/012027

Qin Y, Musa DNS, Lin S, Huang X (2022) Deep peat fire persistently smouldering for weeks: a laboratory demonstration. Int J Wildl Fire 32:86–98. https://doi.org/10.1071/wf22143

Rahman MM, McDermid GJ, Strack M, Lovitt J (2017) A new method to map groundwater table in peatlands using unmanned aerial vehicles. Remote Sens 9:1–14. https://doi.org/10.3390/rs9101057

Ribeiro K, Pacheco FS, Ferreira JW et al (2021) Tropical peatlands and their contribution to the global carbon cycle and climate change. Glob Chang Biol 27:489–505. https://doi.org/10.1111/gcb.15408

Silviana SH, Saharjo BH, Sutikno S (2020) Fire risk analysis based on groundwater level in rewetting peatland, Sungaitohor village, kepulauan Meranti district, Riau province. IOP Conf Ser Mater Sci Eng 796:012041. https://doi.org/10.1088/1757-899X/796/1/012041

Stocks BJ, Lawson BD, Alexander ME et al (1989) The Canadian Forest fire danger rating system: an overview. For Chron 65:258–265

Takakai F, Morishita T, Hashidoko Y et al (2006) Effects of agricultural land-use change and forest fire on N_2O emission from tropical peatlands, Central Kalimantan, Indonesia. Soil Sci Plant Nutr 52:662–674. https://doi.org/10.1111/j.1747-0765.2006.00084.x

Tan ZD, Sutikno S, Carrasco LR, Taylor D (2023) Local community representations of tropical peatlands and implications for restoration in Riau, Indonesia. Restor Ecol. https://doi.org/10.1111/rec.13900

Tata HL, Nuroniah HS, Ahsania DA et al (2022) Flooding tolerance of four tropical peatland tree species in a nursery trial. PLoS One 17:1–20. https://doi.org/10.1371/journal.pone.0262375
UNEP (2022) Global peatlands assessment: the state of the World's peatlands and sustainable management of peatlands
Usup A, Hayasaka H (2023) Peatland fire weather conditions in Central Kalimantan. Fire 6:182
Usup A, Hashimoto Y, Takahashi H, Hayasaka H (2004) The principal types of vegetation in the. Tropics 14:1–19
Vernimmen R, Hooijer A, Akmalia R et al (2020) Mapping deep peat carbon stock from a LiDAR based DTM and field measurements, with application to eastern Sumatra. Carbon Balance Manag 15:1–18. https://doi.org/10.1186/s13021-020-00139-2
Volkova L, Krisnawati H, Qirom MA et al (2023) Fire and tree species diversity in tropical peat swamp forests. For Ecol Manag 529:120704. https://doi.org/10.1016/j.foreco.2022.120704
Wijedasa LS, Jauhiainen J, Könönen M et al (2017) Denial of long-term issues with agriculture on tropical peatlands will have devastating consequences. Glob Chang Biol 23:977–982. https://doi.org/10.1111/gcb.13516
Wong CWY, Wong CY, Boon-Itt S, Tang AKY (2021) Strategies for building environmental transparency and accountability. Sustainability 13:9116. https://doi.org/10.3390/su13169116
Worden J, Jiang Z, Jones DBA et al (2013) El Niño, the 2006 Indonesian peat fires, and the distribution of atmospheric methane. Geophys Res Lett 40:4938–4943. https://doi.org/10.1002/grl.50937
World Bank (2016) The cost of fire. An economic analysis of Indonesia's 2015 fire crisis Indonesia sustainable landscapes knowledge note 1. The World Bank Group, p 9
Wösten JHM, Clymans E, Page SE et al (2008) Peat-water interrelationships in a tropical peatland ecosystem in Southeast Asia. Catena 73:212–224. https://doi.org/10.1016/j.catena.2007.07.010
Yu Z, Loisel J, Brosseau DP et al (2010) Global peatland dynamics since the last glacial maximum. Geophys Res Lett 37:1–5. https://doi.org/10.1029/2010GL043584

Chapter 3
Net Ecosystem Exchange (NEE) Mapping in Tropical Peatland

Albert Sulaiman, Arief Darmawan, Yuta Izumi, Awaluddin, Nobuyuki Tsuji, and Mitsuru Osaki

Abstract This chapter provides various methods for assessing net ecosystem exchange (NEE) in tropical peatlands, such as the use of satellite imagery. Eco-evaluations related to carbon trade and environmental quality are provided. The discussion begins with the concept of NEE and its relationship with the groundwater level (GWL) based on in situ measurements. NEE mapping is conducted with both radar and optical satellites. NEE mapping is performed based on two approaches: GWL and peat subsidence mapping. Based on the maps, carbon valuations are carried out, as carbon is a form of natural capital. Finally, a discussion on carbon trading and environmental quality improvement is given.

Keywords Carbon trading · Groundwater level · Net ecosystem exchange · Peat subsidence

3.1 Introduction

Tropical peat forests, without a doubt, play a crucial role in the world's carbon cycle, which is key to global warming. Approximately 20.3% (36.9 million ha) of tropical peatlands are located in Asia. This peat represents a large amount of global carbon (C), and 88 tC ha^{-1} of this C is released into the atmosphere. The environmental

A. Sulaiman (✉) · A. Darmawan · Awaluddin
Research Center for Climate and Atmosphere, Badan Riset dan Inovasi Nasional (BRIN), Bandung, West Java, Indonesia
e-mail: albe002@brin.go.id

Y. Izumi
Muroran Institute of Technology, Muroran, Japan

N. Tsuji
Japan Peatland Society, Fukuoka, Japan

M. Osaki
Research Faculty of Agriculture, Hokkaido University, and Japan Peatland Society, Sapporo, Japan

M. Osaki et al. (eds.), *Tropical Peatland Eco-evaluation*,
https://doi.org/10.1007/978-981-99-6790-2_3

factors related to climate change influencing tropical peatlands include changes in temperature, rainfall, the composition of the atmosphere, the occurrence of fires, and smoke haze (Leng et al. 2019). Furthermore, reality shows that tropical forests and peatlands are efficient carbon sinks. However, on the other hand, they can be vast carbon emitters if damaged (through land use changes (LUCs), degradation, or burning). This enormous carbon emission causes climate warming to occur more rapidly (Hapsari et al. 2022). Analyses have shown that the rising sea level due to global warming causes coastal peat forests to dry up, making them vulnerable to peat forest fires. These fires increase global warming, so global warming and tropical forest fires have a mutual feedback mechanism.

The dynamics and quantization of how much carbon (C) is released by forests and peatlands can be studied through the net ecosystem exchange (NEE), which is a measure of the net exchange of C between an ecosystem and the atmosphere (per unit of ground area). The NEE is measured by eddy covariance (EC) instruments and is a primary gauge of the carbon sink strength of an ecosystem (Kremer et al. 2002). NEE is released into the atmosphere when peatlands dry, i.e., when the water level drops, causing peat decomposition. The results of NEE measurements obtained with EC instruments in Palangkaraya, Central Kalimantan, Indonesia, in 2004 and 2005 (non-ENSO) showed that the annual amount of net ecosystem CO_2 exchange between May 2004 and May 2005 was positive. The study stated that the peatland ecosystem was a source of atmospheric CO_2. The annual NEE of peatlands ranges between 100 and 800 gC m^{-2} $year^{-1}$ and is largest for peatlands that have been used and smallest for pristine peatland forests (Hirano et al. 2016).

In situ measurements performed with EC instruments provide knowledge of how the NEE varies in a time series. We still need to understand the spatial variation in NEE. Some of the developed technologies, such as satellites, allow us to do this. NEE information is usually not directly obtained by satellite but is derived based on the parameter information measured by the satellite. In other words, satellite-derived NEE data represent a form of estimation. For example, Park et al. 2020 used rainfall global satellite mapping and ground surface temperature data alongside the moderate-resolution imaging spectroradiometer (MODIS) satellite, which produces active photosynthetic radiation, and burn area data. All these input variables were satellite-based and used for estimating the groundwater level (GWL) by combining soil moisture and rainfall into the Keetch–Byram drought index (KBDI). The GWL was then used to estimate respiration (Re) based on the empirical relationship between the GWL and Re. Fire, biomass, and gross primary productivity emissions were estimated using MODIS. Combinations of many of the above parameters can be used to estimate the NEE.

The problems that occur in forests and peatlands are not only natural but are also related to socioeconomic problems, which are currently becoming an increasing problem associated with peatland reductions globally, especially in tropical peat areas in Asia such as Indonesia. In the last three decades, there has been a massive change from tropical forest peatlands to crop production systems, such as oil palm and pulpwood plantations, which cause large greenhouse gas (GHG) emissions. The estimated total carbon loss resulting from the conversion of peat swamp forests to oil palm plantations is 59.4 ± 10.2 Mg CO_2 per hectare per year for 25 years (Murdiyarso et al. 2010). The high annual C emissions, which are indicated by changes in land use, show that peatlands are still considered idle lands with no value, so communities and local governments can easily convert local peatlands into plantations. This means that forests and peatlands must come to be recognized as a bioapparatus serving as natural capital (Osaki et al. 2021).

3.2 Concept of NEE Mapping for Tropical Peatlands

One way to monitor NEE spatially is remote sensing. Through remote sensing, an area's characteristics can be detected and monitored by measuring the reflected and emitted physical parameters at a certain distance. This technique usually uses satellites, airplanes, and unmanned aerial vehicles. Remote sensing does not directly measure NEE but instead measures physical parameters, so the corresponding NEE values must be estimated; therefore, knowledge about how the NEE interacts with environmental parameters is very important.

3.2.1 In Situ Measurements

The NEE is the sum of the net assimilation rate per unit ground area, nighttime leaf respiration rate per unit ground area, respiration of other aerial plant tissues, respiration of the root tissues, and heterotrophic respiration. In general, the NEE can be classified into two categories: the sum of the rate of change in the plant carbon content and the rate of change in the soil carbon content (gCm^{-2} ground d^{-1}). In practice, we do not count these exchanges one-by-one, but by the total of all exchanges, namely, the NEE emitted above the canopy of a forest. An instrument used for this is the eddy-covariance instrument, which consists of a 3D ultrasonic anemometer combined with a fast-infrared gas analyzer installed above the plant canopy. Due to the precision of the sensor, it can record fluctuations or turbulence in the atmosphere above the forest. The turbulent flux of the covariant between the wind and the fluctuating CO_2 concentration is what becomes NEE carbon dioxide.

EC is a popular micrometeorological method for directly observing gas, energy, and momentum exchanges between peat forest ecosystems and the atmosphere. The sensor of the EC instrument is placed over the canopy and measures the area of interest. The CO_2 flux from a forest is calculated as the covariance of the vertical wind speed and CO_2 concentration. Following micrometeorological conventions, a negative value indicates the flux from the atmosphere into the forest, while a positive value denotes the release of CO_2 from the forest into the atmosphere. NEE measurements recorded in the Taman Nasional Sebangau peatland forest are depicted in Fig. 3.1.

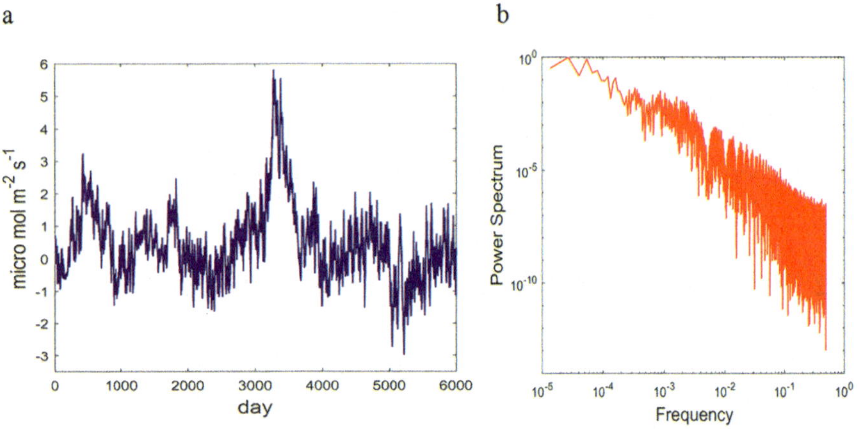

Fig. 3.1 (**a**) Time series of NEE observed by an eddy covariance instrument in Taman Nasional Sebangau and (**b**) its power spectrum

Sebagau National Park is a pristine peat forest where NEE values are alternately positive and negative. In general, NEE is negative during the rainy season (November–December–January) and positive during the dry season (June–August–September). This NEE variation is related to the variability in the GWL, indicating that the carbon balance in tropical peatlands is strongly affected by local hydrology. During the dry season, the GWL decreases, resulting in peat decomposition, which produces a large amount of CO_2 in the atmosphere (Sundari et al. 2012). This condition is the opposite in the rainy season, where the GWL increases and the peat soil does not decompose. The trees above absorb a large amount of CO_2 from the atmosphere, so the NEE is negative.

It is well known that the GWL is a principal parameter used to evaluate peat ecosystems. If the GWL becomes excessively low, then oxygen enters and oxidizes the peat through microorganisms, resulting in the peat emitting CO_2 and becoming vulnerable to fire. There is a positive correlation between the surface peat water content and GWL and a negative correlation between the NEE and GWL (Tsuji et al. 2019, Hirano et al. 2012; Mezbahuddin et al. 2014). The lowering of the GWL tends to increase soil respiration and finally increase carbon emissions (Hirano et al. 2016; Sundari et al. 2012). The relationship between the NEE and GWL is depicted in Fig. 3.2. The relationship between the NEE and GWL is complicated and nonlinear. However, it can be roughly approximated by a linear correlation where NEE generally increases as the GWL decreases. This relationship may be unique for each peat hydrological unit, and Sebangau National Park has the relationship $NEE = -2.7GWL + 0.046$, where the GWL is in meters and the NEE is in micromoles m^{-2}.

The variability of NEE is also seasonally dependent, as shown in Fig. 3.2. There was high variability in NEE during the period between 2004 and 2008; there were two El Niño events (moderate) during this time and two La Niña events. The El Niño

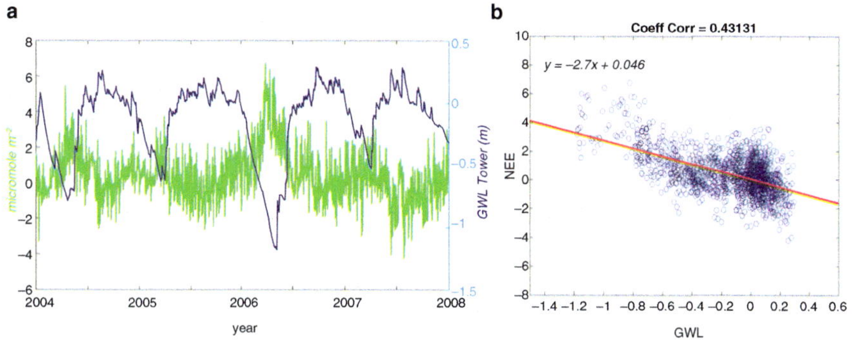

Fig. 3.2 Empirical relation between the NEE and GWL at Taman National Sebangau. NEE in micromoles m^{-2} and GWL in meters

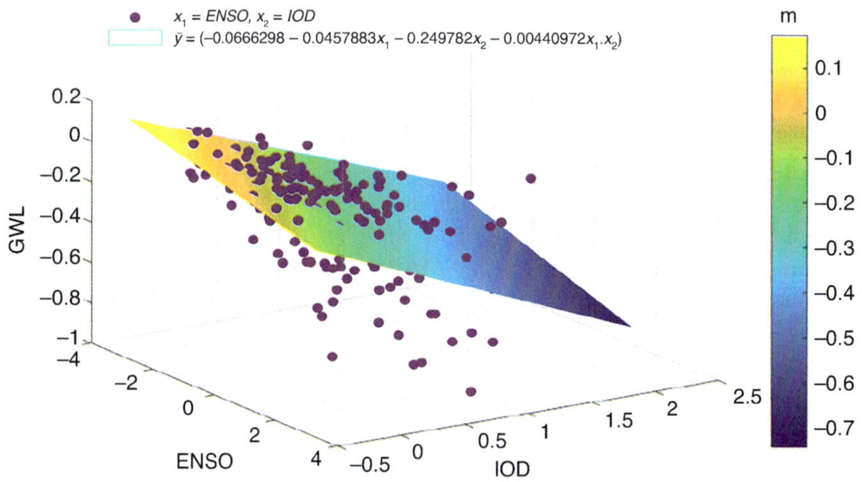

Fig. 3.3 Correlations between the GWL and El Niño and the IoD

events occurred between August 2004 and October 2005 and from June 2006 to July 2007, with amplitudes of approximately 0.8 °C and 1.2 °C, respectively. The other events were La Niña events, with signals at approximately −0.8 °C and −1.7 °C, respectively. In the first El Niño event, the GWL dropped to −0.75 m. Then, during the second El Niño event, the GWL dropped to −1.3 m. During these El Niño events, the NEE increased significantly by approximately 4 micromoles m^{-2} and 6 micromoles m^{-2}, respectively. The GWL is strongly correlated with El Niño and was recently found to be correlated with the Indian Ocean dipole (IOD) mode, as shown in Fig. 3.3.

3.2.2 NEE Mapping Based on Remote Sensing

To date, spatial NEE mapping can be performed only through remote sensing with various vehicles. However, we must remember that such vehicles cannot directly measure the NEE value but instead measure the physical parameters of the Earth's surface, so a model or formula for estimating the NEE is needed with these parameters measured by a remote sensing image. Two models can be developed: models based on fundamental equations and empirical models. Models based on fundamental equations relate electromagnetic waves (from active EM sensors) to the CO_2 molecules emitted from the forest surface. In such cases, the models deal with electromagnetic scattering. In general, this is difficult to do and impractical. The thing that makes more sense and is simpler is using an empirical model, namely, making a model based on the relationships between physical parameters and NEE, such as the relation between the NEE and GWL. Thus, we are dealing with empirical science.

3.2.2.1 Soil Moisture Satellite

One of the technologies that is the cheapest and most accessible for estimating NEE in peatlands is the Soil Moisture Active Passive (SMAP) satellite. This satellite has the mission of measuring the amount of water on Earth's ground surface. Since beginning operation in April 2015, SMAP has been designed to measure soil moisture once every 2–3 days. The SMAP image in the form of soil moisture from the surface to a depth of 5 cm has a GeoTIFF format. The data type used to estimate NEE is SMAP Enhanced L3 Radiometer Global Daily 9-km EASE-Grid Soil Moisture V001 data. The resolution can be downscaled to 3 km using the schematic enhanced-resolution SMAP TB products on both 3-km SMAP project grids. The process is depicted in Fig. 3.4. The detailed SMAP image data processing methods can be seen in Osaki et al. 2021, Chap. 5.

3.2.2.2 Synthetic Aperture Radar

Peatland subsidence monitoring leads to NEE estimations because the microbial decomposition of peat results in peat loss. Spaceborne synthetic aperture radar (SAR) allows regional-scale surface deformation measurements. SAR is known to be an active microwave sensor enabling the Earth's surface to be sensed under almost all weather conditions, as microwaves can penetrate clouds and, to some extent, rain; this is an appropriate property for obtaining measurements in tropical regions (Ulaby and Long 2014). This subsection demonstrates tropical peatland deformation monitoring by spaceborne SAR remote sensing (Izumi et al. 2022a, b).

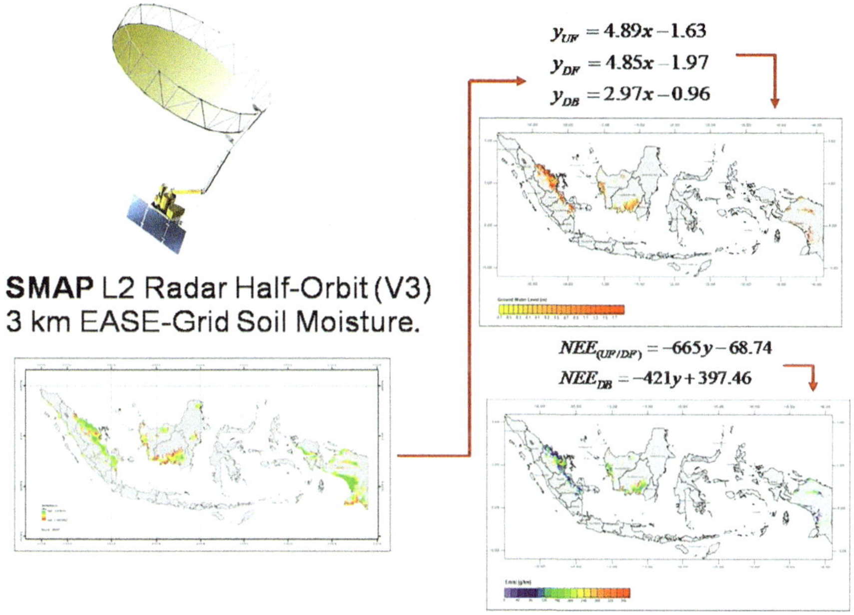

Fig. 3.4 An algorithm for estimating NEE in peatlands from SMAP imagery

Time-Series InSAR

SAR generates 2D complex images of the observed scene consisting of amplitude and phase information. The phase difference between two SAR images contains the target's in-depth information. The interferometric SAR (InSAR) technique accounts for the phase difference to estimate the topography and land deformation of the Earth's surface (Graham 1974). For the deformation map, a digital elevation model (DEM) is employed to minimize the topographic effect on the phase difference, known as differential InSAR (DInSAR). In addition to the topographic phase, nuisance terms should be dealt with to ensure precise deformation estimations. In general, the unwrapped phase difference $\Delta\phi$can be expressed as follows (Graham 1974):

$$\Delta\phi = \Delta\phi_{\text{defo}} + \Delta\phi_{\text{atm}} + \Delta\phi_{\text{topo}} + \Delta\phi_{\text{ion}} + \Delta\phi_{\text{decorr}}, \tag{3.1}$$

where $\Delta\phi_{\text{defo}}$ denotes the phase difference due to deformation, $\Delta\phi_{\text{atm}}$ denotes the atmospheric phase term, known as the atmospheric phase screen (APS) caused by temporal changes in tropospheric refractivity, $\Delta\phi_{\text{topo}}$ denotes the residual topographic phase due to DEM error, $\Delta\phi_{\text{ion}}$ denotes the ionospheric phase term caused by the temporal change in the number of total electrons within the ionosphere, and $\Delta\phi_{\text{decorr}}$ denotes the decorrelation noise caused by several error sources (small correlation length) (Hanssen 2001).

To estimate $\Delta\phi_{defo}$ precisely, the time-series InSAR (TInSAR) technique has been studied. This method requires a large number of SAR images representing the same scene at different times, while DInSAR generally employs two-, three-, and four-pass interferometry. TInSAR can avoid decorrelation sources by taking into account only stable pixels in terms of their temporal scattering properties, named coherent scatter (CS) in this study.

Several TInSAR algorithms are available, categorized by the CS selection criterion, deformation estimation mathematical approach, and interferogram pair selection. The small baseline subsets (SBAS) approach forms interferograms characterized by small spatiotemporal baselines and estimates deformation by least-squares (LS) estimation of a series of unwrapped phases (Berardino et al. 2002; Yunjun et al. 2019). This algorithm is considered an efficient method for rapidly decorrelated pixels such as vegetated pixels; thus, it is appropriate for tropical peatland regions. In particular, we employed the SBAS routine workflow proposed by Yunjun et al. for tropical peatland deformation monitoring (Yunjun et al. 2019). This algorithm was implemented as open-source software, named Miami InSAR time-series software in Python (MintPy) (https://github.com/insarlab/MintPy ((accessed on Nov. 2022)).

Data Processing

We herein demonstrate the deformation estimation result obtained across central Kalimantan and south Kalimantan in Indonesia from freely available Sentinel-1A SAR data using the C-band (5.405 GHz) central frequency, acquired by the Copernicus Programme satellite constellation conducted by the European Space Agency (ESA). The observation mode is interferometric wide swath mode. Sentinel-1A is capable of observing scenes with a 12-day revisit time. Three years of Sentinel-1A vertical-vertical-polarization SLC data obtained from January 2018 to January 2021 were processed.

We used the stack sentinel processor within ISCE2 open-source software (https://github.com/isce-framework/isce2 (accessed on Nov. 2022)) for the coregistered and phase-unwrapped interferogram stack generation. Then, the interferograms were further processed by MintPy to produce a deformation time series. We applied multilooking to each interferogram by 23 in the range (100 m) and 7 in the azimuth direction (100 m) and applied the Goldstein filter with a strength of 0.5. The interferograms were paired with their three nearest neighbors forward in time to limit the maximum temporal baseline to 36 days. The weighted LS inversion approach was applied to a stack of unwrapped interferograms for the deformation estimation; in this method, the weights were defined by the inverse of the phase variance (Yunjun et al. 2019). Finally, APS was compensated by Global Atmospheric Models in MintPy. The ionospheric phase term was omitted in our processing because C-band data have a relatively low ionospheric effect.

We divided the three-year SAR images into three single-year datasets, i.e., 2018, 2019, and 2020 datasets, and estimated the displacement time series for each year.

For TInSAR processing, a reference point indicating the position with known deformation was set. In our case, we used Global Navigation Satellite System (GNSS) observation positions installed by the Geospatial Information Agency (BIG) of Indonesia (Aditiya et al. 2014).

Deformation Results

The vertical deformation velocities estimated across central Kalimantan and southern Kalimantan for 2018, 2019, and 2020 are displayed in Fig. 3.5a, b, c. We used three Sentinel-1A frames to construct these maps. Notably, only the deformation velocity results obtained over CSs are shown in Fig. 3.5. In the figures, a negative deformation sign corresponds to subsidence. From the results, the estimated deformation velocity maps reveal overall subsidence throughout the observed area, especially in the peatlands indicated by the orange hatched lines. Furthermore, the deformation results show larger subsidence in 2019 than in other years. This may be caused by the severe drought recorded in 2019 due to the positive of IOD. A

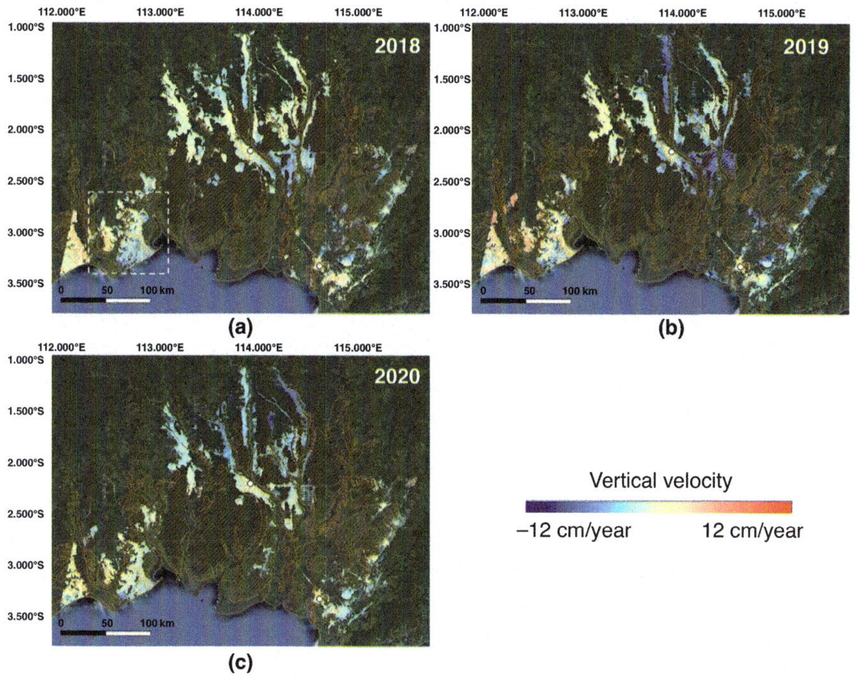

Fig. 3.5 Vertical deformation velocity images of central Kalimantan and south Kalimantan. Three Sentinel-1A frames were processed to construct these maps. The white circles indicate the GNSS locations used as reference points in the TInSAR processing. The orange hatched lines correspond to the peatland distribution. (**a**) Results in 2018. (**b**) Results in 2019. (**c**) Results in 2020

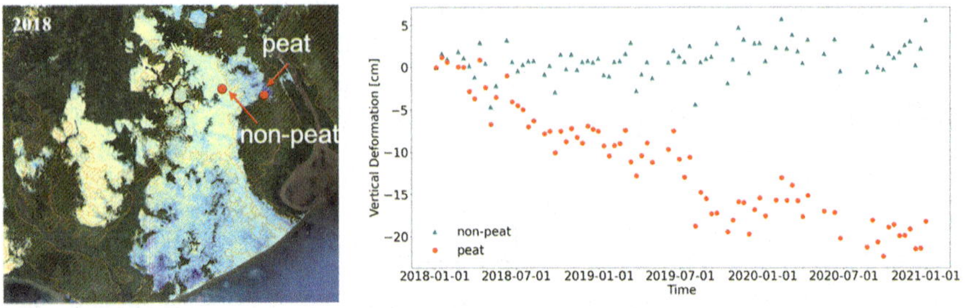

(a) (b)

Fig. 3.6 Deformation time-series in peatland and nonpeatland areas. (**a**) Enlarged view of Fig. 3.5. (**a**) indicated by the dashed rectangle. The two investigated points are shown as red circles. (**b**) Deformation time-series results obtained for the two points

significant GWL decrease in 2019 led to peat shrinkage and accelerated peat microbial decomposition; thus, the peatlands showed larger subsidence that year.

Deformation time series were investigated in both nonpeat and peat areas. Two investigation points are shown as red circles in Fig. 3.6a. The two investigation points were located close to each other, one on the peatlands and the other on a nonpeat area. The deformation time series at both points are shown in Fig. 3.6b. The peatland results reveal large subsidence (~20 cm in 3 years), while the nonpeatland results do not show any subsidence. A sudden large subsidence in the dry season of 2019 can also be seen in the peatland time-series results.

The relationship between peat subsidence and the NEE was obtained not directly but by utilizing the relationship between the NEE and GWL. We already know that a decrease in the peat surface accompanies a decrease in GWL over long time periods. In other words, with peat subsidence mapping, we can estimate the NEE released into the atmosphere. The linear correlation regression performed between the subsidence rates and the mean water table depth of a drained forest with a water depth of 0–0.7 m is $S = 0.41–6.04 \times \text{GWL}$, where the GWL is in meters and S is in centimeters with an R^2 value of ~0.8 (Hooijer et al. 2012). Based on the above relationships, we obtained NEE as follows:

$$\text{NEE} = 0.43\text{S} - 0.18 \quad (3.2)$$

where NEE is in micromoles m^{-2}. The NEE estimation method based on peat subsidence is depicted in Fig. 3.7.

Optical Satellite

Optical satellites are very useful for determining LUCs, especially forest changes, where changes in land cover can be used to estimate net ecosystem production (NEP), which in turn helps estimate carbon stocks and carbon emissions. The

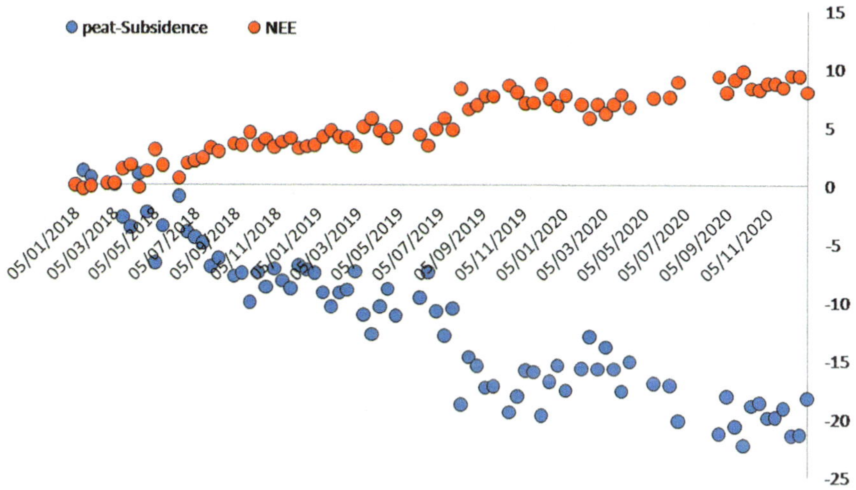

Fig. 3.7 NEE calculation based on peatland subsidence. The red dot is the NEE, the blue dot is the peat subsidence with a position at 112.90494500, −2.88473500, and the pixel resolution is 100 m × 100 m

changes in carbon sequestration and emissions affected by the deforestation of peat forests and LUCs can be quantified by comparing the NEP among different types of land cover in an ecosystem, including tropical peat (Basuki et al. 2021). Many studies have reported soil respiration differences due to land use. For example, soil respiration was higher in intact peat forests than in palm oil and sago estates. The degradation and LUCs of peat swamp forests significantly affect the NPP, and tropical peatland landscapes are now considered net carbon sources with four times more degraded peat forests than intact forest (for example Fig. 3.8). The carbon loss can be calculated using the formula given by the IPCC.

NEE was also estimated with optical satellites indirectly through a GWL estimation by applying a modified KBDI (mKBDI), where the empirical relation is expressed as GWL = −0.0045 × mKBDI (Park et al. 2020). With GWL knowledge, an empirical model can be constructed between the GWL and NEE obtained from the EC, and NEE estimates can be obtained.

3.3 Carbon Valuation of Tropical Peatlands Based on NEE Mapping

Peatlands are not seen as a gift from nature, so they can easily be exploited (low price). However, peatlands must be seen as entities with high prices or known as natural capital. Natural capital provides sustainable goods and services, but accurate calculations are needed for this benefit to be better understood. Peat forests function

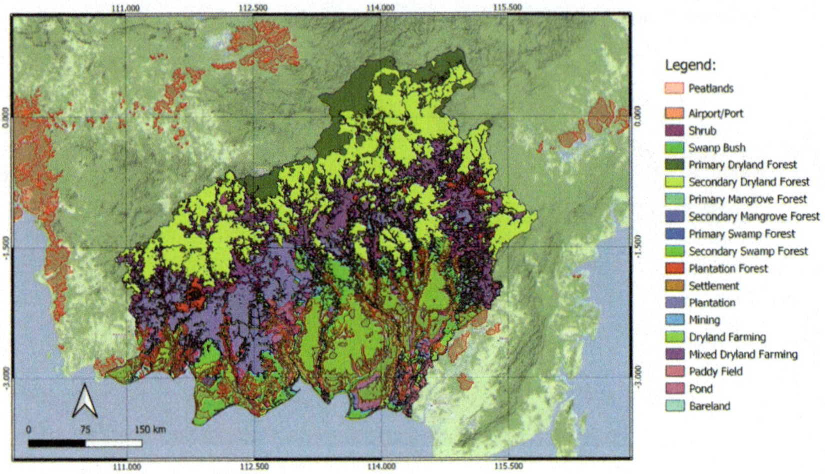

Fig. 3.8 Changes in land use from peatlands to plantations cause peatlands to dry up and burn. The map shows an example of land use in Central Kalimantan, Indonesia

in the water–carbon–tree relationship as a bioapparatus that can be used as natural capital. Simply put, a tree can function like a solar panel (leaves), battery (stems), air conditioning (evapotranspiration), carbon sequestration (biochar), fertilizer production (N_2 fixation), and water dam (water reserves in the soil) (Osaki et al. 2021). Apart from the trees, the function of an ecosystem can also be as natural capital. The next problem is how to quantify natural capacity economically. This is challenging. For example, short-term economic calculations based on the gross domestic product fail to assign accurate natural capital values (Osaki and Kasuga 2022).

There are several ways to evaluate peatlands: the first is by analyzing the nonuse value. Although this is not easy, it is one of the most effective ways to determine natural capital. In this regard, calculating the carbon price is one method in which the accuracy of the peatland carbon calculation largely determines the accuracy of the results. To date, peatland growth has reached a steady state with a dome shape, so that the stored carbon is in top condition. Thus, the estimated carbon capacity is related to the volume of peat in the dome. The study results show that the accumulated volume of peat confined between two canals is proportional to the square of the parcel area times a scale-independent factor (Cobb et al. 2020). This factor describes the shape of the parcel boundary at an earlier time. Generally, the carbon capacity per acre is proportional to the peat area. The carbon capacity can be estimated in near real time using the SMAP with empirical models (Osaki et al. 2021). The NEE map obtained based on the SMAP is depicted in Fig. 3.9.

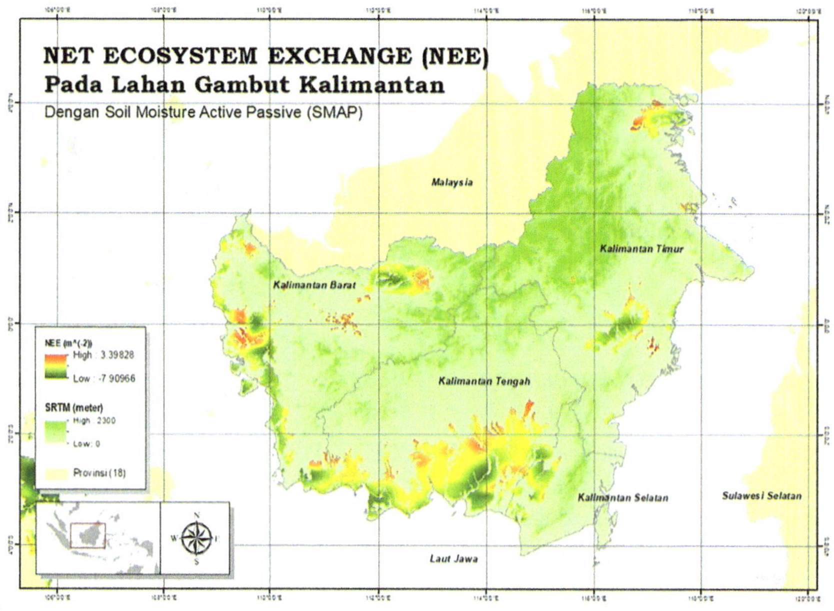

Fig. 3.9 NEE map constructed based on groundwater on Kalimantan Island

3.4 Carbon Trading and Environmental Quality Improvement

The carbon tax is a tax levied on carbon emissions that harm the environment. These mechanisms financially promote emission reductions but do not have the flexibility or guarantees of emission reductions. Another mechanism is the cap and trade method. Caps are defined as limits on the number of permitted emissions. To set emission limits per sector, countries must calculate and report how much GHG they emit each year to be evaluated against baseline emission levels and country-level targets. If an emitter does not use all of its permits, it can sell the unused permits to other countries that want to emit more carbon dioxide than is allowed.

The idea of carbon trading is expected to be one of the solutions to tackling climate change at local and global scales. Of the many ideas about efforts to reduce carbon emissions, carbon trading is a mechanism designed to provide economic incentives to limit GHG. This mechanism is generally referred to as carbon trading because carbon dioxide (CO_2) is the dominant GHG. Carbon trading was coined for the first time at the 1997 Conference of States Parties (COP) III to the United Nations Framework Convention on Climate Change (UNFCCC), which forms the Kyoto Protocol.

The Kyoto Protocol is a United Nations agreement that has set goals for reducing global carbon emissions and mitigating climate change since 2005. The idea is to incentivize each country to reduce its carbon emissions in order to maintain the remaining sales licenses. Relatively large and rich countries are effectively subsidizing the efforts of poorer and more polluted countries by buying their credits. Some argue that carbon trading is a cost-effective partial solution to the problem of climate change and promotes the adoption of innovative technologies. However, some criticisms have arisen regarding the trading of carbon emissions; according to these criticisms, carbon trading is seen as a noncomprehensive solution to widespread and urgent global warming. Despite these criticisms, carbon trading remains a central concept in many proposals to reduce or mitigate climate change and global warming.

3.4.1 The Paris Agreement

The Paris Agreement adopted at the UNFCCC COP 21 in December 2015 marked a new chapter in dealing with climate change. The States Parties are committed to taking action to limit the global temperature rise to below 2 °C. To implement this commitment to control the increase in the global temperature, State Parties can cooperate in fulfilling their emission reductions as stipulated in Article 6 of the Paris Agreement. Cooperation between countries can be achieved through the trading of emission reductions among countries. The Paris Agreement does not explicitly mention a market mechanism or carbon market in the agreement but allows contracting parties to use a "cooperative approach" and voluntarily use "international transfers of mitigation results (ITMO)" to help them achieve GHG emission reduction targets. Before the adoption of the Paris Agreement, carbon trading was implemented through various mechanisms, including the Clean Development Mechanism (CDM) under the Kyoto Protocol and the Joint Credit Mechanism (JCM).

The CDM is a mechanism that can be implemented between Annex I countries and non-Annex I countries with an "offset" (the buyer obtains Certified Emission Reduction/CER credit from the CDM project). Moreover, the JCM is a bilateral carbon trading mechanism, including aspects of technology transfer. The JCM is not solely a carbon-trading mechanism but is also a green investment and low-emission development scheme.

3.4.2 Glasgow Climate Change Conference COP26

Since carbon trading was initiated at the Kyoto Protocol forum, carbon emissions have never decreased. More recently, the implementation of carbon trading has been agreed upon by country leaders at the Glasgow COP26 climate change conference in November 2021. COP26 has established global carbon trading rules that are enforced through a globally integrated approach that was first established by the Paris Climate Agreement in 2015. The agreed-upon framework, known as Article

6, will consist of a centralized and separate bilateral system. Centralized systems are for the public and private sectors, whereas bilateral systems are designed for countries to trade their carbon offset credits, thereby helping them meet their emission targets. This framework is expected to create financial incentives for countries and companies to create emission-reduction technologies and initiatives, such as mechanical carbon capture systems and forest-planting goals, all of which will help reduce atmospheric carbon.

3.4.3 The Economic Benefits of Carbon Trading

Carbon trading can benefit countries that still have forests, especially tropical forests such as Indonesia. Forests that meet the requirements will be compensated at a certain value that has been agreed upon internationally. Regardless of the nominal value offered, carbon trading still has pros and cons, but at least financially, it can benefit countries with well-maintained forests. The important thing that needs to be observed is the extent to which the value of these benefits can be obtained while considering the rate of forest destruction, forest fires, transportation, degradation of peatlands, and other factors. For this reason, comprehensive efforts are needed in various sectors to obtain economic benefits from the environmental services of this natural resource. For Indonesia, as one of the countries contributing 75–80% of the world's carbon credits, carbon trading can contribute billions of US dollars to the Indonesian economy. Therefore, Indonesia needs to move quickly to play an important role in mitigating the risk of global warming and leading the carbon-trading market on an international scale.

3.4.4 The Role of Technology in Carbon Trading

Peatland water level monitoring systems are one way to prevent damage to peatlands due to drought. Such systems use various types of sensors and satellite imagery data that can monitor moisture and the degradation of peatland and carbon stocks. These systems monitor the carbon stocks stored in peat soils to prevent peatland degradation. Peatlands are wetland habitats that can sequester and store (sink) large amounts of carbon to prevent GHG (mainly CO_2) from escaping into the Earth's atmosphere, which can have an impact on climate change. Of course, compensation for preventing the degradation of peatlands is a substantial economic benefit that can be obtained through carbon trading.

On the other hand, mangrove forests also have great potential as carbon trading assets. Mangroves have many leaves, so they have the potential to absorb more carbon than other plants. Research has shown that one hectare of mangrove forest absorbs 110 kg of carbon, while one-third of the carbon is released from organic deposits in the mud. The rates of mangrove deforestation and reforestation can be

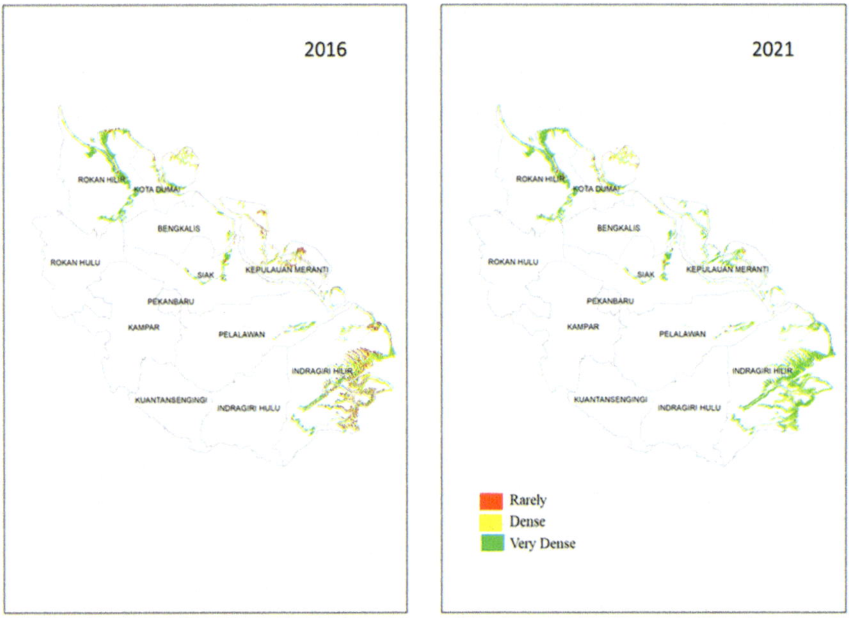

Fig. 3.10 Mangrove density changes from 2016 to 2021 in Riau Province, Indonesia

determined through remote sensing technology. To date, many methods have been developed to monitor the conditions of mangrove forests. One method involves calculating vegetation index values from remote sensing data. Mangrove spectral information has a unique pattern effect compared to other objects on land. This effect is very closely related to the spectral characteristics of the mangrove ecosystem, so spectral analyses involving vegetation index approaches are needed. In general, the spectral vegetation index (VI) is a mathematical combination of different spectral bands, mainly in the visible and near-infrared regions of the electromagnetic spectrum (Tucker 1979; Vina et al. 2011).

Figure 3.10 shows the results of an analysis of the MSAVI2 vegetation index indicating the mangrove forest density in 2016 and 2021; the figure shows that significant changes in the mangrove density have occurred. From this information, the extent of the increased carbon stock and carbon sequestration can be calculated. Furthermore, of course, the results can be used as the basis for calculating tradable carbon

3.4.5 Carbon Trading Scheme

The carbon trading scheme is also known as the baseline-and-crediting system or carbon offset scheme. This scheme does not require quotas (allowances) at the beginning of the period because what is used as a commodity (referred to as carbon credits) is the result of carbon emission reduction certifications resulting from the implementation of projects that reduce carbon emissions. One unit of carbon credit is usually equivalent to one ton of reduced CO_2 emissions. In the carbon credit scheme, the value of the credit that can be sold and used by participants to meet emission reduction targets or position participants as carbon neutral or zero emissions is obtained at the end of a given period. For the emission trade system scheme, the credit value is determined at the beginning of a period, such that new credits can be traded depending on the emitter's business activities.

3.5 Conclusion

We have shown that NEE maps of tropical peatlands can be constructed using an indirect approach based on active and passive (radar) satellite imagery. SMAP and SAR images that map soil moisture and peat subsidence provide GWL data through an empirical formula. Alternatively, optical satellites can map the GWL with the mKBDI approach. Finally, the NEE was estimated by using regression between the GWL and NEE data observed by EC. We have to see peatlands not as a gift but as a source of natural capital, so peatlands contain a must-have value. Assessments of this natural capital can be conducted by looking at the carbon capacity that can be estimated through NEE mapping, which ultimately can lead to carbon trading, and by quantifying the environmental effects that can be obtained from the peatlands.

References

Aditiya A, Efendi J, Syafii A (2014) InaCORS: infrastructure of GNSS CORS in Indonesia, FIG Congress 2014, Engaging the Challenges, Enhancing the Relevance Kuala Lumpur, Malaysia, 16–21 June 2014

Basuki I, Kauffman JB, Peterson JT, Anshari GZ, Murdiyarso D (2021) Land cover and land use change decreases net ecosystem production in tropical peatlands of West Kalimantan, Indonesia. Forests 12:1587. https://doi.org/10.3390/f12111587

Berardino P, Fornaro G, Lanari R, Sansosti E (2002) A new algorithm for surface deformation monitoring based on small baseline differential SAR interferograms. IEEE Trans Geosci Remote Sens 40(11):2375–2383. https://doi.org/10.1109/TGRS.2002.803792

Cobb AR, Dommain R, Tan F, Heng NHE, Harvey CF (2020) Carbon storage capacity of tropical peatlands in natural and artificial drainage networks. Environ Res Lett 15:114009

Ulaby F, Long D (2014) Microwave radar and radiometric remote sensing. The University of Michigan Press, Ann Arbor, MI

Graham LC (1974) Synthetic interferometer radar for topographic mapping. Proc IEEE 62(6): 763–768. https://doi.org/10.1109/PROC.1974.9516

Hanssen RF (2001) Radar interferometry: data interpretation and error analysis. Kluwer, Dordrecht

Hapsari KA, Jennerjahm T, Nugroho SH, Ulianto E, Behling H (2022) Sea level rise and climate change acting as interactive stressors on development and dynamics of tropical peatlands in coastal Sumatra and South Borneo since the last glacial maximum. *Glob Chang Biol 28(10):3459–3479*. https://doi.org/10.1111/gcb.16131

Hirano T et al (2012) Effects of disturbances on the carbon balance of tropical peat swamp forest. Glob Chang Biol 18:3410–3422. https://doi.org/10.1111/j.1365-2486.2012.02793.x

Hirano T, Sundari S, Yamada H (2016) CO2 balance of tropical peat ecosystem. In: Osaki M, Tsuji N (eds) Tropical peatland ecosystem. Springer, Tokyo

Hooijer A, Page S, Jauhiainen J, Leem WA, Lu XX, Idris A, Anshari G (2012) Subsidence and carbon loss in drained tropical peatlands. Biogeosciences 9:1053–1071. https://doi.org/10.5194/bg-9-1053-2012

Izumi Y, Takekuchi W, Widodo J. et al. (2022a) A 3-year tropical peatland subsidence time-series derived by Sentinel- 1: a case study of the Kalimantan, Indonesia. IGARSS 2022–2022 IEEE International Geoscience and Remote Sensing Symposium, 7863–7866, https://doi.org/10.1109/IGARSS46834.2022.9883791

Izumi Y, Takekuchi W, Widodo J et al (2022b) Temporal subset SBAS InSAR approach for tropical peatland surface deformation monitoring using Sentinel-1 data. Remote Sens 14(22): 5825. https://doi.org/10.3390/rs14225825

Kremer K, Leinonen I, Bartelink HH, Berbigier P, Borgeti M et al (2002) Evaluation of six process-based forest growth models using eddy-covariance measurement of CO2 and H2O fluxes at six forest sites in Europe. Glob Chang Biol 8:213–230

Leng LY, Ahmed OH, Jalloh MB (2019) Brief review on climate change and tropical peatlands. Geosci Front 10:373–380. https://doi.org/10.1016/j.gsf.2017.12.018

Mezbahuddin M, Grant RF, Hirano T (2014) Modeling effect of seasonal variation in water table depth on net ecosystem CO2 exchange of a tropical peatland. Biogeosciences 11:577–599

Murdiyarso D, Hergoualch K, Verchot LV (2010) Opportunities for reducing greenhouse gas emissions in tropical peatlands. PNAS 107(46):19655–19661

Osaki M, Kasuga T (2022) Nature as an asset: natural capital in 'FutureCity' Shimokawa, research outreach. https://doi.org/10.32907/RO-129-2508888606

Osaki M, Tsuji N, Kato T et al (2021) Natural capital-based societies in the tropics. In: Osaki M, Tsuji N, Foead N, Rieley J (eds) Tropical peatland eco-management. Springer, Singapore. https://doi.org/10.1007/978-981-33-4654-3

(2015) United Nations

Park H, Takeuchi W, Ichii K (2020) Satellite-based estimation of carbon dioxide budget in tropical peatland ecosystems. Remote Sens 12:250. https://doi.org/10.3390/rs12020250

Tsuji N, Kimura K, Sulaiman A, Hirano T, Takahashi H, Osaki M (2019) Carbon and water interaction model in tropical peatland. Int J Environ Sci Nat Resour 22(5):1–8

Sundari S, Hirano T, Yamada H, Kusin K, Limin S (2012) Effect of groundwater level on soil respiration in tropical peat swamp forests. J Agric Meteorol 68:121–134

Tucker CJ (1979) Red and photographic infrared linear combinations for monitoring vegetation. Remote Sens Environ 8:127–150

Vina A, Gitelson AA, Nguy-Robertson AL, Peng Y (2011) Comparison of different vegetation indices for the remote assessment of green leaf area index of crops. Remote Sens Environ 115: 3468–3478

Yunjun Z, Fattahi H, Amelung A (2019) Small baseline InSAR time series analysis: unwrapping error correction and noise reduction. Comput Geosci 133:104331. https://doi.org/10.1016/j.cageo.2019.104331

Chapter 4
Carbon Accounting System in Tropical Peatlands

Haruni Krisnawati, Wahyu C. Adinugroho, Rinaldi Imanuddin, and Budiharto

Abstract Tropical peatlands are areas of high carbon density and provide numerous important ecosystem services. This ecosystem has traditionally been considered a significant contributor to global greenhouse gas (GHG) emissions due to its high rates of degradation and land use change for agriculture and plantation development. Estimates of the resulting emissions and their contribution to the global total, however, remain uncertain. Understanding the source and size of historical emissions from tropical peatlands is important for planning effective and efficient mitigation actions as well as assessing the potential impact of land management options on future emissions. Carbon accounting is the first prerequisite for measurement, reporting, and verification (MRV) of mitigation and emission reduction actions. The availability of carbon accounting systems for tropical peatlands is important as part of efforts to reduce uncertainty in emission estimation by incorporating better data, improving methodologies, and thus increasing estimation accuracy, consistency, and data processing efficiency. The development of a carbon accounting system for reporting GHG emissions and removal in tropical peatlands requires a combination of data (both remote sensing and ground-based observations) from various sources, and the use of an integration framework would be important for combining data to generate estimates for reporting GHG emissions/removal. The number of studies providing scientific information on emission factors and addressing a significant knowledge gap in the main parameters to improve the accuracy of emission estimation due to disturbance on

H. Krisnawati (✉)
Research Center for Ecology and Ethnobiology, National Research and Innovation Agency, Cibinong, Bogor, Indonesia

Ministry of Environment and Forestry, Jl. Gatot Subroto, Jakarta, Indonesia
e-mail: haru005@brin.go.id

W. C. Adinugroho · R. Imanuddin
Research Center for Ecology and Ethnobiology, National Research and Innovation Agency, Cibinong, Bogor, Indonesia

Budiharto
Directorate of Greenhouse Gas Inventory and Monitoring, Reporting and Verification, Ministry of Environment and Forestry, Jakarta, Indonesia

M. Osaki et al. (eds.), *Tropical Peatland Eco-evaluation*,
https://doi.org/10.1007/978-981-99-6790-2_4

tropical peatlands has increased. Efforts, from simple to complex approaches, have also been implemented to develop carbon accounting tools, combining emissions from peat forest/vegetation and emissions from peat soil. This chapter will discuss the development of a carbon accounting system for tropical peatlands, with particular reference to Indonesia.

Keywords Tropical peatland · Carbon accounting · GHG emission · Uncertainty · Monitoring

4.1 Introduction

Tropical peatlands are areas with large amounts of carbon storage and contain more carbon (C) than other terrestrial vegetation types, estimated at approximately 105 Pg C (Page et al. 2022). In comparison to other ecosystems, this ecosystem stores a larger proportion of carbon in peat or organic soil, estimated at above 80% (Page et al. 2011; Saragi-Sasmito et al. 2018). The soil carbon pool contained in tropical peatlands accounts for 5–6% of the global total, with Indonesian peatlands accounting for approximately 3% of global soil carbon. Large areas of peatlands support forests across the tropics, and Indonesia has approximately 13.4 million ha of tropical peatlands, with exceptionally deep peatlands in Kalimantan on Borneo Island (Anda et al. 2021). This peatland ecosystem plays critical roles in providing vital ecosystem functions and services that adapt to and mitigate the effects of climate change, protect biodiversity and the environment, minimize the risks of flooding and drought, supply food and water, and improve community livelihoods.

Despite the importance of this ecosystem, tropical peatlands are exposed to various disturbance risks. This ecosystem has traditionally been considered a significant contributor to global greenhouse gas (GHG) emissions due to its high rates of degradation and land cover change for agriculture and plantation development (Koh et al. 2009; Turetsky et al. 2014). Drainage of peatlands for agriculture and plantation development can potentially result in a loss of the carbon stored in peat (IPCC 2014). The total ecosystem C stocks of tropical peatlands could reach as high as 3000 Mg C ha^{-1}; however, the rate of loss has been remarkable, causing large GHG emissions of more than 20 Mg C ha^{-1} year^{-1} (Murdiyarso et al. 2019). The growing human population that requires land, the advancing agricultural frontier, and extractive economic and social pressures, and infrastructure development make this ecosystem susceptible to damage with the consequence of increasing GHG emissions and loss of biodiversity (Posa et al. 2011; Giam et al. 2012).

Biomass burning from tropical peatlands is also reported to be a major source of global anthropogenic GHG emissions (Huijnen et al. 2016; Stockwell et al. 2016), with carbon dioxide (CO_2) emissions from wildfires averaging 2.2 Pg C year^{-1} (van der Werf et al. 2017). The lost carbon can be released into the atmosphere from peat drainage due to oxidation and dissolution, which may trigger fires (Ramdani and Hino 2013; Carlson et al. 2015; Warren et al. 2017). Fires generally occur on degraded peatlands and are also likely to continue to dry surface peat over extensive areas, making them susceptible to recurring fires, especially during the long dry season (Cole et al. 2019). Repeated and extensive fires, following drainage and logging, coupled with global warming, affect both natural forests and those subjected to conversion to plantations, and these events resulted in the loss of peat swamp forests during 1973–2005 (Hoscilo et al. 2011). Maintaining peatlands under wet conditions is thus important, as wet peatlands with lower ambient temperatures in surrounding areas are less likely to burn during wildfires. Rewetting and restoring degraded peatlands and ensuring current intact peatlands remain wet are needed to balance the loss of carbon from degraded peatlands to meet the net carbon sink needs for the land sector (IPCC 2018; Humpenöder et al. 2020).

Tropical peatlands are important to global efforts to combat climate change. International organizations, such as the Food and Agriculture Organization (FAO) of the United Nations (UN) and the UN Environment Programme (UNEP), have recommended that countries include peatlands in their national strategies and their global commitments, such as the Paris Agreement on Climate Change and Sustainable Development Goals. To achieve low emissions and build climate-resilient pathways, countries must identify how climate actions will be implemented. Nature-based solutions, such as improved peatland management and peatland conservation and restoration, have been identified as key means for achieving climate and other environmental and development goals. Including this ecosystem in these solutions can help reduce the rate and magnitude of global climate change due to peatland's capacity to sequester carbon. The capacity of tropical peatlands to enhance and sequester carbon and the resulting anthropogenic emissions as well as their contribution to the global total emissions, however, remain uncertain. Thus, improving assessments of carbon and GHG emissions from peatlands is a global strategic priority (FAO and Wetlands International 2012); hence, a carbon accounting system for tropical peatlands is needed.

4.2 Why a Carbon Accounting System Is Important for Tropical Peatlands

The ability to accurately measure, monitor, report, and verify GHG emissions is the foundation of any effective policy for mitigating climate change. Accounting for carbon is the first prerequisite for the measurement, reporting, and verification (MRV) of emissions reduction at any level, from the site or project level to the regional and national levels.

4.2.1 Need for a Carbon Accounting System

The performance-based climate mitigation framework of the United Nations Framework Convention on Climate Change (UNFCCC) encourages emission reductions related to deforestation and forest degradation and removal through enhanced forest carbon stocks, conservation, and sustainable forest management. These activities, including the verification process, should be measured and reported to the UNFCCC. This requires methodologies for estimating GHG emissions and removal and for establishing the reference emission level or reference level (REL or RL), which can be used as a benchmark for assessing performance in implementing mitigation activities. The UNFCCC also encourages countries to build robust and transparent national forest monitoring systems to facilitate GHG emission-related MRV, following the guidelines from the Intergovernmental Panel on Climate Change (IPCC). Thus, the development of carbon accounting systems is necessary for accurately measuring emissions and responsibly reducing emissions. The system is the basis for national GHG inventories and reporting.

For tropical peatlands, accurate assessments of peat forest carbon stocks and the impact of disturbances on carbon are required to determine the magnitude of emissions to the atmosphere (Volkova et al. 2021). Understanding the source and size of historical emissions from tropical peatlands is important for developing strategies for planning climate change mitigation actions as well as assessing the potential impact of peatland management options on future emissions (Krisnawati et al. 2021b). A good and comprehensive understanding of carbon pools and carbon cycles in peatland ecosystems is required to support a carbon accounting system for estimating GHG emissions and removal from tropical peatlands. A system is needed that accurately monitors the resulting emissions from land management actions based on controlling the carrying capacity of peatlands as part of the sustainable peatland management strategy and supports the transparency of peatland emission reduction (Krisnawati et al. 2021a). The development of carbon accounting systems for tropical peatlands is also important as part of efforts to reduce uncertainty by incorporating better and more complete data, improving methodologies, and thus increasing estimation accuracy, consistency, and data processing efficiency.

4.2.2 Carbon Accounting Principles

The estimates of GHG emission reduction resulting from climate change mitigation actions must be measurable, reportable, and verifiable. Thus, appropriate calculation methods are needed. The IPCC provides guidance for producing unbiased estimations that need to be reported in a transparent way and that are accurate, complete, comparable, and consistent over time (referred to as the TACCC principles), as required by the UNFCCC for the MRV of climate change mitigation actions. These principles should be followed and applied in carbon accounting systems (Table 4.1).

In addition to the above principles, when completeness or accuracy of the estimates cannot be achieved, the reduction of emissions should not be overestimated, or at least the risk of overestimation should be minimized (conservativeness). Additionally, the development of a carbon accounting system is based on human resource capacities, which may be involved in the GHG inventory and/or MRV development process. This capacity could be gradually improved while progressing toward its implementation (a learning-by-doing approach).

Table 4.1 Key principles for carbon accounting

Principles	Description
Transparency	All data, methodologies, and results should be clearly explained and appropriately documented, allowing verification of their correctness. This requires sufficient and clear documentation of the accounting process to be available so that the credibility and reliability of estimates can be assessed
Accuracy	Accurate and reliable measurements of carbon (stocks and changes) are needed to report emissions/removal. Accuracy is how close estimates are to the true value (lack bias and systematic error). Precision is the level of agreement between repeated measurements (lower random error). To give confidence in the emission/removal estimates, both accuracy and precision are desirable and can be increased by removing bias and decreasing uncertainty as much as possible
Consistency	A carbon accounting system should provide estimates that are consistent across years. Accounting estimates for different years, gases, and categories should reflect differences in carbon rather than differences in methods
Completeness	Accounting should be inclusive of all relevant categories of sources, sinks, and gases. If carbon pools or gases are excluded, then documentation and justification for their omission must be provided
Comparability	Data, methods, and assumptions used in the accounting process must be based on broad consensus to allow meaningful and valid comparisons between areas

4.3 Advances in Science and Technology Related to Carbon Accounting for Tropical Peatlands

The role of science and technology in supporting climate actions is very important. Accurate data and information, as well as appropriate methodologies, are required for justifying planning and policy implementation to achieve emission reduction goals. Applying an appropriate carbon accounting system for tropical peatlands will increase the accuracy of GHG emission estimation and the reliability of national GHG reporting.

4.3.1 IPCC Guidelines on Peatland Carbon Emissions

Carbon accounting is an important component of MRV, so the quantification of how tropical peatlands contribute to mitigating climate change needs to be measured, reported, and verified. Science to improve data, parameters, and methods for estimating carbon emissions and removal from peatlands continues to evolve. The 2006 IPCC Guidelines for National Greenhouse Gas Inventories provide methodological guidelines for estimating CO_2 emissions caused by land use changes due to drainage (Chap. 4, Volume 4) and peat extraction activities (Chap. 7, Volume 4), with the sources of GHG emissions including CO_2 and N_2O gases (IPCC 2006). To extend the content and address the knowledge gap of the 2006 IPCC Guidelines, the IPCC published the 2013 Supplement to the 2006 IPCC Guidelines for National Greenhouse Gas Inventories: Wetlands, also called the Wetlands Supplement (IPCC

2014), which provides updated information reflecting scientific advances, including updating emission factors.

The 2013 Wetlands Supplement (IPCC 2014) provides guidance on carbon accounting for peatland ecosystems, which cover both drained inland organic soils (Chap. 2) and rewetted organic soils (Chap. 3). Chapter 2 of the Wetlands Supplement provides guidance on data, parameters, and methods for estimating emissions from forests, agricultural land, grassland, and settlements on drained inland organic soils across climate zones. Chapter 3 of the Wetlands Supplement provides guidance on data, parameters, and methods for estimating emissions from any land use category on rewetted organic soils occurring in boreal, temperate, and tropical wetlands. The Wetlands Supplement (IPCC 2014) presents a number of GHG emission sources, which also includes off-site carbon dioxide (CO_2) emissions from waterborne carbon losses. This comprises methane (CH_4) emissions from rewetted organic soils, canals on drained inland organic soils, and CO_2, CH_4, and carbon monoxide (CO) emissions from peat fires.

In the 2019 Refinement to the 2006 IPCC Guidelines for National Greenhouse Gas Inventories (IPCC 2019), there are no updated data, parameters, and methods for estimating carbon emissions from the organic soil pool. However, a guide for estimating emissions from sources of CO_2 and CH_4 gases from drainage activities was added, in which in the previous publication of the Wetlands Supplement (IPCC 2014), only CH_4 gas was included.

In general, carbon accounting for peatland is based on the existing formulas following the IPCC Guidelines (IPCC 2006, 2014) as follows:

$$CO_2 - C_{organicdrained} = CO_2 - C_{on-site} + CO_2 - C_{DOC} + L_{fire} - CO_2 - C \quad (4.1)$$

where

CO_2-$C_{organicdrained}$ = CO_2-C emissions/removals by drained organic soils (t C year^{-1}).

CO_2-$C_{on\text{-}site}$ = on-site CO_2-C emissions/removals by drained organic soils (t C year^{-1}).

CO_2-C_{DOC} = CO_2-C emissions from dissolved organic carbon exported from drained organic soils (t C year^{-1}).

L_{fire}-CO_2-C = CO_2-C emissions from burning of drained organic soils (t C year^{-1}).

$$L_{fire} = A \times M_B \times C_f \times G_{ef} \times 10^{-3} \quad (4.2)$$

where

L_{fire} = amount of CO_2 or non-CO_2 emissions from organic soil fire (t = tons).
A = total area burnt annually (ha).
M_B = mass of fuel available for combustion (t ha^{-1}).
C_f = combustion factor (dimensionless).
G_{ef} = emission factor for each gas, g/kg dry matter burnt.

$$CH_{4\_organic} = \sum_{c,n,p} \left(A_{c,n,p} x\left((1 - Frac_{ditch}) x EF_{CH4_{land}c,n} + Frac_{ditch} x EF_{CH4_{ditch}c,p}\right)\right) \quad (4.3)$$

$$\Delta C_{rewetted\ org\ soil} = CO_2 - C_{rewetted\ org\ soil} + CH_4 - C_{rewetted\ org\ soil} \quad (4.4)$$

$$CO_2 - C_{rewetted\ org\ soil} = CO_2 - C_{composite} + CO_2 - C_{DOC} + L_{fire} - CO_2 - C \quad (4.5)$$

$$CH_4 - C_{rewetted\ org\ soil} = CH_4 - C_{soil} + L_{fire} - CH_4 - C \quad (4.6)$$

4.3.2 Science to Improve Parameters for Estimating Carbon Emissions

Uncontrolled drainage is considered one of the drivers of peat degradation, making peat prone to fire, particularly when the land is covered by shrubs. Emissions that occur as a result of peatland drainage and peat fires have become significant contributors to GHG emissions. The 2020s National Greenhouse Gas Inventory (Directorate of GHG Inventory and MRV 2021) report revealed that peat decomposition is a key source of emissions in the forestry and other land use (FOLU) sector in Indonesia, with an emission contribution of 46%. Peat-related GHG emissions are significant and are estimated to account for approximately 10% of the total emissions from the agriculture, forestry, and other land use (AFOLU) sector and 5% of the global emissions (UN 2017), yet the transparency in these emission estimates is limited.

Efforts to improve emission estimates from tropical peatlands are ongoing (Krisnawati et al. 2021b; Volkova et al. 2021). A number of studies have been conducted in recent years to provide information on carbon stocks and address a significant knowledge gap related to the main parameters to improve the accuracy of emission estimation for tropical peatlands. Studies conducted in several peat swamp forests in Indonesia that included measurements of five C pools (i.e., aboveground biomass, belowground biomass, deadwood, litter, and soil carbon pools) reported that primary (intact) swamp forest, secondary swamp forest, oil palm plantation, and shrub on peatlands tend to have higher total peat ecosystem carbon stocks than those in agricultural land, settlements, and bare land (Dharmawan 2012; Rosalina et al. 2013; Handoyo 2016; Basuki et al. 2017; Qirom et al. 2018, 2021; Novita et al. 2020; Krisnawati et al. 2021a,b; Volkova et al. 2021). Based on these studies, primary peat swamp forest has higher total ecosystem carbon stocks with a range of 1032–5591 Mg C ha^{-1}, while the range of total ecosystem carbon stocks of secondary peat swamp forest, oil palm plantations, and shrubs is 257–4684 Mg C ha^{-1}, 590–4389 Mg C ha^{-1}, and 319–2244 Mg C ha^{-1}, respectively. Primary peat swamp forest has the highest total ecosystem carbon stock, with an estimated mean ± std. error of 3489 ± 543 Mg C ha^{-1} (uncertainty = 31%), followed by secondary peat swamp forest (2751 ± 269 Mg C ha^{-1}, uncertainty = 19%), oil palm plantations (2704 ± 451 Mg C ha^{-1}, uncertainty = 33%), and shrubs (1281 ± 963 Mg C ha^{-1}, uncertainty >100%).

The larger proportion of carbon stored in peat ranges from 62 to 99% depending on the peat characteristics: peat depth, bulk density, and C fraction. A review by Page et al. (2011) estimated an average peat bulk density of 0.09 g cm^{-3}, an average peat C fraction of 0.56, and an average peat depth of 5.5 m. At the same time, studies on Indonesian peatlands (Dharmawan 2012; Rosalina et al. 2013; Handoyo 2016; Basuki et al. 2017; Qirom et al. 2018, 2021; Novita et al. 2020; Volkova et al. 2021) estimated that the average peat bulk density is 0.13 ± 0.003 g cm^{-3} with a range of 0.05–0.21 g cm^{-3}, the average peat C fraction is 0.51 ± 0.01 with a range of 0.36–0.61, and the average peat depth is 2.8 ± 0.2 m with a range of 0.5–8.5 m. In general, there is an increasing trend of peat bulk density with increasing peat depth, but in contrast, there is a decrease in C-org value.

Another source of emissions in tropical peatlands is peat fires, which can be induced by drainage and land-clearing activities for agriculture and plantation development. Although these fires have been reported to be one of the major sources of global anthropogenic emissions (Huijnen et al. 2016; Stockwell et al. 2016; Kumar et al. 2020), a review by Volkova et al. (2021) revealed a lack of published studies to support peat fire emission estimation. Following the guidelines of the IPCC (2006, 2014), GHG emissions from peat fires are estimated as the sum of emissions from the burning of aboveground carbon stocks and the combustion of peat. The combustion factor (CF), the proportion of fuel mass consumed in fire, is one of the significant parameters used to estimate peat fire emissions for both aboveground and peat pools. The IPCC provides default values for this parameter for aboveground CF as 0.50 and for peat CF as 1.0 (assuming 100% of the fuel is combusted). A study by Krisnawati et al. (2021b) improved the estimates of CFs where the CF for the aboveground pool is 0.56, comparable to the default value of the IPCC guidelines, while the CF for peat is 0.4–0.7 (the value increases as the depth of peat burn increases), or 30–60% lower than the IPCC default value (CF = 1.0), which is commonly used in the international reporting of peat fire emissions. A comparison of emissions from peat fires calculated using these improved parameters will reduce uncertainties and result in emission estimates that are 2–4 times lower than the default.

4.3.3 Spatial Information to Improve Peatland Carbon Emissions

Peat maps are one of the key factors used to monitor peat swamp forests. Although several peat maps exist (e.g., Anda et al. 2021), accurate information on peat thickness is still lacking. Spatial information on peat depth is an important component for improving the accuracy of emission estimates from peatlands. Modeling peat characteristics has been developed for peatland mapping. The map is based on the results of a field survey, which is then modeled by following the characteristics of the surrounding peat (Young et al. 2018; Gatis et al. 2019). Multisource remote sensing data, such as digital elevation models (DEM) from the Shuttle Radar Topography Mission (SRTM), geographic information, radar images (Sentinel and ALOS PALSAR), and field measurements using machine-learning models, have been used to map peat thickness (Rudiyanto et al. 2016). Efforts have also been implemented to improve the accuracy of Indonesia's peatland mapping. In 2018, the Indonesian Peat Prize was launched to identify the best method for mapping the extent and depth of peat in Indonesian peatlands. The winning team deployed satellite-based technologies and recommended the worldDEM to model the terrain at a resolution of 10 m, as well as images from the Sentinel satellite, combined with airborne LiDAR surface topography and on-the-ground measurements for mapping the extent and depth of peat that combine accuracy, speed, and affordability.

Another parameter used to monitor emissions/removal generating activities on peatlands is groundwater level (GWL), which is defined as the relative elevation of the groundwater table to the ground surface. The difference in GWL is used as an estimator parameter in calculating emissions/removal resulting from restoration activities through canal blocking (Directorate of GHG Inventory and MRV 2021), where the carbon flux model as a function of GWL is used to estimate the CO_2 emission rate. Measurements of GWL can be performed manually using dipped PVC wells or automatically using a device referred to as a "Diver" (e.g., Model HOBO™ U20–001-02-Ti, 0.3–0.6 cm). To support the availability of GWL data as a monitoring parameter for preventing peat degradation, Indonesia has issued a policy through the Minister of Environment and Forestry Regulation No. 15 of 2017 concerning Procedures for Measuring Groundwater Levels at Peat Ecosystem Compliance Points. There are two GWL monitoring systems that have been developed in Indonesia, i.e., SIPALAGA and SiMATAG-0.4 m. SiMATAG-0.4 m provides a GWL database in the concession area, including industrial plantation forests and oil palm plantations, while SIPALAGA provides a GWL database outside the concession area (Fig. 4.1).

GWL data for Indonesian peatlands are often only available at certain measurement locations and often have problems in terms of quality and high cost. Many measurement stations are located in relatively accessible locations to protect and maintain equipment and may not be an optimal location from a scientific point of view to represent the GWL in different land covers and land use types. These factors create challenges and gaps when using field data for estimating the GWL over a broader landscape and for assessing peatland status. A change in GWL can cause a redox (reduction–oxidation) potential effect due to biogeochemical processes. When the GWL is low, most organic matter decomposes aerobically to CO_2 (Itoh et al. 2017), thereby gradually increasing CO_2 emissions and limiting CH_4 production (Wang et al. 2018). Increasing the GWL is an option to reduce peat vulnerability to fires and CO_2 emissions due to oxidation processes; however, an increase in the GWL will increase microbial processes that can increase the emission of other gases, such as CH_4. A study by Hirano et al. (2014) reported that the depth of the groundwater table is closely related to soil moisture. Based on this relationship, soil moisture data are often used as a proxy for estimating the depth of the

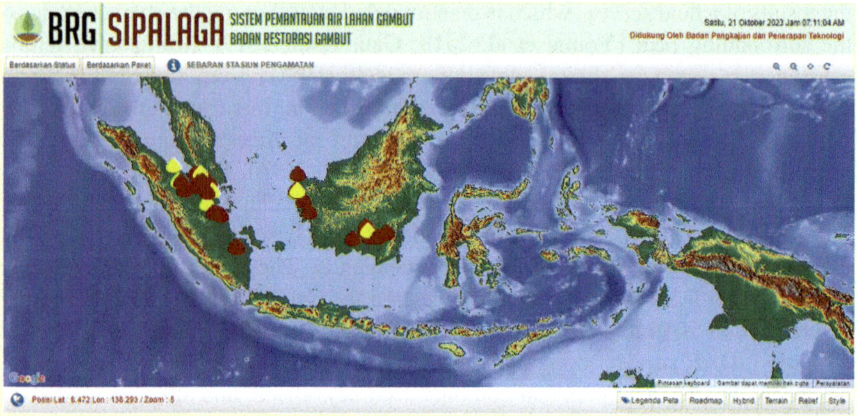

Fig. 4.1 SIPALAGA. (Source: https://sipalaga.brg.go.id/)

groundwater table. Estimation of soil moisture using microwave-based radar sensors such as Sentinel-1 in remote sensing is an indirect method that can potentially be used to overcome the challenges arising from GWL measurements in the field, as in the case studies by FAO (2020) and Adinugroho et al. (2021). The potential of soil moisture maps as a proxy for groundwater levels in peat has also been used for monitoring peatlands in Indonesia (Fig. 4.2).

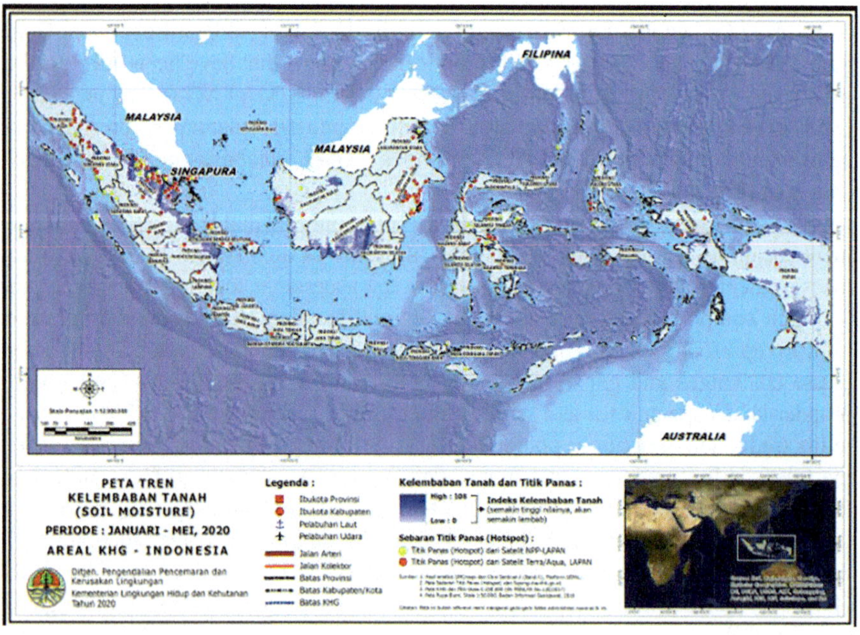

Fig. 4.2 Soil moisture map. (Source: http://pkgppkl.menlhk.go.id/)

Some empirical studies have used GWL as a predictor variable to estimate the CO_2 emissions rate (CO_2 emissions equivalent to peat subsidence losses caused by drainage). For example:

- $C_L = 21 - 69GWL$ (Hooijer et al. 2012)
- $C_L = (1.1 + 0.28GWL) \times 3.67$ (Carlson et al. 2015)
- $C_L = -0.723\ \Delta GWL$ (adopted from Wakhid et al. (2017), which emitted 0.723 Mg CO_2eq ha^{-1} year^{-1} for every 1 cm decrease in GWL

where

C_L = total carbon loss/soil respiration (Mg CO_2eq ha^{-1} year^{-1}).

GWL = depth of groundwater level (cm).

4.4 Development of a Carbon Accounting System for Tropical Peatlands

4.4.1 Integration Framework

Estimating GHG emissions/removal requires the integration of activity data (AD) (i.e., areas of land use change with data related to land management) and emission factor (EF) (i.e. pre- and postdisturbance carbon pools (i.e., biomass (aboveground and belowground biomass), dead organic matter (deadwood and litter), and soil carbon stock pools)) (IPCC 2006, 2019). Following this approach, the development of a carbon accounting system for reporting GHG emissions/removal requires the combination of data from different sources integrating both remote sensing and ground-based measurement data. The use of an integration framework would be important for facilitating the data combination to generate estimates and simplify reporting by assigning land uses and emissions to the required classes based on rules consistent with national circumstances. The integration framework may vary, from a simple to a fully integrated framework, depending on the data, assumptions, and models used as well as the approach and tier applied, as described in the IPCC Guidelines.

The simple integration framework integrates AD x EF by applying a Tier 1 (using global data) or Tier 2 (using national data) approach. The fully integrated framework integrates data with spatially referenced models and/or spatially explicit methods, which track individual units of land (polygons or pixels) by applying the Tier 3 approach. This simple framework has been used by many countries to integrate activity data and emission factors for reporting GHG emissions/removal (e.g., MOEF 2016, 2022). This method is more suitable for landscapes with a few sequential changes over time. The fully integrated framework generally more easily accommodates changes and disturbances in land management and has a greater ability to analyze the effects of management on emissions and project future emissions. The estimation using the Tier 3 approach generally provides estimates with a lower level of uncertainty, as it may not need to simplify assumptions inherent in emission/removal factors as in the Tier 1 or Tier 2 approach (GFOI 2016).

Selecting an integration framework for the carbon accounting system for the MRV requires consideration of practical and scientific issues, including (i) reporting requirements, (ii) data availability, (iii) technical capacity, (iv) standards for assessing the system and its outputs, (v) availability of integration frameworks, and (vi) available resources (cost-effectiveness) (Krisnawati et al. 2021a). The IPCC recommends using a higher tier in measuring GHG emissions and removal. However, it is reasonable to implement a system progressively, starting with a simple integration framework to meet short-term needs and then continuously improving it.

4.4.2 *Integrated System for Carbon Accounting and Monitoring in Tropical Peatlands*

Carbon accounting for GHG emissions/removal is needed for monitoring the impacts of land management activities and assessing the effectiveness of the mitigation outcomes. A few systems have been developed for tropical peatlands. The Excel-based Peat-GHG tool and the INCAS-peatland model are examples of very limited studies on carbon accounting systems developed for peatlands in Indonesia.

4.4.2.1 Peat-GHG Tool

The Peat-GHG tool was developed based on the Ex-Ante Carbon-balance Tool (EX-ACT), a scoring system based on Excel, which was initially used to estimate the impact of agricultural, forestry, and fishery development projects, programs, and policies on carbon balance (Bernoux et al. 2010). The Peat-GHG tool is a scoring system that provides ex ante estimates of anthropogenic GHG emissions/removal resulting from peatland management practices in Indonesia (Fig. 4.3), providing estimates of how the carbon balance is affected by peatland management activities. The Peat-GHG tool calculates and estimates changes in carbon stock per unit of land and CH_4 and N_2O emissions in ton CO_{2eq} per hectare per year.

The Peat-GHG tool was developed by the FAO to support key national stakeholders and answer questions such as what activities and practices should be undertaken and where based on a rapid analysis of potential GHG outcomes. The ultimate goal of using this tool is to measure the climate change mitigation potential of peatland management practices and activities carried out and coordinated by the Indonesian Peatland Restoration Agency. Overall, this tool aims to help those with an interest in the future development of peatland landscapes, such as policy-makers, planners, and other stakeholders, make informed decisions about peatland management. In addition, the tool aims to provide data and information needed by

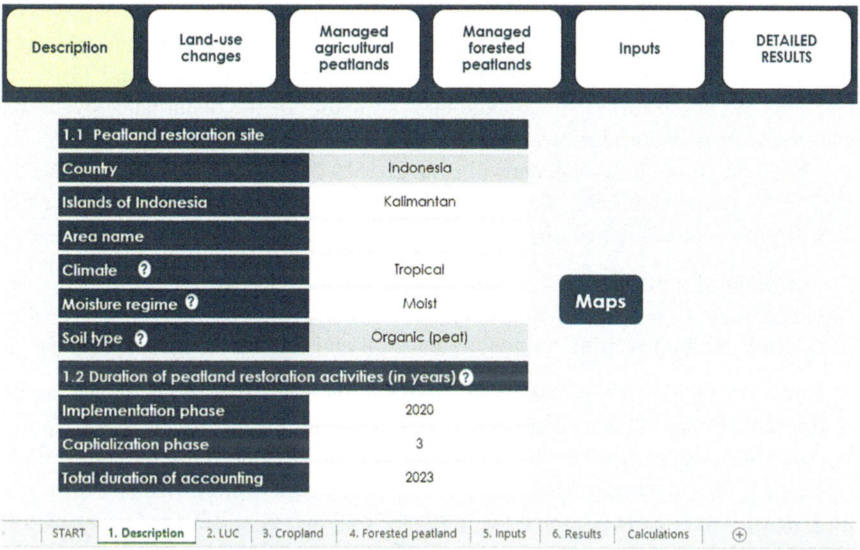

Fig. 4.3 Screenshot of the Peat-GHG tool

stakeholders to better understand the main sources and sinks of GHG emissions related to peatlands and participate in decision making.

The Peat-GHG tool includes five carbon pools: (1) aboveground biomass, (2) belowground biomass, (3) dead wood, (4) litter, and (5) soil carbon in peatlands. Data and information used in the Peat-GHG tool were based on the default values contained in the Wetlands Supplement (IPCC 2014) and the 2019 Refinement to the 2006 IPCC Guidelines for National Greenhouse Gas Inventories (IPCC 2019). In addition, information reported in the National Forest Reference Emission Level for deforestation and forest degradation in Indonesia (MOEF 2016) was included. This tool was designed to be aligned with internationally recognized methodologies and to be applied to specific national needs.

4.4.2.2 INCAS-Peatland Model

Empirical models based on the formulas contained in the IPCC guidelines (Wetlands Supplement) are also used as the basis for developing a carbon accounting system for peatland in the INCAS (Indonesian National Carbon Accounting System). A combination of spatial data (forest vs. nonforest, land cover, peat, fire, etc.) is used as input in determining activity data and emission factors. The system has been developed to estimate annual GHG emissions from forests and peatlands, which include five C pools and apply as detailed an approach (tier) as possible. The event-driven process is used to model and quantify the progressive impact of disturbance events or land management, including peat swamp forest, from which GHG emissions/removal are derived. This approach tracks the flow of carbon between the different carbon pools in the landscape and estimates the net GHG emissions released into the atmosphere (Krisnawati et al. 2015a). This allows for GHGs to be estimated based on the net change and the nature of the disturbance event that caused the peat swamp forest to change.

The standard methods (Krisnawati et al. 2015b) describe the process used by the INCAS to model GHG emission/removal from peatlands in Indonesia. GHG emissions from peatlands are estimated annually for the following sources and gases:

(i) Biological oxidation from peat drainage: CO_2-C, CO_2-e
(ii) Peat fires: CO_2-C, CO_2, CO, CH_4
(iii) Direct emissions from drained organic soil and canal drainage: N_2O, CH_4

Emission factors for peat biological oxidation were developed based on the Kalimantan Forest Climate Partnership (KFCP) study, as documented by Hooijer et al. (2014). Emission factors for peat fires were also developed through this project, which used the information on peat depth burnt in the first and subsequent fires, peat bulk density, and carbon content, as reported by Page et al. (2014) but adapted to meet reporting requirements following the IPCC guidelines. These emission factors are considered to be more representative under normal fire conditions in Indonesia (Hooijer et al. 2014). A combustion factor of 1 (100% fuel combusted) was used in the peat fire emission estimation due to the lack of data on this parameter at the time of system development (Krisnawati et al. 2015a, 2016). For that reason, the new CF of 0.54 found in the study by Krisnawati et al. (2021a,b) could be used to improve the estimate of emissions from peat fires, as was also used in the second National Forest Reference Level reporting (MOEF 2022). Direct emissions from drained organic soils and canal drainage are estimated by multiplying the annual area of drained peatland in a land use category by the default emission factor of the IPCC Wetland Supplement. Total GHG emissions are estimated by multiplying the area affected by drainage and/or fires by activity-specific emission factors. This approach for estimating peat-GHG emissions is consistent with the approach used by the INCAS for modeling GHG emissions from biomass and dead organic matter, where emissions are triggered by events in land management (Fig. 4.4).

In applying the Tier 3 approach with the modeling basis, the use of assumptions and model parameterization greatly determines accuracy. Therefore, efforts are needed to continuously update the assumptions and models used based on recent data and the results of the latest studies specific to Indonesia. Research support is needed to expand key modeling parameters, such as developing a model that describes the accumulation of carbon over time in various management activities and the impact of disturbances or management events on carbon loss and recovery. The process of integrating data from land cover change maps, biophysical conditions, and GHG emission and removal models is a challenge in managing large national data. The fully integrated framework that uses full spatial modeling could improve data processing efficiency and reduce the processing time for calculating GHG emissions and removal that involve large datasets.

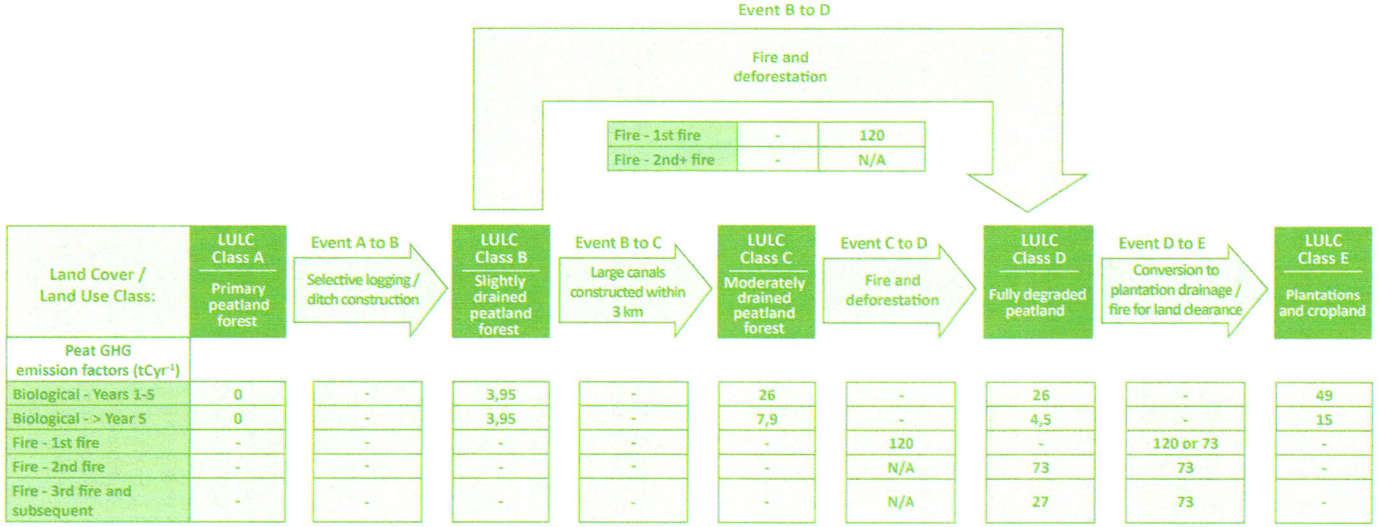

Fig. 4.4 Pathway for peatland land use change and emission factors

4.5 Conclusion

The development of a national carbon accounting system that includes peatland ecosystems is important for supporting the MRV of climate change mitigation actions for tropical peatlands. Applying an appropriate carbon accounting system for tropical peatlands will increase the accuracy of GHG emission/removal estimation and the reliability of the national GHG inventory and reporting. Strengthening science and technology through research and innovation should be continuously conducted to address data and knowledge gaps, improving the calculation of emission/removal estimates. The important parameters representing peatland conditions should be continuously updated and incorporated into the carbon accounting system to support accurate estimations of the resulting GHG emissions/removal from mitigation actions and to assess the effectiveness of the actions. Attribution and spatial analysis of changes in annual peat swamp forest cover need to be continuously expanded to improve the identification of the causes of changes and disturbance events more specifically. Accurate peatland mapping is also needed to accurately monitor peatland degradation and identify peatland areas that need to be protected, conserved, or restored.

References

Adinugroho WC, Imanuddin R, Krisnawati H, Syaugi A, Santosa PB, Qirom MA, Prasetyo LB (2021) Exploring the potential of soil moisture maps using sentinel imagery as a proxy for groundwater levels in peat. IOP Conf Ser: Earth Environ Sci 874:012011. https://doi.org/10.1088/1755-1315/874/1/012011

Anda M, Ritung S, Suryani E, Hikmat M, Yatno E, Mulyani A, Subandiono RE (2021) Revisiting tropical peatlands in Indonesia: semi-detailed mapping, extent and depth distribution assessment. Geoderma 402:115235. https://doi.org/10.1016/j.geoderma.2021.115235

Basuki I, Kauffman JB, Anshari GZ, Peterson J, Myrold D (2017) Carbon dynamics in response to land cover change in tropical peatlands, Kalimantan, Indonesia. Dissertation, Oregon State University

Bernoux M, Branca G, Carro A, Lipper L, Smith G, Bockel L (2010) Ex-ante greenhouse gas balance of agriculture and forestry development programs. Sci Agric 67:31–40. https://doi.org/10.1590/S0103-90162010000100005

Carlson KM, Goodman LK, May-Tobin CC (2015) Modeling relationships between water table depth and peat soil carbon loss in southeast Asian plantations. Environ Res Let 10:74006. https://doi.org/10.1088/1748-9326/10/7/074006

Cole LES, Bhagwat SA, Willis KJ (2019) Fire in the swamp forest: palaeoecological insights into natural and human-induced burning in intact tropical peatlands. Front For Glob Change 2:48. https://doi.org/10.3389/ffgc.2019.00048

Dharmawan IWS (2012) Evaluasi dinamika cadangan karbon tetap pada hutan gambut primer dan bekas terbakar di Hampangen dan Kalampangan Kalimantan Tengah. Disertasi, Institut Pertanian Bogor (in Indonesian)

Directorate of GHG Inventory and MRV (2021) Laporan Inventarisasi Gas Rumah Kaca, Monitoring, Pelaporan, dan Verifikasi Nasional Tahun 2021. Kementerian Lingkungan Hidup dan Kehutanan, Jakarta (in Indonesian)

FAO (2020) Peatland mapping and monitoring recommendations and technical overview. FAO, Rome

FAO, Wetlands International (2012) Peatlands - guidance for climate change mitigation through conservation, rehabilitation and sustainable use, 2nd edn. FAO, Rome. http://www.fao.org/3/a-an762e.pdf

Gatis N, Luscombe DJ, Carless D, Parry LE, Fyfe RM, Harrod TR, Brazier RE, Anderson K (2019) Mapping upland peat depth using airborne radiometric and lidar survey data. Geoderma 335:78–87. https://doi.org/10.1016/j.geoderma.2018.07.041

GFOI (2016) Integrating remote-sensing and ground based observations for estimation of emission and removals of greenhouse gases in forests: methods and guidance from the global Forest observations initiative. Group on Earth Observations, Geneva

Giam X, Koh LP, Tan HH, Miettinen J, Tan HTW, Ng PKL (2012) Global extinctions of freshwater fishes follow peatland conversion in Sundaland. Front Ecol Env 10:465–470. https://doi.org/10.1890/110182

Hoscilo A, Page SE, Tansey KJ, Rieley JO (2011) Effect of repeated fires on land-cover change on peatland in southern Central Kalimantan, Indonesia, from 1973 to 2005. Int J Wildland Fire 20: 578–588. https://doi.org/10.1071/WF10029

Handoyo GC (2016) Cadangan karbon dan emisi gas rumah kaca dari input budidaya pada perkebunan kelapa sawit (*Elaeis guineensis* Jacq.) di lahan gambut. Disertasi, Institut Pertanian Bogor (in Indonesian)

Hirano T, Kusin K, Limin S, Osaki M (2014) Carbon dioxide emissions through oxidative peat decomposition on a burnt tropical peatland. Glob Change Biol 20:555–565. https://doi.org/10.1111/gcb.12296

Hooijer A, Page S, Navratil P, Vernimmen R, der Vat MV, Tansey K, Konecny K, Siegert F, Ballhorn U, Mawdsley N (2014) Carbon emissions from drained and degraded peatland in Indonesia and emission factors for measurement, reporting and verification (MRV) of peatland greenhouse gas emissions – a summary of KFCP research results for practitioners. IAFCP, Jakarta

Hooijer A, Page S, Jauhiainen J, Lee WA, Lu XX, Idris A, Anshari G (2012) Subsidence and carbon loss in drained tropical peatlands. Biogeosciences 9:1053–1071. https://doi.org/10.5194/bg-9-1053-2012

Huijnen V, Wooster MJ, Kaiser JW, Gaveau DLA, Flemming J, Inness A, Murdiyarso D, Main B, van Weele M (2016) Fire carbon emission over maritime Southeast Asia in 2015 largest since 1997. Sci Rep 6:26886. https://doi.org/10.1038/srep26886

Humpenöder F, Karstens K, Lotze-Campen H, Leifeld J, Menichetti L, Barthelmes A, Popp A (2020) Peatland protection and restoration are key for climate change mitigation. Environ Res Lett 15:104093. https://doi.org/10.1016/j.clet.2021.100305

IPCC (2006) In: Eggleston HS, Buendia L, Miwa K, Ngara T, Tanabe K (eds) IPCC guidelines for national greenhouse gas inventories. National Greenhouse Gas Inventories Programme. IGES, Japan

IPCC (2014) In: Hiraishi T, Krug T, Tanabe K, Srivastava N, Baasansuren J, Fukuda M, Troxler TG (eds) 2013 Supplement to the 2006 IPCC guidelines for National Greenhouse gas Inventories: wetlands: methodological guidance on lands with wet and drained soils, and constructed wetlands for wastewater treatment. IPCC, Geneva

IPCC (2018) In: Masson-Delmotte V, Zhai P, Pörtner H-O, Roberts D, Skea J, Shukla PR, Pirani A, Moufouma-Okia W, Péan C, Pidcock R, Connors S, JBR M, Chen Y, Zhou X, Gomis MI, Lonnoy E, Maycock T, Tignor M, Waterfield T (eds) Global warming of 1.5°C. IPCC, Geneva

IPCC (2019) In: Buendia E, Tanabe K, Kranjc A, Baasansuren J, Fukuda M, Ngarize S, Osako A, Pyrozhenko Y, Shermanau P, Federici S (eds) 2019 refinement to the 2006 IPCC guidelines for National Greenhouse gas Inventories. IPCC, Geneva

Itoh M, Okimoto Y, Hirano T, Kusin K (2017) Factors affecting oxidative peat decomposition due to land-use in tropical peat swamp forests in Indonesia. Sci Total Environ 609:906–915. https://doi.org/10.1016/j.scitotenv.2017.07.132

Koh LP, Butler RA, Bradshaw CJA (2009) Conversion of Indonesia's peatlands. Fron Ecol Environ 7:238–238. https://doi.org/10.1890/09.WB.013

Krisnawati H, Imanuddin R, Adinugroho WC, Hutabarat S (2015a) National inventory of greenhouse gas emission and removals on Indonesia's forests and peatlands. Research, development and innovation Agency of the Ministry of Environment and Forestry, Bogor. ISBN: 978–979-8452-64-2

Krisnawati H, Imanuddin R, Adinugroho WC, Hutabarat S (2015b) Standard methods for estimating greenhouse gas emission from forests and peatlands in Indonesia (version 2). Research, development and innovation Agency of the Ministry of Environment and Forestry, Bogor. ISBN: 978–979-8452-65-9

Krisnawati H, Imanuddin R, Adinugroho WC, Hutabarat S (2016) Estimating greenhouse gas emission from peatland in Indonesia: an annual time series analysis using INCAS framework. Proceeding of 15th International Peat Congress:308–315

Krisnawati H, Adinugroho WC, Imanuddin R (2021a) Climate-responsible management of tropical peatlands: the need for integrated MRV for tropical peatland ecosystem. In: Osaki M, Tsuji N, Foead N, Rieley J (eds) Tropical peatland eco-management. Springer, Singapore. https://doi.org/10.1007/978-981-33-4654-3_15

Krisnawati H, Adinugroho WC, Imanuddin R, Suyoko, Weston CJ, Volkova L (2021b) Carbon balance of tropical peat forests at different fire history and implications for carbon emissions. Sci Total Environ 779:146365. https://doi.org/10.1016/j.scitotenv.2021.146365

Kumar P, Adelodum AA, Khan MF, Krisnawati H, Garcia-Menendez F (2020) Towards an improved understanding of greenhouse gas emission and fluxes in tropical peatlands of Southeast Asia. Sustain Cities Soc 53:101881. https://doi.org/10.1016/j.scs.2019.101881

MOEF (2016) National Forest Reference Emission Level for deforestation and forest degradation: in the context of decision 1/CP.16 Para 70 UNFCCC (encourages developing country parties to contribute to mitigation actions in the forest sector): post technical assessment by UNFCCC. Directorate general of climate change of the Ministry of Environment and Forestry, Jakarta

MOEF (2022) National Forest Reference Level for deforestation, forest degradation and enhancement of forest carbon stock: in the context of decision 12/CP.17 Para 12 UNFCCC (encourages developing country parties to update the forest reference emission level and/or forest reference level periodically). Directorate general of climate change of the Ministry of Environment and Forestry, Jakarta

Murdiyarso D, Lilleskov E, Kolka R (2019) Tropical peatlands under siege: the need for evidence-based policies and strategies. Mitigat Adapt Strat Global Chang 24:493–505. https://doi.org/10.1007/s11027-019-9844-1

Novita N, Kauffman JB, Hergoualc'h K, Murdiyarso D, Tryanto DH, Jupesta J (2020) Carbon stock from peat swamp forest and oil palm plantation in Central Kalimantan, Indonesia. In: Djalante R, Jupesta J, Aldrian E (eds) Climate change research, policy and actions in Indonesia. Springer Climate. https://doi.org/10.1007/978-3-030-55536-8_10

Page SE, Rieley JO, Banks CJ (2011) Global and regional importance of the tropical peatland carbon pool. Glob Change Biol 17:798–818. https://doi.org/10.1111/j.1365-2486.2010.02279.x

Page S, Tansey K, Navratil P, Hooijer A, Mawdsley N (2014) Measuring emission from peat fire: commentary on a proposed methodology for Indonesia. Report prepared for the Indonesia-Australia Forest Carbon Partnership. IAFCP, Jakarta

Page S, Mishra S, Agus F, Anshari G, Dargie G, Evers S, Jauhiainen J, Jaya A, Jovani-Sancho AJ, Laurén A, Sjögersten S, Suspense IA, Wijedasa LS, Evans CD (2022) Anthropogenic impacts on lowland tropical peatland biogeochemistry. Nat Rev Earth Environ 3:426–443. https://doi.org/10.1038/s43017-022-00289-6

Posa MRC, Wijedasa LS, Corlett RT (2011) Biodiversity and conservation of tropical peat swamp forests. Bioscience 61:49–57. https://doi.org/10.1525/bio.2011.61.1.10

Qirom MA, Yuwati TW, Santosa PB, Halwany W, Rachmanadi D (2018) Carbon stocks potential of peatland forests typologies in Central Kalimantan. Jurnal Ilmu Kehutanan 12:196. https://doi.org/10.22146/jik.40150

Qirom MA, Windawati TA, Kissinger FA (2021) Carbon stock potential on various land covers in heath forest in Liang Anggang, South Kalimantan. Galam 1:61–78. https://doi.org/10.20886/GLM.2021.2.1.61-78
Ramdani F, Hino M (2013) Land use changes and GHGs emission from tropical forest conversion by oil palm plantations in Riau Province, Indonesia. PLoS One 8:1–6. https://doi.org/10.1371/journal.pone.0070323
Rosalina Y, Kartawinata K, Nisyawati NE, Supriatna J (2013) Kandungan karbon di hutan rawa gambut kawasan konservasi PT. National Sago Prima Kepulauan Meranti, Riau. Buletin Kebun Raya 16:115–130. https://publikasikr.lipi.go.id/index.php/buletin/article/view/102 (in Indonesian)
Rudiyanto, Minasny B, Setiawan BI, Arif C, Saptomo SK, Chadirin Y (2016) Digital mapping for cost-effective and accurate prediction of the depth and carbon stocks in Indonesian peatlands. Geoderma 272:20–31. https://doi.org/10.1016/j.geoderma.2016.02.026
Saragi-Sasmito MF, Murdiyarso D, June T, Sasmito SD (2018) Carbon stocks, emission, and aboveground productivity in restored secondary tropical peat swamp forests. Mitig Adapt Strateg Glob Chang 24:521–533. https://doi.org/10.1007/s11027-018-9793-0
Stockwell CE, Jayarathne T, Cochrane MA, Ryan KC, Putra EI, Saharjo BH, Nurhayati AD, Albar I, Blake DR, Simpson IJ, Stone EA, Yokelson RJ (2016) Field measurements of trace gases and aerosols emitted by peat fires in Central Kalimantan, Indonesia, during the 2015 El Niño. Atmos Chem Phys 16:11711–11732. https://doi.org/10.5194/acp-16-11711-2016
Turetsky MR, Benscoter B, Page S, Rein G, van der Werf GR, Watts A (2014) Global vulnerability of peatlands to fire and carbon loss. Nat Geosci 8:11–14. https://doi.org/10.1038/ngeo2325
UN (2017) CO_2 emissions from peatlands degradation. http://www.grida.no/resources/12529
van der Werf GR, Randerson JT, Giglio L, van Leeuwen TT, Chen Y, Rogers BM, Mu M, van Marle MJE, Morton DC, Collatz GJ, Yokelson RJ, Kasibhatla PS (2017) Global fire emissions estimates during 1997-2016. Earth Syst Sci Data 9:697–720. https://doi.org/10.5194/essd-9-697-2017
Volkova L, Krisnawati H, Adinugroho WC, Imanuddin R, Qirom MA, Santosa PB, Halwany W, Weston CJ (2021) Identifying and addressing knowledge gaps for improving greenhouse gas emissions estimates from tropical peat forest fires. Sci Total Environ 763:142933. https://doi.org/10.1016/j.scitotenv.2020.142933
Wakhid N, Hirano T, Okimoto Y, Nurzakiah S, Nursyamsi D (2017) Soil carbon dioxide emissions from a rubber plantation on tropical peat. Sci Total Environ 581-582:857–865. https://doi.org/10.1016/j.scitotenv.2017.01.035
Wang J, Bogena HR, Vereecken H, Brüggemann N (2018) Characterizing redox potential effects on greenhouse gas emissions induced by water-level changes. Vadose Zone J 17:170152. https://doi.org/10.2136/vzj2017.08.0152
Warren M, Frolking S, Dai Z, Kurnianto S (2017) Impacts of land use, restoration, and climate change on tropical peat carbon stocks in the twenty-first century: implications for climate mitigation. Mitig Adapt Strateg Glob Change 22:1041–1061. https://doi.org/10.1007/s11027-016-9712-1
Young DM, Parry LE, Lee D, Ray S (2018) Spatial models with covariates improve estimates of peat depth in blanket peatlands. PLoS One 13:e0202691. https://doi.org/10.1371/journal.pone.0202691

Chapter 5
An Innovative Restoration Technology for Tropical Peatlands: AeroHydro Culture (AHC)

Maman Turjaman, Chairil Anwar Siregar, Tien Wahyuni, Sisva Silsigia, Asep Hidayat, Aryanto, Laras Murni Rahayu, Niken Andika Putri, Tsuyoshi Kato, Nobuyuki Tsuji, and Mitsuru Osaki

Abstract The major drawback of managing peat swamp forest ecosystems in Indonesia for production and conservation purposes is the intricate natural characteristics of peat soils. Peat soils naturally have limited oxygen availability because of the high water table and low nutrient concentrations. This unfavorable condition always leads to poor or stunted growth of planted native species and even death. AeroHydro Culture (AHC) is the practice of managing water and air exchange in peat soils to promote the growth of crops or plants. This type of cultivation is of significant importance in peatlands, as the unique characteristics of these ecosystems require specific management practices to ensure their stability and productivity. This particular culture is a way of overcoming the challenges of plant roots to adapt to

M. Turjaman (✉) · A. Hidayat · Aryanto · L. M. Rahayu
Research Center for Applied Microbiology, National Research and Innovation Agency (BRIN), Cibinong, West Java, Indonesia
e-mail: maman.turjaman@brin.go.id; asep066@brin.go.id; arya013@brin.go.id; laras002@brin.go.id

C. A. Siregar · T. Wahyuni
Research Center for Ecology and Ethnobiology, National Research and Innovation Agency (BRIN), Cibinong, West Java, Indonesia
e-mail: chai008@brin.go.id; tien003@brin.go.id

S. Silsigia
PT Wana Subur Lestari/PT Mayangkara Tanaman Industri, Jakarta, Indonesia
e-mail: sisva@wanasl.com

N. A. Putri · T. Kato
Sumitomo Forestry Co. Ltd., Tokyo, Japan
e-mail: niken_AP@star.sfc.co.jp; KATO_tsuyoshi5@star.sfc.co.jp

N. Tsuji
Japan Peatland Society, Fukuoka, Japan

M. Osaki
Research Faculty of Agriculture, Hokkaido University, Sapporo, Japan
e-mail: mosaki@agr.hokudai.ac.jp

M. Osaki et al. (eds.), *Tropical Peatland Eco-evaluation*,
https://doi.org/10.1007/978-981-99-6790-2_5

taking up oxygen and essential nutrients by stimulating root respiration in peatland under relatively high-water table conditions. AHC is still being developed and adapted to peatland ecosystem conditions as a forest conservation and smart farming-based agricultural strategy for application and dissemination at the site level. The basic concept of AHC involves cultivation using high groundwater levels through a storage-oriented water management system. With the addition of complete nutrients, the presence of oxygen, natural compost, and root stimulants that are symbiotic with plant growth-promoting rhizobacteria and mycorrhizal fungi, the plant roots will grow structures that resemble aerial roots. This conservation-type culture contrasts with conventional methods focusing on drainage-oriented water management systems. In experimental fields in Riau and Central Kalimantan, AHC considerably improved the early growth and biomass of indicator plants such as oil palm (*Elaeis guineensis*), balangeran (*Shorea balangeran*), and sago palm (*Metroxylon sagu*). The new perspective on AHC applications is focused on how local materials can be tailored to produce complete AHC packages that are more readily available, cheaper, practical, and can be applied at the farm level in the form of smart farming. This chapter discusses the AHC knowledge gained from field trials and espouses new perspectives on AHC to be adopted and disseminated by policy-makers at the central and site levels for improving the rehabilitation success of degraded peat swamp forest ecosystems.

Keywords Tropical peatland · AeroHydro Culture (AHC) · Aerial roots · Mound roots · Adoption

5.1 Introduction

Climate change conditions are causing degraded peat swamp forests in Indonesia to become inundated for long periods (Osaki et al. 2021a; Yuwati et al. 2021). This makes it difficult to successfully carry out mass-scale rehabilitation activities, especially if the height of native forest seedlings is also submerged in water, which causes the seedlings to stagnate and die. As a result, implementers must replant the dead seedlings, as the survival rates of the planted tree seedlings are very low. This situation causes the peat swamp ecosystem to accumulate more organic matter and endure a lack of oxygen. Therefore, peat swamp forests are known as ecosystems with very large carbon stocks and thus are a priority for maintaining a better climate for life on Earth (Anshari et al. 2021). Biomass carbon productivity is very high in tropical peat swamp forest ecosystems compared with forests growing on mineral soils. Managing peatland requires an understanding of its key components, such as oxygen and nutrients in the peat material. These components—oxygen and nutrients—determine vegetation growth, and both are limiting factors (Girkin et al. 2022). Lack of oxygen and nutrients inhibits the growth of forest vegetation, and this phenomenon is usually introduced in forest rehabilitation activities. One of the reasons why rehabilitated forest plants experience stagnant growth is due to the solubility of oxygen in water is very low; the roots of forest plants cannot absorb nutrients because nutrient absorption in the active uptake process depends on energy supplied by O_2 and on respiration (Wu et al. 2022). In addition, plant nodulation and N_2 fixation in legume species are highly dependent on the oxygen supply. If water issues and nutrient concerns are mismanaged in peat swamp forest ecosystems, devastating peat land processes will occur at an alarming rate, and restoration activity will become more demanding.

Sustainable management of tropical peatlands needs to be augmented with innovations, such as landscape restoration programs, which are still under development (Indrajaya et al. 2022). Almost all programs related to tropical peatland management are still based on the "drainage systems" method, which causes damage and decomposition of tropical peatlands, and thus far still use only physical development mechanisms (Saputra 2019). This approach does not yet touch on the rudimentary aspects of peat research involving basic biological and chemical mechanisms related to vegetation, peat characteristics, and microorganisms interacting in conditions of oxygen deficiency. AeroHydro Culture (AHC), embracing multidisciplinary technology, is a cultivation or sylviculture technique for various native tree species, including agricultural crops, carried out on peatland with a relatively high groundwater level (GWL) environment. Thus, AHC, as a newly emerged innovation technology and designed to increase the productivity of tropical peatlands, is essential for peatland restoration activities that target sustainable management outputs. Under this technology, peat management is regulated by the hydrological conditions at a high GWL while promoting forest plants to grow well and flourishing roots to grow on the soil surface. This good root growth performance is possible because more oxygen is available in the rhizosphere adjacent to the soil surface, where plant nutrients are also introduced for surface applications in plantations. Of course, the target of successful peatland restoration is to produce large amounts of biomass to achieve high carbon sequestration and carbon-negative systems (Wetadewi et al. 2021).

According to previous research, forest tree biomass productivity is very high in native tropical peat swamp forest ecosystems with high GWLs (Osaki et al. 2021b). This contradicts the prevailing theory that forest plant growth in peatlands is usually insignificant under high GWL conditions (low oxygen supply) and nutrient-poor conditions. A critical restoration method for adequately managing tropical peat swamp forests should mimic the conditions of a naturally flooded peat swamp ecosystem without altering soil moisture through the construction of a drainage system. A good lesson can be gained from the sago palm, as it is naturally well adapted to permanently flooded conditions such as peat swamp ecosystems. Sago palm has natural AHC capacity, perceived as a plant adaptation mechanism in a saturated soil atmosphere, in addition to the ability of its roots to fix free nitrogen (N_2) from the air and symbiotically associate with arbuscular mycorrhizal (AM) fungi (Asano et al. 2021; Tampubolon et al. 2021). Thus, the "environmental management of tropical peatlands" roadmap and the actual implementation on the ground have already been developed, following the implementation of an innovative technology called the natural capital wrapping mechanism based on the water–tree–

carbon relationship, referred to as the "Water-Tree-Carbon Relationship Model" (Osaki and Tsuji 2016).

We discuss the prospects and perspectives of AHC applications from a multidisciplinary point of view. Our knowledge of soil sciences, including the physical–chemical–biological characteristics of tropical peat, peat hydrology, vegetation–water–nutrient–microbial interactions, sylviculture, and ecosystem restoration principles, is the basis for a way forward in the mass-scale application of AHC. Similarly, a new perspective on AHC should be put forward as a proven cultivation technique to rehabilitate devastated peatland ecosystems and, in turn, become a specific sylviculture system for adoption and implementation at the local level. Alternative AHC content packages readily available at the site level should be discussed, as well as how the forest community can produce local AHC to develop food production in peatland agriculture based on smart farming.

5.2 Current Knowledge of AHC

The AHC concept was first practiced on oil palm plantations in Riau Province. Subsequent experiments were conducted on field plantations of oil palm (*Elaeis guineensis*), sago palm (*Metroxylon sagu*), and balangeran (*Shorea balangeran*). Some interesting outcomes from AHC activities have been discussed in the 2021 book, *Tropical Peatland Eco-management* (Osaki et al. 2021a; Wetadewi et al. 2021). The application of the AHC technology as an effort to improve the environment must be disseminated in detail among the scientific community and field practitioners. AHC is formulated through serial field trials carried out by joint research and innovation teams across multiple disciplines. The diverse expertise of researchers who have performed fundamental research on peatland ecosystems has resulted in tremendously essential observations, and the results have been published in international journals. Considerable expertise in various fields of study is involved in AHC innovation, including knowledge of soil sciences, symbiotic fungi and bacteria, plant–nutrient interactions, sylviculture, compost making, climatology, carbon measurement, seed technology, and forest hydrology. The main problematic points in peatland rehabilitation activities have been determined through the knowledge of all prominent researchers who have researched peatland ecosystems for approximately 30 years. As a result, the inherent problems are overcome promptly by implementing the lessons learned.

A fundamental tool for peatland rehabilitation managers is knowledge of the principles of the physical, chemical, and biological nature of peatland as a natural body. Peatlands are ecosystems characterized by the accumulation of organic matter in water-logged environments. The high water-holding capacity of peat soils and the low oxygen levels create a challenging environment for plant growth. Other important characteristics of peat soil include a low fertility level due to low macro- and micronutrient concentrations and a low soil and water pH. For instance, when rehabilitation managers decide to plant 30-cm-tall forest seedlings in saturated open peatland without knowing the nature of peat soils and the climate behavior, large mistakes are made (Dohong et al. 2018; Graham et al. 2016). In fact, peatland field managers must be aware of daily, monthly, and annual fluctuations in GWLs. Knowing the GWL of a given peatland can be a basis for determining the optimal height of native forest seedlings ready to be planted at the field level. Field experience indicates that the average height of forest plant seedlings required for planting is approximately 100 cm. After understanding how to deal with peatland work, of course, this should change the mindset of field-level implementers, and hence the cost of seedling production would have to be revised accordingly. Very low nutrient concentrations coupled with low oxygen limitations have caused many native trees planted for rehabilitation to grow very poorly and die in the long term (Hayasaka et al. 2020). AHC, which involves cultivation techniques using high GWLs through storage-oriented water management systems, with supplementation of nutrients, soil microbial stimulants, and oxygen through natural compost and aerial roots, is being developed to resolve this problem. In other words, the implementation of AHC can help promote good plant growth and development, leading to better soil and water conservation practices. This is true because the surface application of nutrient fertilizer and effective soil microbial inoculation help to improve the formation of the tree rhizosphere near the peatland surface, leading to improved nutrient absorption, regardless of the high GWLs in the peat soil (Chaddy et al. 2021). The above concept, by its very nature, can be classified as a conservation tillage technique, which contrasts with conventional methods focusing on drainage-oriented water management systems where peat degradation is likely to occur over time through the increase in organic matter decomposition as soil temperature increases while soil moisture decreases (Cárceles et al. 2022).

The idea of developing AHC has become a reality based on several real field examples in peat swamp forest ecosystems (Wetadewi et al. 2021). Four native tree species have aerial roots that can adapt to long periods of stagnant water and oxygen-depleted peatland conditions (Fig. 5.1): *S. balangeran*, *Dyera polyphylla*, *Combretocarpus rotundatus*, and *Calophyllum macrocarpum* are examples of native species with aerial roots and aero-hydro behavior that have the adaptive capacity to cope with the waterlogged atmosphere in peat swamp ecosystems. In addition, *D. polyphylla* and *Calophyllum hosei*, native to peatland, can form symbiotic associations with AM fungi in nursery conditions (Turjaman et al. 2006, 2008). This means that the good adaptability of local tree species is evident not only from the formation of adventitious roots but also from their distinct association with mycorrhizal fungi to streamline the absorption process of macro- and micronutrients in peatland conditions exhibiting poor physical, chemical, and biological merit (Turjaman and Osaki 2021). Accordingly, the formation of this type of aerial root model must be stimulated in planting activities so that plant seedlings are armed to seek adequate plant nutrients in oxygen-deficit conditions. The AHC technique basically mimics the natural nutrient circulation mechanism. In short, the aero-hydro capacity of some native species should be manipulated such that the surface application of added soil nutrients will stimulate root growth, which is normally inhibited due to waterlogging, and in turn, root growth is accompanied by the

Fig. 5.1 Four native tropical peat swamp tree species with natural aero-hydro roots and ability to adapt to long periods of inundation and oxygen deprivation. (**a**) kapur naga (*Calophyllum macrocarpum*), (**b**) balangeran (*Shorea balangeran*), (**c**) jelutung (*Dyera polyphylla*), and (**d**) tumih (*Combretocarpus rotundatus*) (Source: Purwanto Budi Santosa, BRIN)

formation of new aerial roots. The tree roots together will absorb adequate amounts of plant nutrients and in turn will provide oxygen for living organisms in the vicinity of the forest tree rhizosphere. The AHC concept is to create a plant nutritional technology package consisting of composted organic materials, a microbial consortium that is symbiotic with tree roots, and materials, including biochar and zeolite, that also provide aeration processes and hence homes for microbes (Maftuah et al. 2021; Santi et al. 2021).

Applying AHC to improve the stagnant growth of peat swamp forest plantations aims to induce root growth by improving the rhizosphere conditions on the peatland surface as a precursor to stimulate bursting root growth and, in turn, the formation of the typical root morphology of aerial roots (Osaki et al. 2021b). The formation of abundant aerial roots will improve the absorption of nutrients from the air, including free oxygen and nitrogen. Litter from the forest floor can be utilized by the aerial roots of trees that begin to spread around the trees through biomass degradation and nutrient leaching even from fresh material, as affected by the driving force of water movement. This succession process will accelerate the nutrient cycle of the plant and peatland surface system. As reported by Tawaraya et al. (2003), the roots of rubber

Fig. 5.2 Various tropical forest tree species in peatland have adapted to address the prolonged lack of oxygen and nutrients in flooded areas through the formation of aerial roots that absorb oxygen and nitrogen directly from the air. The rubber tree (*Hevea brasiliensis*) has adapted to survive in peatland ecosystems

plants (*Hevea brasiliensis*) were found to be naturally symbiotic with AM fungi. This means that *H. brasiliensis* adapts not only via formation of adventitious root but also via colonization of its root hairs with AM fungi (Fig. 5.2). Thus, the dynamics of nutrient cycles and food chains connected among tree roots will spread rapidly to increase the plant metabolism rate and simultaneously make use of the photosynthesis process of each tree. The spread of photosynthate belowground causes the population of symbiotic microbes and decomposers to share their rations to carry out their functions as part of the food network in the peatland. In the meantime, the biodiversity of tree species in peatland ecosystems is very high, but mastering sylviculture to produce forest plant seeds requires much work. Damage to the peat swamp forest ecosystem causes the reproductive systems of forest plants to become irregular, the flowering season to become unpredictable, and pollinating insect populations to disappear or become extinct. As a result, all these unfavorable conditions put pressure on the source of forest plant seeds in peatland, which are challenging to obtain, especially to meet the target of producing seedlings on the scale of millions of trees per permanent nursery. Currently, the supply of forest plant seeds in peatland is mostly from wilding in remaining forests or national parks bordering degraded peat forest areas. This pattern of forest plant seedling provision is not sustainable. Peatland rehabilitation policy-makers must create a road map to provide forest seedlings to restore forest function in peatlands (Budiman et al. 2020).

The role of mycorrhizal fungi in AHC technology has been discussed previously in *Tropical Peatland Eco-management* (Turjaman and Osaki 2021; Tawaraya and Turjaman 2014). Some researchers, however, are still questioning the role of mycorrhizae during flooding and how much mycorrhizae can benefit hosts by absorbing nutrients under flooded conditions in peatlands. In short, the scientific community has questioned the existence of mycorrhizal fungi in stagnant water. Especially in the case of ectomycorrhizal fungi from the Basidiomycota group, the debate is about how they grow their fruiting bodies and spread spores under a water-saturated atmosphere. Mycorrhizal fungi in limited conditions, such as a lack of oxygen, apparently form vegetative hyphae and mycelia that survive in still water. Hyphae are similar to feeding tubes that continue to perform their function by connecting their network to tree roots. In studies carried out in mineral soils, *Nauclea orientalis* seedlings inoculated with AM fungi under flooded conditions experienced increased plant development, especially in regard to the number of leaves, compared to control plants (Tuheteru et al. 2022). A study conducted in dry peatland conditions, when the water table drops, showed that the fruiting bodies of ectomycorrhizal fungi emerge in the dry peat and then spread their spores to other places to form symbiotic associations with other host trees (Helgason and Fitter 2009). The effectiveness of mycorrhizal fungi inoculated at the seedling stage of native peatland forest plants in the nursery was evident and significant. Mycorrhizal forest trees had superior height, stem diameter, and leaf density than control plants (Turjaman et al. 2006, 2008, 2011). The higher the richness of mycorrhizal forest plant seedlings, the better the forest plant survival rate after transplant to the field. Ectomycorrhizal and AM fungi in conjunction below ground. They utilize the photosynthesis products of forest plants. A rich forest plant community can increase carbon storage in biomass, and the microbial community will aid in balanced nutrient cycling. Furthermore, the pelawan tree (*Tristaniopsis obovata*) is a native tree species in the family Myrtaceae that forms a dual symbiosis with ectomycorrhizal and AM fungi and is able to adapt well to conditions of thin and thick peat substrate. Additionally, a specific ectomycorrhizal fungus symbiotic with *T. obovata* has been identified, namely, *Hemioporus retisporus*, an edible mushroom commercialized on the island of Bangka Belitung (Helbert et al. 2019; Turjaman et al. 2019).

Forest trees in peat swamp ecosystems absorb water and nutrients mostly through their roots, a process regulated by leaf mechanisms in the plant–soil–water continuum system (Cusack et al. 2021). Actual examples are the water-saving function of stomata stimulated by solar energy, the promotion of root function by carbohydrate application, and root interaction with the rhizosphere (10 μm around the roots) and soil microorganisms on the forest floor (Fig. 5.3). It is conceivable that peat swamp forest ecosystems with abundant tree species biodiversity have larger-scale transdisciplinary functions. These above- and belowground functions globally include interconnections between ecosystems, modulating the water–carbon cycle, soil fertility, peatland protection, natural balance, and biodiversity conservation in the biosphere and geosphere. Forest trees associated with understory vegetation are the basis of the solar energy, CO_2, water, and nutrient cycle and the essential investment for building peat swamp forest ecosystems that form the main currency of natural capital (Osaki et al. 2021b).

In undisturbed native tropical peatlands, biomass productivity is very high compared with that in typical mineral soils, and forest trees and understory vegetation have nutrient supplies (Silvianingsih et al. 2021). To increase the productivity of peatland forests, forest tree communities and other vegetation have developed

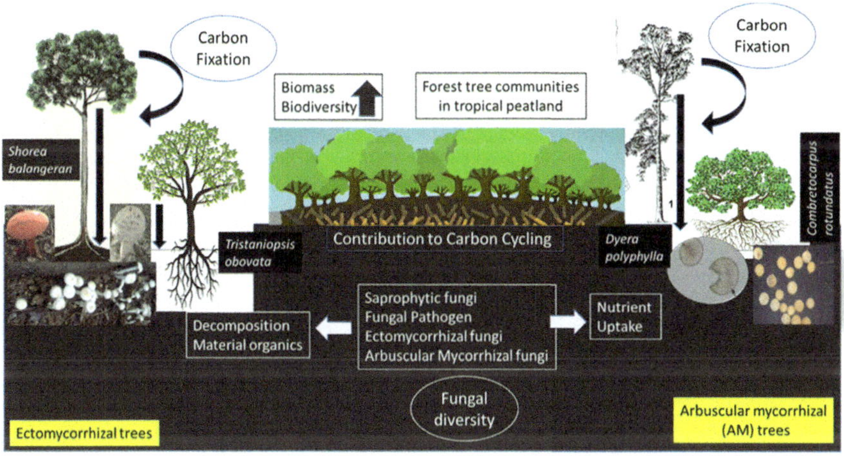

Fig. 5.3 Role of fungal diversity interacting with host trees in peatland ecosystems in enhancing biomass productivity and accumulation and carbon cycling and storage

environmental adaptation strategies to maximize nutrient utilization under oxygen-deficient conditions (Antala et al. 2022). Forest trees in peatlands have evolved mechanisms to recycle nutrients from litter via symbiosis with organic matter-degrading microorganisms on the peatland surface, nutrient absorption and nitrogen (N_2) fixation by tree roots with the help of ectomycorrhizal and AM fungi, nitrogen (N_2) fixation by free-living N_2-fixing bacteria in the rhizosphere (the thin layer of soil immediately adjacent to the roots), plant growth-promoting rhizobacteria (PGPR) association, endophytic fungal association, and oxygen uptake through the development of aerial roots and taproots, with each tree species having its own coping strategy (Antonius et al. 2021; Maulana et al. 2018). Potential forest diseases from fungal pathogens may occur in peatlands. However, the percentage of damage caused by fungal diseases will likely be insignificant if forest tree diversity is high. Supporting the role of symbiotic microbes in maintaining nutrient balance is essential to prevent outbreaks of fungal pathogens (Fig. 5.4).

Based on field observations, it is very clear that *S. balangeran* exhibits good adaptation to various long inundation conditions (4–6 months) by producing adventitious roots at the base of the rootstock. The function of roots other than anchorage to the ground is to absorb water and nutrients. In addition, *S. balangeran* also forms a symbiotic association with a native ectomycorrhizal fungus, so in nonflooded peat conditions, the ectomycorrhizal network can be observed on the peat surface in the form of litter and ectomycorrhizal formation (Graham et al. 2013; Turjaman et al. 2011). The root growth and development of *S. balangeran* is better than that of *Cratoxylum arborescens*, as it has a 13-week inundation tolerance (Tata et al. 2022). Knowledge of the ability of native peat tree species to develop their own aero-hydro mechanism in nature should be well documented. Afterward, this knowledge can be used to formulate effective and robust AHC technology that can play a critical role in

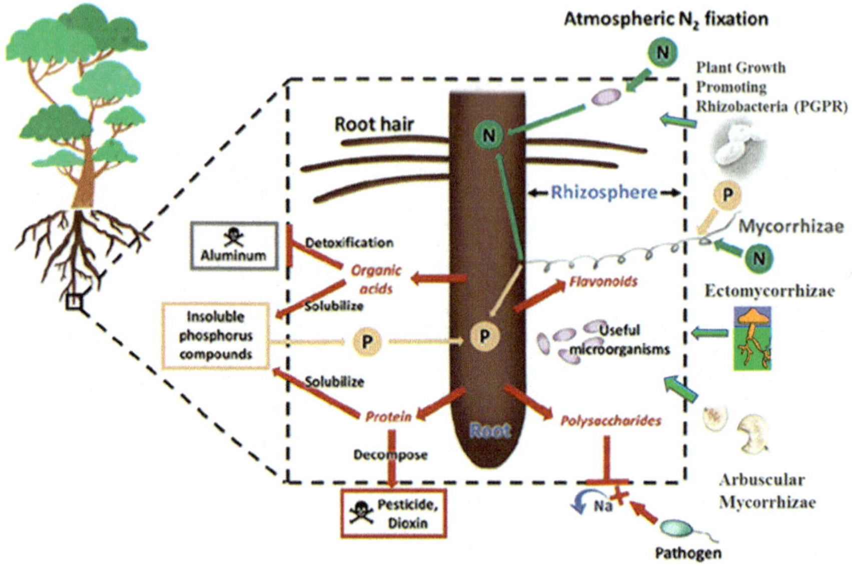

Fig. 5.4 The rhizosphere of trees in peat swamp forest ecosystems has limited nutrient absorption capacity. The oxygen and nutrient deficit conditions mean that tree roots must form a symbiotic relationship with microbes to adapt to long inundation periods

the establishment of peat swamp forest plantations for production and conservation purposes under future conditions of prolonged inundation.

Tadon, also called biochar or woody charcoal, is an essential component for growth stimulation, toxin neutralization, and microorganism (bacteria and fungi) habitat provision in peatland forests. Tadon stands for "empty fruit bunch", which is the raw material used for biochar production from oil palm. Tadon is one of the media applied in AHC. The technology of "tadon" production was first adopted in Japan, in which a hardened scrap of woody charcoal is produced during biochar production, which was used as household fuel in the past (Osaki et al. 2021b). The primary material of tadon is biochar, a porous substance that is excellent at retaining water and oxygen. As with other AHCs, the key idea of tadon is to support the growth of trees in peatland by increasing the supply of oxygen and nutrients from the land surface, which is known to be very limited. Biochar is a pyrolyzed woody charcoal product and has been found to be the best material for improving peatland quality, mainly by functioning as a soil conditioner. The characteristics of biochar depend on the raw materials used and the conditions of the pyrolysis process, such as temperature, pyrolysis time, and oxygen or air supply. Biochar plays a very important role in improving the physical, chemical, and biological properties of soils in suboptimal lands such as peatlands. Biochar application to peatlands can improve and maintain peat soil fertility, especially by increasing the cation exchange capacity (CEC), soil pH, and microorganism population and increasing the capacity of substrates to hold water and nutrients and release them when needed by plants (Ahmad

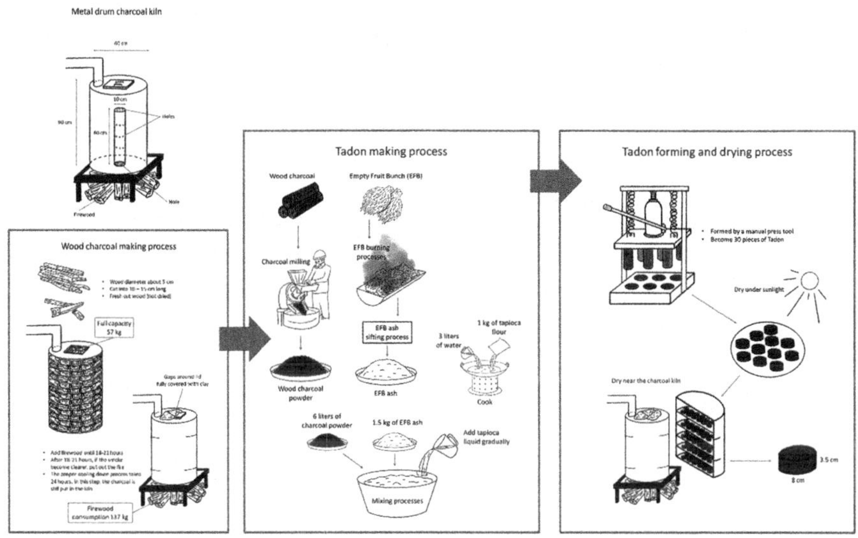

Fig. 5.5 Flowchart of the tadon production process

Bhat et al. 2022). Biochar can also increase peatland productivity, and fluctuations in land productivity due to biochar application depend on soil type, biochar type, and biochar dosage. Biochar houses symbiotic microbes and can sequester carbon in the soil for hundreds to thousands of years, making it an ideal component to slow global warming. Numerous organic material wastes can be good feedstock for biochar production, such as waste biomass from weed plants, waste wood, and any part of unwanted tree biomass. The application of biochar for peat swamp ecosystem rehabilitation activities is expected to help preserve peat in the future (Alkharabsheh et al. 2021).

Figure 5.5 illustrates the completed step of the tadon production process and its application in the field. PT. Wana Subur Lestari (WSL) and PT. Mayangkara Tanaman Industri (MTI) developed tadon as a substance that activates rhizobia and promotes the formation of lateral roots in *Acacia crassicarpa* species, which support the absorption of nutrients in peatlands with relatively high GWLs. Tadon is primarily composed of natural ingredients such as wood charcoal and an empty oil palm fruit bunch mashed and blended with water and tapioca flour to make it sticky and formable (Laurén et al. 2021). Using a manual press tool, the tadon is formed into a cylindrical shape with a diameter of 8 cm and a height of 3.5 cm and then dried in direct sunlight or near a biochar-making kiln.

Two tadon pieces (approximately 400 g) were placed on the ground surface at the center of four *A. crassicarpa* trees. Five sampling plots were selected, one in WSL and four in MTI. Each sampling plot was divided into two treatments: the control and tadon application plots. In the study, tadons were applied near plants older than 2 years (Fig. 5.6). In the first year, the root growth of the *A. crassicarpa* tree is concentrated on the primary root to support the plant and prevent it from collapsing.

Fig. 5.6 Application of tadon in the field. Company technician digging a 20-cm-deep hole (**a**) and embedding the tadon in the hole (**b**)

Acacia crassicarpa is considered a fast-growing species; thus, not only must the primary roots be robust, but lateral roots are also necessary for plants to grow optimally to uptake oxygen and nutrients at high GWLs. The diameter at breast height and survival rate data from the control and tadon plots were collected monthly during the monitoring period.

AHC packages contain the primary raw material, organic matter, in the form of mature compost that is ready to be absorbed by forest plants. The composting process is based on decomposers from PGPR bacteria and fungi, which are very useful in stimulating the formation of aerial roots in field plantations (Antonius et al. 2021). In the experimental stage, composting was conducted using a common standardized method. Starter products for compost development have been tested in mineral soil and peatland agricultural areas. This compost product is one of the mainstay ready materials in promoting AHC packaging to improve plant growth in degraded peatland areas. Based on previous AHC experiments, applying AHC increased the productivity of oil palms with stagnant growth and without fruits into healthy and fruitful plants. Likewise, *S. balangeran* (balangeran) and *M. sagu* (sago palm) plants were initially stagnant in growth in flooded peatland areas but later became plants that grew well when compared to untreated control plants, finally exhibiting the emergence of leaf buds and new aerial roots on the peatland surface (Fig. 5.7).

Preliminary research results on application of AHC have been discussed in detail and depth in the *Tropical Peatland Eco-management* book (Wetadewi et al. 2021). The application of AHC to oil palm plants grown on flooded peatlands led to the development of new aerial roots compared with the control plants (Fig. 5.8).

Fig. 5.7 Stages of production of AHC packages based on biological organic fertilizers that can be produced by forest communities at the site level of peat swamp forest ecosystems. (**a**) collecting raw materials from agricultural waste; (**b**) cutting raw materials; (**c**) mixing and turning raw materials with a compost starter; (**d**) adding dolomite to raise the pH; (**e**) the fermentation process for compost production; (**f**) compost was available and ready for AHC packing; (**g**) experimenting with the application AHC on plots of plant in peatland; (**h**) the AHC research team conducting experimental activities

Similarly, the height of *S. balangeran* and *M. sagu* plants showed a significant increase compared with control plants in flooded peatland conditions (Fig. 5.9).

5.3 Challenges and New Perspectives of AHC Application

There are still many challenges and new insights needed to improve for effective use of AHC technology. Technically, all AHC materials must be prepared for AHC mass production conditions. The AHC conceptmay be good, but at the local level, materials must always be available on the market, at low prices, and be practical for use. The provision of symbiotic microbial feedstock for mass-scale AHCs must also be considered for wide-scale and practical applications. The aim is that the production cost of AHC should be low and affordable for all growers. One way to achieve this is to plant as many bamboo species as possible that can grow well in peatland. It has been reported that 17 bamboo species grow well in peat swamp forest ecosystems in Riau Province, Sumatra (Fitmawati et al. 2020). Bamboo roots and leaf litter harbor symbiotic mycorrhizae, especially a group of bacteria that can promote plant growth, often called PGPR. In the bamboo species (*Phyllostachis edulis*), 182 rhizosphere bacterial isolates were found, dominated by *Lysinibacillus*, *Bacillus*, and *Burkholderia* (Han et al. 2009). All three have the potential as PGPR to

Fig. 5.8 Application of AHC to oil palm plants in peatland can stimulate new hair roots and the initial formation of aerial roots compared to control plants

be integrated into the AHC package. However, field trials using PGPR originating from the rhizosphere of bamboo still need to be carried out to verify their effectiveness. AHC application also needs further clarification and replication covering several experimental sites that have variations in terms of peatland depth, forest vegetation types, and duration of waterlogging. For interdisciplinary studies focusing on AHC applications at the field scale, AHC researchers still need support in terms of facilities and funding from sponsors interested in developing AHC in the future. Research collaboration is the key pathway to implementing AHC technique by sharing knowledge, facilities, suitable research designs, and funding accordingly.

Next, the very intricate question is about how to provide ectomycorrhizal inoculants and AM fungi to support the production of AHC application packages in mass production lines. Inoculant material for mycorrhizal fungi does not always have to come from the laboratory. On a field scale, materials can be taken directly from the forest community area. As long as communities are willing to plant trees as hosts for mycorrhizal fungi, then their availability will be abundant. One practical way to provide mycorrhizal fungi inoculants is to plant native tree species that are adapted to peat swamps, namely, *S. balangeran*, *C. rotundatus*, *Melaleuca cajuputi*, and *D. polyphylla*, around peat swamp ecosystems near the forest community settlements (Turjaman and Osaki 2021). These tree species can proliferate in 2–3 years, and the roots on the peat surface evidently grow very fast. Planting can be done near

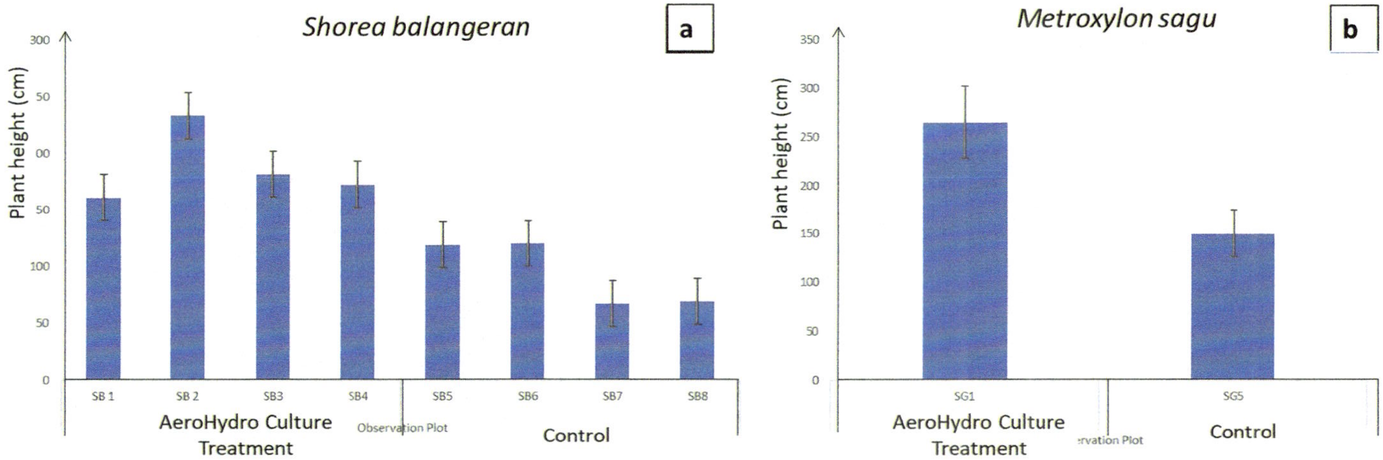

Fig. 5.9 The plant height of *Shorea balangeran* (left) and *Metroxylon sagu* (right) increased significantly with AHC application compared to that of control plants under long periods of flooded peatland conditions

people's yards and gardens. To date, the Ministry of Environment and Forestry has distributed free forest plant seeds throughout Indonesia. One grower can receive 25 pots of forest plant seeds that are available at the local nursery. For example, if 200 forest residents in a village are interested in planting these trees, then the movement to plant 5000 forest tree seedlings in gardens and yards in residential areas can be accelerated, and at the same time, the village will be greened. If the availability of tree species that can grow well near residential areas is there, then the community can collect surface peat soil as a source of mycorrhizal inoculants. Thus, the microbial supply package for AHC applications is no longer an obstacle (Asano et al. 2021; Churchland and Grayston 2014).

Another AHC material is zeolite. Zeolite is a type of aluminosilicate crystal, a porous material containing alkaline cations such as sodium, calcium, and magnesium that are classified as essential plant nutrients. The application of zeolite is important because it also functions as a shelter for symbiotic microbes and a good adsorption surface for retaining toxic material (Santi et al. 2021). Zeolite is actually a mining material and is widely used as a good supporting nursery material in agricultural fields. Today, zeolite is available and easily found in fertilizer stores. However, with the development of online shopping, zeolite can be purchased even more easily through online markets. Generally, agricultural growers use only a small amount of zeolite in their nursery medium to produce good seedling growth. Moreover, good growers normally use a large amount of compost to enrich their nursery medium, and some growers also use compost in field plantations. If zeolite is not available, it can be replaced by biochar. Biochar can serve as a home for beneficial microorganisms for the growth of the target trees.

AHC applications in peat swamp forest ecosystems are not done solely for peatland rehabilitation (Wetadewi et al. 2021; Osaki et al. 2021a). Restoring the forest function of peat swamp forest ecosystems to address climate change and carbon mitigation is a long-term endeavor of AHC application. AHC should be used to increase the productivity of plantations and food crops produced daily by forest communities so that they obtain a monthly source of income from AHC-based peat management. The results from several AHC experiments indicated that oil palm plantation productivity increased significantly. In the future, we should try the approach on other crops favored by the community, such as rubber, pineapple, chili, corn, and vegetable crops. Of course, AHC modifications need to be made for smart farming applications in peatlands. AHC applications for smart farming will be the mainstay in peat swamp ecosystems for the production of various types of agricultural food crops, in agroforestry systems with zero burning and environmental friendliness, and in other high-productivity farming systems (Silvianingsih et al. 2021). In the scale-up of AHC applications, it is necessary to ascertain the need for developing specific AHCs based on the forest tree species, plant height, peat thickness, length of inundation, and macronutrient content if the peatland area will be managed extensively.

5.4 How the AHC Strategy Can Be Implemented and Adopted by Stakeholders

AHC is a breakthrough in planting technology for the management of peatland (Osaki et al. 2021a; Wetadewi et al. 2021; Putra et al. 2020). As an innovative plant culture system, AHC provides an arrangement of plant–peat–water structures such that oxygen supply and nutrient inputs through the surface application are available to growing plants under high GWL conditions. By implementing this system, plant productivity remains high even though the peatland is saturated with excessive water, as this is the original status of natural old growth in peat swamp forest ecosystems. This innovative management system is in line with the government regulations regarding a GWL of 40 cm.

Many stakeholders, such as farmers, companies, and agribusinesses, will invest and implement AHC technology and innovative management practices if this cultivation system works well. They expect the investment to be profitable. However, the adoption of AHC technology by various stakeholders will take time to ensure that the benefits and profits continue to flow. However, farmers may also be reluctant to invest in uncertain climates with more constraints in relation to technical and economic reasons (Ong et al. 2022).

Adopting AHC technology will have a fundamental impact on responsible eco-management of tropical peatland utilization (Osaki et al. 2021a). This innovative technology can also be applied to other ecosystems by creating various modifications. Furthermore, this innovative management system will support agriculture production management in a broader sense in a sustainable manner (Dudek and Wrzaszcz 2020).

Although there are still challenges and emotional issues in convincing many parties, mainly palm oil companies, similar to other new innovations, true endeavors to disseminate this system widely apparently require a longer period of time. A strategy is needed to convince many stakeholders (farmers, companies, extension services, agribusinesses, and policy-makers) (Fig. 5.10). AHC technology can improve productivity in an economically efficient, profitable, and socially and environmentally responsible manner. This specific technology should warrant high productivity in planting systems implemented in the agricultural and forestry sectors, and at the same time, it embraces public concern at the national and global levels about improving and conserving environmental quality (Murphy and Goggin 2021).

5.4.1 Stakeholder Attitude and Behavior Toward AHC Innovation

The adoption of innovation is a mental process in which people experience behavioral change in the form of knowledge systems, attitudes, or skills when encountering something new. An example is the hybrid corn innovation case in the United

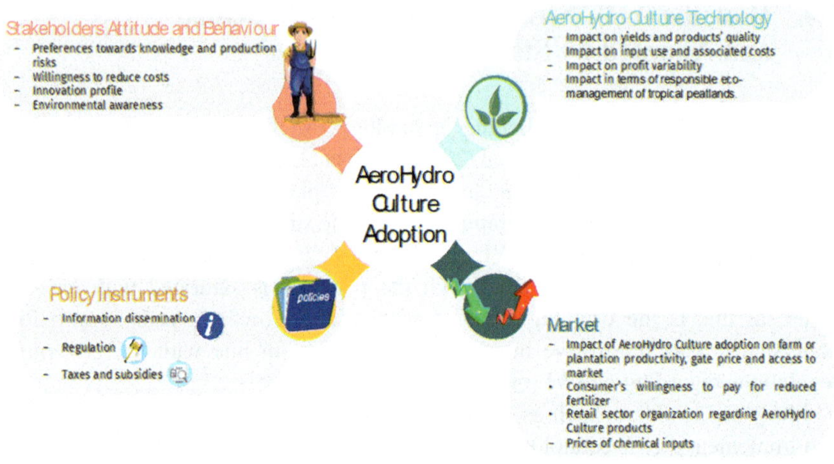

Fig. 5.10 The AHC strategy can be implemented and adopted by stakeholders

States, which took more than 10 years for articulate farmers to fully accept the innovation since it was established. In short, the adoption of a new technology by smart people requires ample time and patience (Rogers and Shoemaker 1995). *Innovation* is an idea, action, or item considered something new by someone. The novelty of innovation is measured subjectively according to the views of the individuals who perceive it.

Furthermore, it is also important to comprehend that the adoption of AHC technology may be affected by the education level, training, advice, and information availability of concerned individuals who are likely to bring about the notable findings to their production line in the business. Intensive training through education accompanied by a good example in field plantations, such as field days and visiting demonstration sites, may help to cut down long adoption periods (Coman et al. 2020).

As a new technology, applying AHC is one way to reduce the use of inputs and associated costs that must be paid by farmers and companies. Smart farmers and qualified profit-oriented companies are likely to consider the positive and negative impacts of adopting the new technology by taking into account all technical and economic benefits that may be gained (Schimmelpfennig 2016). While time passes, it is interesting to witness profitable investments in AHC technology by fellow farmers and companies. Likewise, it often helps decision-making institutions with sophisticated facilities and enough resources to push for the ultimate adoption and broad implementation of AHC in the peatland management process.

As an innovation, the characteristics of AHC can be evaluated, and a series of indicators to measure fitness are as follows (Schiffman and Kanuk 2010):

1. Relative advantages. AHC can be considered an innovation that is better than previous innovations and economically profitable.

2. Compatibility is the extent to which an innovation's past is considered consistent with existing values, past experiences, and adopters' needs (recipients). Therefore, innovations incompatible with salient social system features will be adopted slower than compatible ideas.
3. Complexity is a level where innovation is considered relatively difficult to understand and use, becoming an obstacle to the speed of the innovation adoption process.
4. The possibility to try (trial ability) is a level where innovation is on a small scale. New ideas that can be tried on a small scale are usually adopted more quickly than innovations that cannot be tried first.
5. Observability is the degree to which innovation results can be easily seen as a technical and economic advantage, thus accelerating the adoption process. Other prospective adopters no longer need to go through the trial phase; they can continue to the adoption stage.

5.4.2 Market

Several cases of good oil palm practices with high GWL show that applying compost nutrients on the peatland surface can increase the productivity of oil palm fresh fruit bunches (Wetadewi et al. 2021). High productivity will affect farmers' income and open up market access. However, the availability of many available technologies will affect the choices of farmers and companies, leading to uncertainty. The government's role, policy support, and a conducive market are needed to encourage the application of appropriate AHC technology to increase the sustainability of peatlands at the site level.

5.4.3 Policy Instruments

The occurrence of environmental changes in the peat ecosystem requires accelerating the adoption of AHC technology by peatland users for cultivation purposes. Fulfillment of these goals depends greatly on attractive information and smart dissemination activities provided by the competent body. In this context, dissemination activities are critical. For disseminating AHC technology, the work of researchers and extension workers is needed. The strategies applied must pay attention to the characteristics of farmers and companies as very diverse users, both in terms of the conditions of their regions and socioeconomic diversity. Efforts must be made to overcome the constraints that may arise. Farmers and companies will be encouraged to adopt AHC technology for sustainable farming/plantation systems if the dissemination of information is efficient (Stringer et al. 2020).

5.5 Conclusion

AHC application is a breakthrough to restore and rehabilitate degraded peat swamp forest ecosystems. The major positive impact of AHC is to provide additional livelihoods for forest communities. Encouraging organic farming in peatland areas can lead to the sustainable production of food and fruit crops and help improve the stability of peat soils. The utilization of AHC has many advantages, namely, improving the peatland ecosystem and ensuring an increase in agricultural productivity managed by the forest community in the nursery. Small-scale businesses based on smart farming can also gain a new source of income for communities whose settlements are in peatland areas. Business units in the form of cooperatives that now exist in oil palm plantations are suitable business units for producing AHC packages. The role of agricultural extension workers in introducing AHC applications in reaching farmers in rural peatlands is highly important. New perspectives for AHC application can facilitate its mass-scale practice and modified AHC content formulations (in terms of nutrients and microbes) can help to make it more practical, effective, and relatively low cost. The dissemination and adoption of AHC application at the central and site levels is an impetus not only to improve forest conservation but also to increase the productivity of plantation land and peatland agriculture based on smart farming. Moreover, the production of AHC packages to increase agricultural crop production can improve the welfare of farmers on the one hand and mitigate climate change on the other. AHC specifications can be formulated based on the various natures of the peatland in question. Different peatland sites show variations in ecological conditions, level of ecosystem damage, length of inundation, and local microclimate factors. Aerial root stimulation caused by AHC application should be monitored periodically. A representative time to show that AHC technology works well is approximately three years. By using a set of important parameters in terms of soil science, microbiology, silviculture, hydrology, carbon measurement estimation, and financial and economic feasibility, scientific judgment is certainly possible. We recommend AHC as an innovation in forest and land rehabilitation programs in peatland because this technology is evidently a good solution to help tree roots adapt and function under high GWLs to restore damaged peatland ecosystems.

References

Ahmad Bhat S, Kuriqi A, Dar MUD, Bhat O, Sammen SS, Towfiqul Islam ARM, Elbeltagi A, Shah O, Ai-Ansari N, Ali R, Heddam S (2022) Application of biochar for improving physical, chemical, and hydrological soil properties: a systematic review. Sustainability 14(17):11104. https://doi.org/10.3390/su141711104

Alkharabsheh HM, Seleiman MF, Battaglia ML, Shami A, Jalal RS, Alhammad BA, Almutairi KF, Al-Saif AM (2021) Biochar and its broad impacts in soil quality and fertility, nutrient leaching and crop productivity: A review. Agronomy 11(5):993. https://doi.org/10.3390/agronomy11050993

Anshari GZ, Gusmayanti E, Novita N (2021) The use of subsidence to estimate carbon loss from deforested and drained tropical peatlands in Indonesia. Forests 12(6):732. https://doi.org/10.3390/f12060732

Antala M, Juszczak R, van der Tol C, Rastogi A (2022) Impact of climate change-induced alterations in peatland vegetation phenology and composition on carbon balance. Sci Total Environ 827(154294):1–15

Antonius S, Agustiyani D, Dewi TK, Laili N, Osaki M (2021) Plant growth-promoting Rhizobacteria (PGPR) and compost materials for AeroHydro culture. Principles of AeroHydro culture. In: Osaki M, Tsuji N, Foead N, Rieley J (eds) Tropical peatland eco-management. Springer, Singapore. https://doi.org/10.1007/978-981-33-4654-3_9

Asano K, Kagong WVA, Mohammad SMB, Sakazaki K, Talip MSA, Sahmat SS, Chan MKY, Isoi T, Kano-Nakata M, Ehara H (2021) Arbuscular mycorrhizal communities in the roots of sago palm in mineral and shallow peat soils. Agriculture 11(11):1161. https://doi.org/10.3390/agriculture11111161

Budiman I, Bastoni, Sari ENN, Hadi EE, Asmaliyah, Siahaan H, Januar R, Hapsari RD (2020) Progress of paludiculture projects in supporting peatland ecosystem restoration in Indonesia. Global Ecol Conserv 23(e01084):1–17. https://doi.org/10.1016/j.gecco.2020.e01084

Cárceles RB, Durán-Zuazo VH, Soriano RM, García-Tejero IF, Gálvez RB, Cuadros TS (2022) Conservation agriculture as a sustainable system for soil health: a review. Soil Syst 6(4):87. https://doi.org/10.3390/soilsystems6040087

Chaddy A, Melling L, Ishikura K, Goh KJ, Toma Y, Hatano R (2021) Effects of long-term nitrogen fertilization and ground water level changes on soil CO_2 fluxes from oil palm plantation on tropical peatland. Atmos 12(10):1340. https://doi.org/10.3390/atmos12101340

Churchland C, Grayston SJ (2014) Specificity of plant-microbe interactions in the tree mycorrhizosphere biome and consequences for soil C cycling. Front Microbiol 5:261. https://doi.org/10.3389/fmicb.2014.00261

Coman C, Țîru LG, Meseșan-Schmitz L, Stanciu C, Bularca MC (2020) Online teaching and learning in higher education during the coronavirus pandemic: students' perspective. Sustainability 12(24):10367. https://doi.org/10.3390/su122410367

Cusack DF, Addo-Danso SD, Agee EA, Andersen KM, Arnaud M, Batterman SA, Brearley FQ, Ciochina MI, Cordeiro AL, Dallstream C, Diaz-Toribio MH, Dietterich LH, Fisher JB, Fleischer K, Fortunel C, Fuchslueger L, Guerrero-Ramírez NR, Kotowska MM, Lugli LF, Marín C, McCulloch LA, Maeght J-L, Metcalfe D, Norby RJ, Oliveira RS, Powers JS, Reichert T, Smith SW, Smith-Martin CM, Soper FM, Toro L, Umaña MN, Valverde-Barrantes-O, Weemstra M, Werden LK, Wong M, Wright CL, Wright SJ, Yaffar D (2021) Tradeoffs and synergies in tropical forest root traits and dynamics for nutrient and water acquisition: field and modeling advances. Front For Glob Change 4:704469. https://doi.org/10.3389/ffgc.2021.704469

Dohong A, Abdul Aziz A, Dargusch P (2018) A review of techniques for effective tropical peatland restoration. Wetlands 38:275–292. https://doi.org/10.1007/s13157-018-1017-6

Dudek M, Wrzaszcz W (2020) On the way to eco-innovations in agriculture: concepts, implementation and effects at national and local level. Case Poland Sustain 12(12):4839. https://doi.org/10.3390/su12124839

Fitmawati F, Saputri NA, Hartanto S, Resida E, Kholifah SN, Kapli H, Sopiyanti N, Wahibah NN, Khairijon (2020) Diversity and utilization of bamboo (Bambusoideae) in five islands around Riau province, Indonesia. Sabrao J Breed Gen 52(2):177–190

Girkin NT, Cooper HV, Ledger MJ, O'Reilly P, Thornton SA, Akesson CM, Cole LES, Hapsari KA, Hawthorne D, Roucoux KH (2022) Tropical peatlands in the anthropocene: the present and the future. Anthropocene 40(100354):1–16

Graham LLB, Turjaman M, Page SE (2013) *Shorea balangeran* and *Dyera polyphylla* (syn. *Dyera lowii*) as tropical peat swamp forest restoration transplant species: effects of mycorrhizae and level of disturbance. Wetland Ecol. Manage 21(5):307–321

Graham LLB, Giesen W, Page SE (2016) A common-sense approach to tropical peat swamp forest restoration in Southeast Asia. Restor Ecol 25(2):312–321

Han J, Xia D, Li L et al (2009) Diversity of culturable bacteria isolated from root domains of moso bamboo (*Phyllostachys edulis*). Microb Ecol 58:363–373. https://doi.org/10.1007/s00248-009-9491-2

Hayasaka H, Usup A, Naito D (2020) New approach evaluating peatland fires in Indonesian factors. Remote Sens 12(12):2055. https://doi.org/10.3390/rs12122055

Helbert, Turjaman M, Nara K (2019) Ectomycorrhizal fungal communities of secondary tropical forests dominated by *Tristaniopsis* in Bangka Island, Indonesia. PLoS One 14(9):1–9

Helgason T, Fitter AH (2009) Natural selection and the evolutionary ecology of the arbuscular mycorrhizal fungi (phylum Glomeromycota). J Exp Bot 60(9):2465–2480. https://doi.org/10.1093/jxb/erp144

Indrajaya Y, Yuwati TW, Lestari S, Winarno B, Narendra BH, Nugroho HY, Rachmanadi D, Pratiwi, Turjaman M, Adi RN, Savitri E (2022) Tropical Forest landscape restoration in Indonesia: a review. Land 11(3):328. https://doi.org/10.3390/land11030328

Laurén A, Palviainen M, Page S, Evans C, Urzainki I, Hökkä H (2021) Nutrient balance as a tool for maintaining yield and mitigating environmental impacts of *acacia* plantation in drained tropical peatland—description of plantation simulator. Forests 12(3):312. https://doi.org/10.3390/f12030312

Maftuah E, Hairani A, Susilawati A, Sosiawan H, Nursyamsi D, Osaki M (2021) Biochar for the improvement of peatland and suboptimal land. In: Osaki M, Tsuji N, Foead N, Rieley J (eds) Tropical peatland eco-management. Springer, Singapore. https://doi.org/10.1007/978-981-33-4654-3_11

Maulana AF, Turjaman M, Sato T, Hashimoto Y, Cheng W, Tawaraya K (2018) Isolation of endophytic fungi from tropical forest in Indonesia. Symbiosis 76(2):151–162

Murphy DJ, Goggin K, Paterson RM (2021) Oil palm in the 2020s and beyond: challenges and solutions. CABI Agric Biosci 2:39. https://doi.org/10.1186/s43170-021-00058-3

Ong JW, Rahim MFA, Lim W, Nizat MNM (2022) Agricultural technology adoption as a journey: proposing the technology adoption journey map. Int J Technol 13(5):1090–1096

Osaki M, Tsuji N (2016) Introduction to tropical peatland. In: Osaki M, Tsuji N (eds) Tropical peatland ecosystems. Springer, Japan

Osaki M, Kato T, Kohyama T, Takahashi H, Haraguchi A, Yabe K, Tsuji N, Shiodera S, Rahajoe JS, Atikah TD, Oide Y, Matsui K, Wetadewi RI, Silsigia S (2021a) Basic information about tropical peatland ecosystems. In: Osaki M, Tsuji N, Foead N, Rieley J (eds) Tropical peatland eco-management. Springer, Singapore. https://doi.org/10.1007/978-981-33-4654-3_1

Osaki M, Tsuji N, Kato T, Yamanaka MD, Sulaiman A, Silsigia S, Wetadewi RI (2021b) Natural capital-based societies in the tropics. In: Osaki M, Tsuji N, Foead N, Rieley J (eds) Tropical peatland eco-management. Springer, Singapore. https://doi.org/10.1007/978-981-33-4654-3_6

Putra FM, Sitanggang IS, Sobir S (2020) Decision support system for evaluation of peatland agroecology suitability in pineapple plants. Sci J Inf 7(1):75–86. https://doi.org/10.15294/sji.v7i1.23819

Rogers EM, Shoemaker FF (1995) Communication of innovation, New edn. The Free Press, New York

Santi LP, Goenadi DH, Osaki M (2021) Zeolites and aggregate-stabilizing microbes for reducing the degradation and carbon emissions in tropical peatlands. In: Osaki M, Tsuji N, Foead N, Rieley J (eds) Tropical peatland eco-management. Springer, Singapore. https://doi.org/10.1007/978-981-33-4654-3_10

Saputra E (2019) Beyond fires and deforestation: tackling land subsidence in peatland areas, a case study from Riau, Indonesia. Land 8(5):76. https://doi.org/10.3390/land8050076

Schiffman L, Kanuk L (2010) Consumer behavior, 10th edn. Global Edition. Prentice-Hall Inc., Upper Saddle River, NJ

Schimmelpfennig D (2016) Farm profits and adoption of precision agriculture. Economic research report number 217. United States Department of Agriculture (USDA). 46. https://www.ers.usda.gov/webdocs/publications/80326/err-217.pdf?v=0

Silvianingsih YA, Hairiah K, Suprayogo D, van Noordwijk M (2021) *Kaleka* agroforest in Central Kalimantan (Indonesia): soil quality, hydrological protection of adjacent peatlands, and sustainability. Land 10(8):856. https://doi.org/10.3390/land10080856

Stringer LC, Fraser EDG, Harris D, Lyon C, Pereira L, Ward CFM, Simelton E (2020) Adaptation and development for different type of farmers. Environ Sci Pol 104:174–189

Tampubolon AP, Turjaman M, Osaki M (2021) Sago palm practice as natural AeroHydro culture. In: Osaki M, Tsuji N, Foead N, Rieley J (eds) Tropical peatland eco-management. Springer, Singapore. https://doi.org/10.1007/978-981-33-4654-3_12

Tata HL, Nuroniah HS, Ahsania DA, Anggunira H, Hidayati SN, Pratama M et al (2022) Flooding tolerance of four tropical peatland tree species in a nursery trial. PLoS One 17(4):e0262375. https://doi.org/10.1371/journal.pone.0262375

Tawaraya K, Turjaman M (2014) Use of arbuscular mycorrhizal fungi for reforestation of degraded tropical forests. In: Solaiman ZM, Abbott LK, Varma A (eds) Mycorrhizal fungi: use in sustainable agriculture and land restoration, soil biology 41. Springer, Japan. https://doi.org/10.1007/978-3-662-45370-4_22

Tawaraya K, Takaya Y, Turjaman M, Tuah SJ, Limin SH, Tamai Y, Cha JY, Wagatsuma T, Osaki M (2003) Arbuscular mycorrhizal colonization of tree species grown in peat swamp forests of Central Kalimantan, Indonesia. For Ecol Manag 182:381–386

Tuheteru FD, Arif A, Mansur I, Turjaman M, Hadijah MH, Rumambi A, Prasetya B, Male AR (2022) Tolerance of lonkida (*Nauclea orientalis* L.) seedlings inoculated with mycorrhizae against drought and waterlogging stress. J Degraded Mining Lands Manag 9(4):3725–3732. https://doi.org/10.15243/jdmlm.2022.094.3725

Turjaman M, Osaki M (2021) The role of mycorrhizal fungi for supporting AeroHydro culture in tropical peatland. In: Osaki M, Tsuji N, Foead N, Rieley J (eds) Tropical peatland eco-management. Springer, Singapore. https://doi.org/10.1007/978-981-33-4654-3_8

Turjaman M, Tamai Y, Santoso E, Osaki M, Tawaraya K (2006) Arbuscular mycorrhizal fungi increased early growth of two non-timber forest product species *Dyera polyphylla* and *Aquilaria filaria* under greenhouse conditions. Mycorrhiza 16:459–464

Turjaman M, Tamai Y, Sitepu IR, Santoso E, Osaki E, Tawaraya K (2008) Improvement of early growth of two tropical peat-swamp forest tree species *Ploiarium alternifolium* and *Calophyllum hosei* by two arbuscular mycorrhizal fungi under greenhouse conditions. New For 36:1–12

Turjaman M, Santoso E, Susanto A, Gaman S, Limin SH, Tamai Y, Osaki M, Tawaraya K (2011) Ectomycorrhizal fungi promote growth of *Shorea balangeran* in degraded peat swamp forests. Wetl Ecol Manag 19:331–339

Turjaman M, Faulina SA, Aryanto N, Yani A, Hidayat A (2019) Isolation, identification and utilization of fungi associated with *Tristaniopsis obovata*. Jurnal Penelitian Hutan dan Konservasi Alam 16(1):73–90. https://doi.org/10.20886/jphka.2019.16.1.73-90

Wetadewi RI, Osaki M, Turjaman M, Antonius S, Goenadi DH, Nursyamsi D, Maswar, Surayah L, Kato T (2021) Principles of AeroHydro culture. In: Osaki M, Tsuji N, Foead N, Rieley J (eds) Tropical peatland eco-management. Springer, Singapore. https://doi.org/10.1007/978-981-33-4654-3_7

Wu J, Wang J, Hui W, Zhao F, Wang P, Su C, Gong W (2022) Physiology of plant responses to water stress and related genes: a review. Forests 13(2):324. https://doi.org/10.3390/f13020324

Yuwati TW, Rachmanadi D, Pratiwi TM, Indrajaya Y, Nugroho HYSH, Qirom MA, Narendra BH, Winarno B, Lestari S, Santosa PB, Adi RN, Savitri E, Putra PB, Wahyuningtyas RS, Prayudyaningsih R, Halwany W, Nasrul B, Bastoni, Mendham D (2021) Restoration of degraded tropical peatland in Indonesia: a review. Land 10(11):1170. https://doi.org/10.3390/land10111170

Chapter 6
Function of Humic Acid

Masahiko Uomori and Tatsuaki Yamaguchi

Abstract Humic materials have been studied since the beginning of the twentieth century, and due to poor evaluation methods, several researchers have been skeptical about the structure of humic acid. The method of extracting humic materials from soils was not established until the establishment of the International Humic Substance Society (IHSS) in 1981, and the IHSS seeks to advance knowledge and research on natural organic materials in soil and water, identify methods to isolate humic acid from soil, and develop a standard. Since the IHSS extraction method was established, studies of humic acid and fulvic acid have been performed in soil and water, especially in peatland and wetland.

Hence, the physical properties and chemical structure of humic acid were characterized (colloidal property, elemental analysis, acidity of humic acid and ion exchanging character, and NMR spectrum), and researchers have focused on the dynamics of chelation between humic acid and iron ions.

Keywords Humic acid · IHSS method · Tropical peat · Iron chelates · Mangrove ecosystem

6.1 Introduction

Humic acid plays an essential role in agriculture, forestry, and fishery fields. However, it was difficult to study humic acid chemistry before the International Humicosol Society (IHSS) method was established because humic acid is a complex compound. Tropical peatlands produce large amounts of humic acid, which forms chelates with Fe^{2+} (Rue and Bruland 1995) and carries iron from peatland trough rivers into mangrove and coastal ecosystems. Due to its high pH, Fe^{2+} deposits in

M. Uomori (✉)
NPO Peat-Greening Association, Atami-shi, Shizuoka, Japan
e-mail: peat-society.jp@nifty.com

T. Yamaguchi
Chiba Institute of Technology, Narashino-shi, Chiba, Japan

M. Osaki et al. (eds.), *Tropical Peatland Eco-evaluation*,
https://doi.org/10.1007/978-981-99-6790-2_6

seawater (Kida et al. 2016); thus, Fe-chelate is a compound that supplies iron nutrients to sea plants. Therefore, it is hypothesized that humic acid from peatland affects coastal and marine ecosystems over long distances.

6.2 History of Studies on Humic Acid

J.G. Wallerius (1761) was a Swedish chemist and mineralogist who studied organic materials in soil, and he named them "Fumus" from the Latin word for soil in 1761. J.J. Berzelius was the first chemist to articulate the differences between inorganic and organic compounds and studied Fumus as well. J.F. Liebig was a German scientist who made major contributions to agricultural and biological chemistry, and in his writing, Leibig stated that "Fumus is soluble in alkaline solvent colored dark brown and insoluble in water." W. Fuchs (1930) discovered that brown coal has the following molecular structure:

$$C60H_{39}O_{15}(COOH)_4(OH)_4(-O-)\ (CH_2CO)\ (CH_3)$$

Based on a study of Fuchs' brown coal, Fumus was characterized as a weakly acidic compound containing carboxyl groups.

The method of extracting humic acid from peat was quickly developed by several researchers (Kononova 1966). Humic substances are complex and heterogeneous mixtures of polydispersed materials that are formed in soils, sediments, and natural waters. These materials form through biochemical and chemical reactions as plant and microbial remains decay and transform (a process called humification). The International Humic Substance Society (IHSS) was founded in 1981 in Denver, Colorado, USA, and aimed to establish sampling, extraction, and analysis methods.

6.3 IHSS Method of Isolating Fulvic Acid and Humic Acids from Soil

Remove the roots and pass the dried soil sample through a 2-mm sieve. Equilibrate the sample to a pH value between 1 and 2 with 1 M HCl at room temperature. Adjust the solution volume with 0.1 M HCl to provide a final concentration with a ratio of 10 mL liquid/1 g dry sample. Shake the suspension for 1 h, and then separate the supernatant from the residue by decantation after the solution settles or by low-speed centrifugation. Save the supernatant (FA Extract 1) for the isolation of fulvic acid using XAD-8. Neutralize the soil with 1 M NaOH to pH = 7, and then 0.1 M NaOH under an atmosphere of N_2 to give a final extractant to soil ratio of 10:1. Extract the suspension under N_2 with intermittent shaking for a minimum of 4 h. Settle the alkaline suspension overnight and collect the supernatant by decantation or centrifugation. Acidify the supernatant with 6 M HCl with constant stirring to pH = 1 and then store it for 12–16 h. Centrifuge the samples to separate the humic acid (precipitate) and fulvic acid (supernatant—FA Extract 2) fractions.

Redissolve the humic acid fraction by adding a minimum volume of 0.1 M KOH under N_2. Add solid KCl to attain a concentration of 0.3 M $[K^+]$ and then centrifuge at high speed to remove the suspended solids. Reprecipitate humic acid by adding 6 M HCl with constant stirring to pH = 1 and let the suspension stand for 12–16 h. Centrifuge the samples and discard the supernatant. Suspend the humic acid precipitate in 0.1 M HCl/0.3 M HF solution in a plastic container and shake overnight at room temperature. Centrifuge and repeat the HCl/HF treatment, if necessary, until the ash content is below 1%. Transfer the precipitate to a Visking dialysis tube by mixing with water and dialyzing against distilled water until the dialysis water gives a negative Cl– test with silver nitrate $AgNO_3$. Freeze-dry the humic acid.

Pass the supernatant designated "FA Extract 1" through a column of XAD-8 (0.15 mL of resin per gram of initial sample dry weight at a flow rate of 15 bed volumes per hour). Discard the effluent and rinse the XAD-8 column containing sorbed fulvic acid with 0.65 column volumes of distilled H_2O. Back elute the XAD-8 column with 1 column volume of 0.1 M NaOH, followed by 2–3 column volumes of distilled H_2O. Immediately acidify the solution with 6 M HCl to pH = 1. Add concentrated HF to a final concentration of 0.3 M HF. The solution volume should be sufficient to maintain fulvic acid in the solution.

Pass the supernatant designated "FA Extract 2" through a column of XAD-8 (1 mL of resin per gram of initial sample dry weight). Repeat the back elution and acidification as for "FA Extract 1" above. Combine the final eluates from each of the fulvic acid extracts and pass them through XAD-8 resin in a glass column (column volume should be one-fifth of the sample volume). Rinse with 0.65 column volumes of distilled H_2O. Back elute with 1 column volume of 0.1 M NaOH followed by two column volumes of distilled H_2O. Pass the eluate through H^+-saturated cation exchange resin [Bio-Rad AG MP^{-5} (Bio-Rad, Richmond, CA) using three times the mole of Na ions in solution]. Freeze-dry the eluate to recover the H^+-saturated fulvic acid (Kuwatsuka et al. 1992).

The product is sometimes difficult to obtain, and it may be necessary to use an alternative resin such as Polyclar, which is a cross-linked poly (vinylpyrrolidone) (De Nobili et al. 1990; Watanabe and Kuwatsuka 1991).

Extensive purification procedures of the resins are necessary before use. These methods and methods used to store the resin were detailed by Thurman and Malcolm in 1981 (Thurman and Malcolm 1981).

6.4 Physical Properties and Chemical Structure of Humic Acid

6.4.1 Colloidal Property

Humic acid does not pass through the colloidal membrane, and the particles of humic acid show Brownian movement, similar to the irregular motion of small particles in fluids. Humic acid is not chemically decomposed by hot water and dispersed by diluted neutral salt. Humic acid is a colloidal material that carries a negative electric charge. Humic acid sol is dark brown and mist slurry or milky turbid. Colorimetric analysis is a conventional method used to evaluate colloidal properties. Humic acid is swelled with a water-miscible organic solvent. Due to ion dissociation, the molecular weight of humic acid salt is smaller than that of humic acid sol by osmolality measurement. The molecular weights of humic acid and fulvic acid are between 3000 and 800, respectively (Malcom 1985). The molecular weight distribution of typical humic acids (Uomori et al. 1998) measured by the gel filtration method is shown in Fig. 6.1.

6.4.2 Elemental Analysis

The results of elemental analysis for several peats are shown in Table 6.1 (Yamaguchi et al. 1998) and Table 6.2 (Uomori et al. 1998).

Tropical peat contains a larger amount of carbon than other peat (Uomori et al. 1995). The fumigation of tropical peat is accelerated over several years under certain weather conditions. C/N is over 15, and that of tropical peat is approximately 35.

6.4.3 Acidity of Humic Acid and Ion Exchange Characteristics

Humic acid contains carboxyl groups and phenolic hydroxyl groups and shows acidic properties. When comparing humic acid extracted from weathered coal and tropical peat, the former contains many carboxyl groups; in contrast, the latter contains fewer carboxyl groups due to their generation processes. For the former, a part of the coal constitution is assumed to be oxidized, and the aromatic ring is cleaved to generate many carboxyl groups. Regarding the latter, flora and fauna that contain lignin, polysaccharides, proteins, and lipids are oxidized and depolymerized under anaerobic conditions and generated many phenolic hydroxyl groups.

When comparing the ion exchange ratio between ammonium humic acid and humic acid from a physical property perspective, each carboxylic group and carboxylate strongly affect the cation exchange reaction. Carboxylate is generated by the

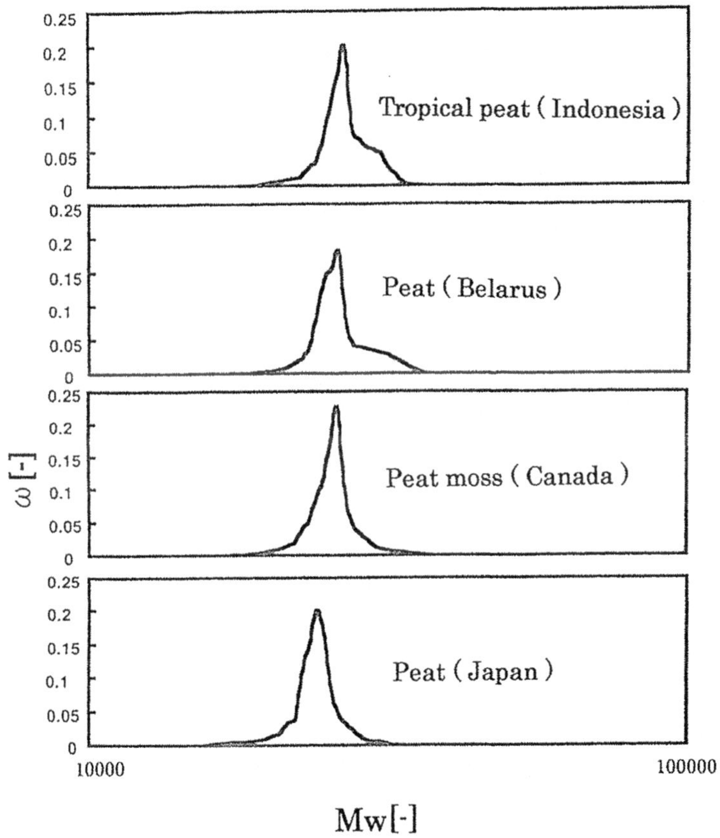

Fig. 6.1 Molecular weight distribution of extracted humic acid

Table 6.1 Organic substances group composition of peats

	Chemical composition of organic substances (wt % to organic mass)			
	Humic acid	Easy-hydrolyzable substances	Hard-hydrolyzable substances	Nonhydrolysable substances
Peat moss (Canada)	14.50	27.65	17.43	12.02
Reed-sedge peat (Belarus)	36.93	12.04	5.02	29.05
Reed peat (Japan)	40.45	5.22	10.88	12.86
Tropical peat (Indonesia)	52.94	8.83	1.69	24.27

Table 6.2 Ionic components of peats

	pH	Ionic concentration (meq/100 g-dried material)								
	(KCl)	Na	NH^{4+}	K^+	Mg^{2+}	Ca^{2+}	SO_4^{2-}	NO^{3-}	Cl	PO_4^{2-}
Peat moss (Canada)	2.65	0.40	2.83	0.54	0.65	1.32	1.08	0.73	5.12	2.77
Reed-sedge peat (Belarus)	2.80	0.37	2.30	0.38	0.63	3.51	2.32	0.66	4.33	2.57
Reed peat (Japan)	3.65	0.98	2.15	0.29	3.19	3.62	6.20	0.54	1.38	2.92
Tropical peat (Indonesia)	2.70	1.27	3.32	1.10	0.61	0.95	5.60	1.55	1.40	9.54

substitution between hydrogen ions of carboxylic groups in humic acid and ammonium ions.

The ion exchange character and/or ability of humic acid to generate chelates affect the fertilizer retention of soil. Humic acid exhibits polyfunctionalized characteristics in several pH ranges.

6.4.4 NMR Spectrum

In the previous year, it was difficult to specify the functional group of organic material, as it is similar to that of a powder substance. A solid material can be specified with the CP/MAS^{13}C-NMR spectrum, and several carbon group elements of humic acid have been identified. As shown in Fig. 6.2, the CP/MAS^{13}C-NMR spectra of four selected kinds of peat, including tropical peat, peat in Belarus, peat in Japan (Hokkaido), and peat moss in Canada, were extracted using a conventional method (0.1 N NaOH).

An aliphatic carbon appeared at 10–50 ppm, especially in the 30 ppm range, and methyl aliphatic, methylene, and methine carbons appeared.

The "50–110 ppm" methoxy group and carbohydrate carbon appeared in the 70–75 ppm range, and the CHOH carbon appeared.

The "110–160 ppm" aromatic carbon appeared in the 130 ppm range, and aromatic carbon bonded with aliphatic polymer chains appeared. 150 ppm range, carbon bonded with an aromatic carbon group.

Regarding each NMR peak, aromatic carbon shows a broad range, and the relative intensity ratio is small. In contrast, that of aliphatic carbon is significantly larger.

The ^{13}C chemical shift is affected by the electric state of the electron nucleus, and the relaxation time of the CP/MAS^{13}C-NMR spectrum is short. The cumulative efficiency is high performance without taking into account the solubility of the sample.

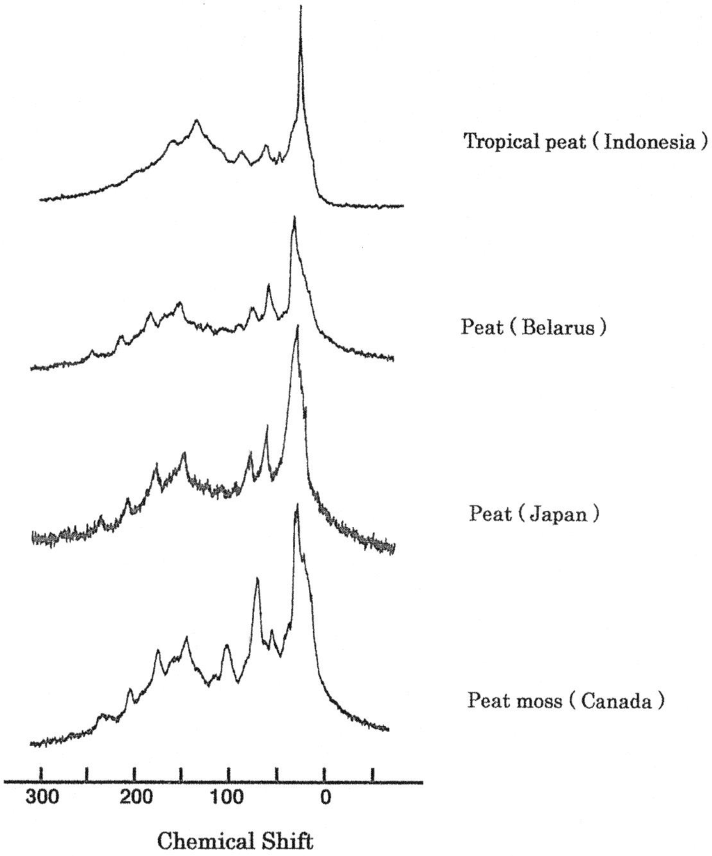

Fig. 6.2 CP/MAS^{13}C-NMR spectrum

6.5 Property of Tropical Peat

Tropical or "woody" peat is generated by mother plants in wooded forests. More than 52.4% of tropical peat in tropical areas is found in Indonesia, and more than 87.3% of tropical peat in southeastern Asia is found in Indonesia. We may say that tropical peat in Indonesia is a typical tropical peat. The effective water content of tropical peat can be estimated by pF curves. The fiber components of tropical peat effectively hold water. Available water (pF 1.5–4.2) for plants comprises capillary water in the fiber components of tropical peat. The water absorption of peat has been studied (Yamaguchi and Tsukagoshi 1991; Uomori and Yamaguchi 1997).

There are two types of peatlands in Indonesia: "topogenous" and "ombrogeneous" (Widjaya-Adhi 1988). The former is recharging groundwater beside lakes and low-level ground, and the latter is gregarious and excavated in hills, in which the aquifers are recharged by rainwater.

Table 6.3 Characteristics of humic acid extracted from tropical peat, weathered coal, and peat moss

	Tropical peat (Kalimantan)		Weathered coal (Uirumqi)		Peat moss (Canada)
Extractant	NaOH	NH_4OH	NaOH	NH_4OH	NaOH
COOH [cmol/kg]	345.0	294.0	303.0	291.0	241.0
Phenolic OH [cmol/kg]	1.9	1.6	5.8	5.6	6.5
Ash content [wt %]	2.5	0.4	3.8	1.3	2.8

The former is a limited area, and almost the whole area is occupied by the latter. Tropical peat has accumulated between rivers that flow through coastal plains and filled in the area. Tropical peat in Indonesia is generated from "topogenous" peat, and the bulk specific gravity is 0.1–0.2 g/cm^3; in addition, the water retention ability is high (150–200% dry weight). However, the peat exhibits water repellency and dries up, has a pH range of 3–4, possesses a low-level ash content, shows a relatively high C/N (24–33) depending on the depth, and exhibits a cation exchange capacity range of 50–100 cmol/kg. The amount of cellulose decreases; hence, the property of soil changes from fibric to sapric, and the decomposition rate is sluggish.

Malaysia is located close to Indonesia, and the properties of tropical peat in Malaysia are similar to those in Indonesia. The ash content is in a low range (1–5%), and the humic acid content is in a high range (49–77%).

The characteristics of humic acid extracted from tropical peat (Indonesia) and weathered coal (China) with an aqueous ammonia solution were compared. Humid acid was extracted using a conventional method (NaOH, $NaOH/Na_4P_2O_7$) and a new process with NH_4OH as an extractant. Regardless of the extractant, humic acid extracted from tropical peat has approximately four times larger average molecular weight than that of weathered coal. For the functional groups, humic acid extracted from weathered coal has a higher quantity of carboxyl groups; however, the quantity of phenolic hydroxyl groups is lower, as shown in Table 6.3. Humic acid extracted from tropical peat contains many aromatics. The yield of humic acid with NaOH was the highest among the three extractants. The humic acid extracted with NH_4OH contained a smaller amount of ash but a larger amount of ammoniacal nitrogen than that extracted with $NaOH/Na_4P_2O_7$.

The cation exchange reactions of humic acid were accelerated by the higher electric charge of metal ions, the higher carboxyl group content, and the lower molecular weight of the humate.

6.6 Dynamics of Iron Ions and Organic Material Containing Humic Acid

The connection between forests, rivers, and oceans is essential for the circulation of materials. Iron transport of humic acid in humic materials is important for coastal water.

Iron is an essential material for the growth of phytoplankton in water. Based on several previous studies on iron biogeochemistry in coastal and freshwater systems, the particular influence of dissolved organic materials containing humic acid and its quality on iron transformation kinetics is remarkable, and iron can affect phytoplankton in coastal waters. The residual rate of iron ions is dependent on the content of humic acid in the dissolved organic material. A number of previous studies, particularly those in the field of ocean science, revealed that in air_saturated water at circumneutral pH (between 6.5 and 7.5), the solubility of ferric iron (Fe^{2+}) is at the subnanomolar level, whereas the presence of dissolved organic material substantially increases iron solubility.

Iron ions (Fe^{2+}) are essential for phytoplankton, and Fe^{2+} is easily oxidized to Fe^{3+}, which is insoluble in water; thus, phytoplankton cannot attain Fe^{2+}. Humic acid contains many carboxyl groups, and humic acid and Fe^{2+} are chemically bonded to form a “Humic Iron Chelate (HIC)” (Rue and Bruland 1995). Fe^{2+} is retained in dissolved conditions (Tipping 2002) at circumneutral pH values (between 6.5 and 7.5). HIC is carried to coastal areas, although it is removed at high salt concentrations and plays an important role in biological productivity.

Under aerobic (high Pε) and circumneutral pH conditions, insoluble $Fe(OH)_3$ is stable; in contrast, under anaerobic (low Pε) conditions, Fe^{2+} and $FeCO_2(s)$ are stable. The iron redox potential is changed with dissolved organic materials, such as humic acid, and the iron redox potential is decreased by coordinate bonding with humic acid. The oxidation process of Fe^{2+} with the Haber–Weiss reaction is as follows:

$$Fe^{2+} + O_2 \rightarrow Fe^{3+} + O_2{}^{-}.$$

$$Fe^{2+} + O_2{}^{-} + 2H^{+} \rightarrow Fe^{3+} + H_2O_2.$$

$$Fe^{2+} + H_2O_2 \rightarrow Fe^{3+} + OH\bullet + OH^{-}.$$

$$Fe^{2+} + OH\bullet \rightarrow Fe^{3+} + OH^{-}.$$

Research on the effectiveness of Fe^{2+} generated from land to coastal areas in Amur has continued (Sholkovitz and Copland 1981). Several projects have been conducted regarding the effectiveness of iron generated from the upper Amur to a coastal area and to oceanic provinces.

The importance of forests and peatlands has been confirmed by these projects.

Regarding the reduction of Fe^{3+}, organic ligands absorb light energy and accumulate electric charge on iron atoms. The reduction of Fe^{3+} is accelerated by humic acid-containing materials (Rakshit et al. 2009).

6.7 Dynamics of Humic Acid in Mangroves

The content percentage of humic materials in the water depends upon the environment and is 10–30% in oceans, 40% in lakes, and 70–90% in colored rivers (Thurman 1985; Kida et al. 2015; Tsuda et al. 2016).

Mangrove forests are tropical forests with carbon-rich soils, such as humic acid. The mangrove carbon cycle has been studied by several researchers (Dittmar et al. 2006; Cawley et al. 2014). Mangrove forests are located next to the coast and are connected with river tributaries. The ecological roles and turnover rate of humic acid contained in materials are dependent on their composition.

Mangrove forests are in the brackish water between rivers and the seashore. Their area covers approximately 60–77% of the tropical seashore. Wild animals live there, and nutrient salts are supplied through mangrove forests.

The soil under mangrove forests contains a large amount of organic carbon, and although its area is 0.7% of the total tropical plants (Giri et al. 2011) in tropical areas worldwide, it absorbs approximately 2.5 times the carbon dioxide emissions (2.5 PgC) throughout the year (Donato et al. 2011), and the reserve volume of organic carbon is 26.2 TgC (Breithaupt et al. 2012). Mangrove forests supply dissolved organic materials to coastal areas. Approximately 10% of dissolved organic material from land area is from mangrove forests, and more than 90% of total carbon generated from mangrove forests (Davis et al. 2001; Dittmar et al. 2006; Sutula et al. 2003) is dissolved organic materials.

The outflow quantity of humic material generated from the humification of withered mangrove roots to the river or coast side depends upon the basicity of the conditions. At a high basicity level, humic materials are precipitated (Sholkovitz 1976) and cannot drain to the river or coast. At a low basicity level, humic materials are dispersed, and dispersed humic materials conjugate the humic material–clay complex; these materials show Brownian movement similar to the irregular motion of small particles in fluids. The dispersion rate depends on the salinity concentration. The dispersion is in the range of electrical conductivity of approximately 77 mS/m.

Humic acid is precipitated even in seawater, although it is composed of humid acid–cation complexes (Fox 1983); in contrast, fulvic acid is not precipitated even in seawater (Kida et al. 2016).

6.8 Conclusion

Humic acid is not a popular material among chemists because the method to analyze humic acid is poor and humic acid exhibits complicated molecular structures. Since the IHSS was founded in the 1980s, studies on humic acid have been rapidly accelerating. Peat contains humic acid, and an iron transport role by humic acid is important in coastal water because iron is an essential material for the growth of phytoplankton. Hence, the role of humic acid in agriculture, forestry, and fisheries will prevail, especially as a carrier of Fe^{2+} from forests to coastal areas, and the growth of phytoplankton in water will soon be the focus.

References

Breithaupt JL, Smoak JM, Smith TJ, Sanders CJ, Hoare A (2012) Organic carbon burial rates in mangrove sediments: strengthening the global budget. Glob Biogeochem Cycles 26. https://doi.org/10.1029/2012GB004375

Cawley KM, Yamashita Y, Maie N, Jaffe R (2014) Using optical properties to quantify fringe mangrove inputs to the dissolved organic matter (DOM) Pool in a subtropical estuary. Estuar Coasts 37:399–410. https://doi.org/10.1007/s12237-013-9681-5

Davis SE, Childers DL, Day JW, Rudnick DT, Sklar FH (2001) Wetland-water column exchanges of carbon, nitrogen, and phosphorus in a southern Everglades dwarf mangrove. Estuaries 24: 610–622

De Nobili M, Bragato G, Alcaniz JM, Puigbo A, Comellas L (1990) Characterization of electrophoretic fractions of humic substances with different electro focusing behavior. Soil Sci 150: 763–770

Dittmar T, Hertkorn N, Kattner G, Lara RJ (2006) Mangroves, a major source of dissolved organic carbon to the oceans. Glob Biogeochem Cycles 20:7p. https://doi.org/10.1029/2005GB002570

Donato DC, Kauffman JB, Murdiyarso D, Kurnianto S, Stidham M, Kanninen M (2011) Mangroves among the most carbon-rich forests in the tropics. Nat Geosci 4:293–297. https://doi.org/10.1038/ngeo1123

Fox LE (1983) The removal of dissolved humic acid during estuarine mixing. Estuar Coast Shelf Sci 16:431–440. https://doi.org/10.1016/0272-7714(83)90104-X

Giri C, Ochieng E, Tieszen LL, Zhu Z, Singh A, Loveland T, Masek J, Duke N (2011) Status and distribution of mangrove forests of the world using earth observation satellite data. Glob Ecol Biogeogr 20:154–159. https://doi.org/10.1111/j.1466-8238.2010.00584.x

Kida M, Maki K, Takata A, Kato T, Tsuda K, Hayakawa K, Sugiyama Y, Fujitake N (2015) Quantitative monitoring of aquatic humic substances in Lake Biwa, Japan, using the DAX-8 batch method based on carbon concentrations. Org Geochem 83–84:153–157. https://doi.org/10.1016/j.Orggeochem.2015.03.015

Kida M, Otsuka T, Kato T, Suzuki T, Fujitake N (2016) Evaluation of salinity effect on quantitative analysis of aquatic humic substances using nonionic DAX-8 resin. Chemosphere 146:129–132

Kononova MM (1966) Soil organic matter. Pergamon Press, Oxford

Kuwatsuka S, Watanabe A, Itoh K, Arai S (1992) Comparison of two methods of preparation of humic and fulvic acids, IHSS method and NAGOYA method. Soil Sci Plant Nutr 38(1):23–30

Malcom RL (1985) Humic substances in soil sediment and water. Wiley International, New York, pp 181–209

Rakshit S, Uchimiya M, Sposito G (2009) Iron (III)bioreduction in soil in the presence of added humic substances. Soil Sci Soc Am J 73:65–71

Rue EL, Bruland KW (1995) Complexation of iron(III)by natural organic ligands in the central North Pacific as determined by a new competitive ligand equilibration/adsorptive cathodic stripping voltammetric method. Mar Chem 50:117–138

Sholkovitz E (1976) Flocculation of dissolved organic and inorganic matter during the mixing of river water and seawater. Geochim Cosmochim Acta 40:831–845

Sholkovitz ER, Copland D (1981) The coagulation, solubility and adsorption properties of Fe, Mn, cu, Ni, cd, co and humic acids in a river water. Geochim Cosmochim Acta 45:181–189

Sutula MA, Perez BC, Reyes E, Childers DL, Davis S, Day Jr. JW, Rudnick D,Sklar F (2003) Factors affecting spatial and temporal variability in material exchange between the Southern Everglades wetlands and Florida Bay (USA). Estuarine, Coastal and Shelf Science, 57:757–781

Thurman EM (1985) Aquatic humic substances. In: Thurman EM (ed) Organic geochemistry of natural waters. Kluwer Academic Publishers Group, Boston, MA, pp 273–362

Thurman EM, Malcolm RL (1981) Preparative isolation of aquatic humic substances. Environ Sci Technol 15:463–466

Tipping E (2002) Cation binding by humic substances. Cambridge University Press, New York, NY, p 448

Tsuda K, Kida M, Aso S, Kato T, Fujitake N, Mamo M, Hayakawa K, Hirota M (2016) Determination of aquatic humic substances in Japanese lakes and wetlands by the carbon concentration-based resin isolation technique. Limnology 17:1–6. https://doi.org/10.1007/sl0201-015-0455-6

Uomori M, Yamaguchi T (1997) The water absorption of peat. Int Peat J 7:41–44

Uomori M, Yazaki F, Yamaguchi T (1995) The research of sapropel (Lake-deposit) in Belarus., journal of Japan society on water. Environment 18(10):745–754

Uomori M, Hayashi H, Yazaki F, Yamaguchi T (1998) Characteristic of Sapropel (Lake-deposit)., Journal of Japan Society on Water. Environment 21(8):540

Watanabe A, Kuwatsuka S (1991) Fractionation of soil fulvic acids using polyvinyl-pyrrolidone and their ionization difference spectra. Soil Sci Plant Nutr 37:611–617

Widjaja-Adhi I P.G (1988) Physical and chemical characteristic of peat soil of Indonesia. IARD J 10:59–64

Yamaguchi T, Tsukagoshi S (1991) Effect of drying on the water absorption of grass peat. Chem Exp 6(7):543–546

Yamaguchi T, Hayashi H, Yazawa Y, Uomori M, Yazaki F, Bambalov NN (1998) Comparison of basic characteristics of humic acids extracted from peat and other sources. Int Peat J 8:87–94

Chapter 7
Camera Traps to Survey Mammals in Peatland

Hidemi Nagano-Tatsuno and Midzuho Tatsuno

Abstract Acquiring information on mammal communities in the field has been challenging due to various factors. These factors include the nocturnal behavior of many mammals, their tendency to evade humans, and the difficulty involved in trapping large mammals. However, mammals, which are usually larger in body size than other animals (e.g., insects, reptiles, and birds) and consist of primary consumers to top predators, can significantly impact the ecosystems of their habitats. Therefore, missing data on mammals lead to a critical lack of information for assessing the biodiversity and conservation value in a given area. In recent years, camera traps have been used to survey ground-dwelling mammals. By detecting animal movement and heat, camera traps can automatically capture animals that appear in front of them at any time of day or night.

Peatlands are established on moist soils with accumulated undecomposed organic matter, and the soils are generally oligotrophic and have a low pH. Accordingly, peatlands are composed of unique tree species, which can influence the mammal community. Their muddy soils can also affect the migration and nest formation of mammals. Nonetheless, there is little information on mammals in peatlands. We installed camera traps in peatlands in the southwestern part of Borneo Island and detected 21 mammal species and 12 bird species. These recorded animals included eight endangered mammal species and two endangered bird species on the IUCN Red List. Notably, porcupines (Hystricidae spp.) and muntjacs (*Muntiacus* spp.), which are common species in non-peatlands, were not recorded in this area. This observation suggests that mammal communities in peatlands may differ significantly from those in non-peatlands. In addition, the Sunda stink badger, a species not previously reported in the area, was recorded in this survey. These results have demonstrated that camera traps provide vital information on mammal communities for evaluating the conservation value of peatlands.

H. Nagano-Tatsuno (✉)
Graduate School of Agriculture, Kyoto University, Kyoto, Japan
e-mail: tatsuno.hidemi@naga-ecolab.com

M. Tatsuno
Na-ga Ecology Laboratory, Nara, Japan

M. Osaki et al. (eds.), *Tropical Peatland Eco-evaluation*,
https://doi.org/10.1007/978-981-99-6790-2_7

Keywords Mammal communities · Camera traps · Peatland · Ground-dwelling mammals · Birds

7.1 Introduction

Declining biodiversity is a global concern. Medium- and large-sized mammals need a wide range of habitats (Ofstad et al. 2016). Therefore, they are particularly vulnerable to human-induced loss and fragmentation of habitats. Hence, the mammalian community structure is a reliable indicator for evaluating forest conditions. However, due to the difficulty of comprehensively surveying mammal communities in a forest, knowledge of these communities has been insufficient. In recent years, these assessments have become possible owing to improvements in the performance of camera traps and their availability at a relatively low cost.

In this chapter, we discuss the effectiveness of camera traps in peatlands, by presenting the findings of a study performed in a peatland in West Kalimantan.

7.1.1 Ecological Significance of Mammals in Peatland

A peatland is a forest on soil composed of organic matter from plant debris under waterlogged conditions. Moist and muddy ground can have a significant impact on ground-dwelling mammals because it can make foraging and nesting more challenging compared to regular ground conditions. Accordingly, the mammalian community in peatland is unique. Nonetheless, researchers have rarely focused on peatlands, and reports on their biota are few (Yule 2010) because peatlands are often hard to access and have been considered to have low animal diversity (World Wildlife Fund n.d.). However, recent studies have revealed that endangered species, such as orangutans and clouded leopards, inhabit these areas (Posa et al. 2011). In addition, peatlands are often left in pristine conditions because humans cannot easily use them, and they are increasingly recognized as important sites that serve as wildlife refugia (Yamada 2006; Nowak 2013). While several studies have concentrated on rare Felidae species and orangutans, it is crucial to obtain data on all the mammalian species in peatlands and understand the distinctive features of their communities to evaluate mammal distribution effectively on a global level.

7.1.2 Biological Conservation Significance in Peatland

Peatlands have been significantly reduced by fires caused by the dryness of El Niño (Langner et al. 2007) as well as by human land use (Yule 2010). It has also been noted that fires are more widespread due to the drying caused by the drainage systems that have been installed to use peatlands (Page and Hooijer 2016). It is estimated that 60% of the peatlands in the tropics are in Southeast Asia, of which 45% are in Kalimantan (Posa et al. 2011). However, approximately 55% of the peatlands in Kalimantan have already been lost, and only 24% of the remaining forests are conserved (Posa et al. 2011). The urgent task is to evaluate the value of peatland as a mammal habitat, encourage more eco-friendly management, and conserve the remaining peatland.

7.2 Camera Trap Technology

7.2.1 *Noninvasive*

Compared to surveys conducted by capturing or attaching tracking devices to mammals, surveys using camera traps have less impact on mammals because camera traps only take images of the mammals. Some mammal species are also sensitive to the sounds and smells produced by human activity in the forest and may change their behavior. Thus, the presence of humans in the forest can significantly affect their behavior. Camera traps allow for the collection of data on mammals with less influence from humans.

7.2.2 *Low Cost*

Camera trap surveys require only two trips to the site for installation and for data collection. Furthermore, camera traps can collect data on mammals that pass in front of the cameras for several months. Although selecting the camera location may require some knowledge and experience, once the camera has been installed, data collection does not require specialized knowledge and can be left to local collaborators. Additionally, the data are recorded 24 h a day, regardless of weather conditions. Therefore, a large amount of data can be obtained with little effort. Although camera traps are expensive and require a large initial investment, in recent years, the cost of camera traps has decreased, making them more affordable for researchers and conservationists.

7.2.3 *Highly Credible Data*

Unlike visual observation, using camera traps allows for greater accuracy and reliability in species identification because image data remain and can be shared with other researchers. In particular, it can be a very effective means of proving the existence of rare species that are encountered infrequently.

7.2.4 Comprehensive Understanding of a Whole Mammal Community

A major obstacle to surveying peatlands is their inaccessibility. In addition, mammals are often nocturnal and escape when they sense a human presence. Therefore, it is difficult to obtain reliable presence/absence data on mammals, especially for rare species. The camera traps continue to record mammals that pass in front of them for several months, 24 h a day, with minimal human impacts. The possibility of recording rare mammals can be greater than with other methods. In addition, by surveying a large number of camera traps over a long period, researchers can acquire an indication of the relative abundance of mammals based on the frequency of imaging from camera traps. However, we must exercise caution when interpreting imaging frequency, because various factors such as the mammal species, temperature, season, and food resources can influence the moving behavior of mammals, leading to differences in their frequency of appearance in front of camera traps. Therefore, the frequency of imaging can vary even in areas in which the same mammal species are present at the same density. Nonetheless, currently, camera trapping can be considered an extensional method to obtain information on the community of ground-dwelling mammals. In recent years, innovative methods for density estimation using camera trap data, such as the REST method (Nakashima et al. 2018), have also been developed.

7.2.5 Detailed Long-Term Observations in Particular Locations

Camera traps are highly effective tools for observing mammalian behavior, especially in specific locations where they frequently visit for specific purposes, such as nests, salt licks, mud baths, and latrines. Placing camera traps in these locations enables detailed and long-term observations of mammalian behaviors, which are often difficult to observe in the presence of humans. Video images taken by camera traps are particularly useful for observing behaviors.

7.3 Case Study of Camera Trap Surveys in Peatlands in Borneo

A camera trap survey was conducted on peatland in West Kalimantan. Twenty camera traps were installed for 2 years. This chapter describes the mammal and bird species recorded during this camera trap survey and discusses the features of the animal community and its conservation value.

From 14,845 camera day data, we obtained 1561 animal images, with 19 species (14 families in 5 orders) of medium- and large-sized mammals, 2 species (2 Sciuridae) of small animals (other small mammals could not be identified), and 12 species (9 families in 5 orders) of birds (Table 7.1 and Fig. 7.1). The number of endangered medium- and large-sized mammals recorded in this survey was 8 (vulnerable + critically endangered species) out of 19 (42%), and the number of birds was 2 (vulnerable species) out of 12 (17%) (Fig. 7.2). The top three medium- and large-sized mammals that were recorded with the greatest number of camera traps were the Malay civet, southern pig-tailed macaque, and common palm civet. The top three bird species were the Bornean crestless fireback, black-capped babbler, and black partridge (Table 7.1).

The species accumulation curve did not reach an asymptote using the camera data with or without the bait (Fig. 7.3). The Fisher's exact test results demonstrated that the bait significantly affected six species, including four families in two orders (Table 7.1).

7.3.1 Survey Methods

The study site was a peatland located in the southwestern part of Borneo Island (ca. 20 m asl). Twenty camera traps were set up from August 2019 to August 2021. We moved all 20 camera traps to other locations in September 2020, and the locations of 6 of the 20 camera traps were changed again in March 2021. Therefore, we set camera traps in 46 locations in total (Fig. 7.4). Baits were hung in front of all the cameras from October 2019 to August 2020. The baits were pineapple, sweet potatoes, mangosteen, rambutan, and cempedak. We set two to four types of baits every 2 weeks.

Camera traps were placed at a height of approximately 50 cm above the ground, and the shooting range was approximately 3–5 square meters. However, the range varied depending on environmental factors such as the tree density and the degree of the slope.

Species accumulation curves on medium- and large-sized mammals were created to evaluate the efforts by camera days. The regression curve was based on the Mao Tau estimate (Colwell et al. 2012). Birds were eliminated from the accumulation curve analysis because of the insufficient number of species.

To evaluate the effect of the bait, we performed Fisher's exact test on each animal species. We made a two-by-two matrix composed of presence/absence and with/without bait to perform the analysis. Each cell indicated the number of camera days.

Table 7.1 Summary of mammals and birds detected in the study site

						With bait		Without bait			
	Order	Family	Species	English name	Threatened states	Frequency of detection/ 1000 camera day	Percentage of detected cameras (%)	Frequency of detection/ 1000 camera day	Percentage of detected cameras (%)	Total no. of detections	Total no. of detected cameras
Medium- and large-sized mammals	Carnivora	Felidae	*Neofelis diardi*	Sunda clouded leopard	VU	0.26	5	0.43	8	5	3
			Pardofelis marmorata	Marbled cat	LC	0	0	0.14	4	1	1
		Mephitidae	*Mydaus javanensis*	Sunda stink-badger	LC	0.13	5	0	0	1	1
		Mustelidae	*Martes flavigula*	Yellow-throated marten**	LC	2.93	40	0.14	4	24	9
			Mustela nudipes	Malay weasel	LC	0.26	5	0	0	2	1
		Prionodontidae	*Prionodon linsang*	Banded linsang**	LC	1.02	20	0	0	8	4
		Ursidae	*Helarctos malayanus*	Sun bear	VU	1.15	30	0.71	15	14	10
		Viverridae	*Viverra tangalunga*	Malay civet	LC	7.27	80	5	58	92	31
			Paradoxurus hermaphroditus	Common palm civet*	LC	8.55	70	1.57	31	78	22
			Arctogalidia trivirgata	Small-toothed palm civet***	LC	4.21	20	0	0	33	4
			Paguma larvata	Masked palm civet	LC	0.13	5	0	0	1	1
	Cetartiodactyla	Cervidae	*Rusa unicolor*	Sambar	VU	0.13	5	0	0	1	1
		Suidae	*Sus barbatus*	Bearded pig	VU	0.26	10	1.14	19	10	7

		Tragulidae	*Tragulus kanchil*	Lesser oriental chevrotain	LC	1.79	30	1.28	15	23	10
	Eulipotyphla	Erinaceidae	*Echinosorex gymnura*	Moonrat	LC	1.02	15	0.14	4	9	4
	Pholidota	Manidae	*Manis javanica*	Sunda pangolin	CR	0.38	15	0.57	15	7	7
	Primates	Cercopithecidae	*Macaca nemestrina*	Southern pig-tailed macaque	VU	2.55	50	3.57	38	45	20
		Hominidae	*Pongo pygmaeus*	Bornean orangutan†	CR	1.66	35	0.57	12	17	10
		Tarsiidae	*Cephalopachus bancanus*	Horsfield's tarsier	VU	0.26	10	0	0	2	2
Small mammals	Rodentia	Sciuridae	*Callosciurus notatus*	Plantain squirrel***	LC	27.94	85	0.71	15	224	21
			Callosciurus prevostii	Prevost's squirrel*	LC	2.55	30	0	0	20	6
Birds	Accipitriformes	Accipitridae	*Spilornis cheela*	Crested serpent-eagle	LC	0.26	10	0	0	2	2
			Accipiter trivirgatus	Crested goshawk	LC	0.13	5	0	0	1	1
	Cuculiformes	Cuculidae	*Carpococcyx radiceus*	Bornean ground-cuckoo	NT	0.13	5	0	0	1	1
	Galliformes	Phasianidae	*Lophura erythrophthalma*	Bornean crestless fireback	VU	1.66	25	1.86	27	26	12
			Melanoperdix nigra	Black partridge	VU	0.13	5	0.57	12	5	4
	Passeriformes	Monarchidae	*Hypothymis azurea*	Black-naped monarch	LC	0.13	5	0	0	1	1
		Muscicapidae	*Copsychus pyrropygus*		NT	0.13	5	0.14	4	2	2

(continued)

Table 7.1 (continued)

	Order	Family	Species	English name	Threatened states	With bait		Without bait		Total no. of detections	Total no. of detected cameras
						Frequency of detection/ 1000 camera day	Percentage of detected cameras (%)	Frequency of detection/ 1000 camera day	Percentage of detected cameras (%)		
				Rufous-tailed shama							
		Pellorneidae	*Pellorneum capistratum*	Black-capped babbler	LC	1.02	20	0.71	12	13	7
			Malacopteron affine	Short-tailed babbler	NT	0.38	10	0.86	8	9	4
		Timaliidae	*Stachyris erythroptera*	Chestnut-winged babbler	LC	0.13	5	0	0	1	1
		Turdidae	*Turdus obscurus*	Eyebrowed thrush	LC	0.13	5	0	0	1	1
	Strigiformes	Strigidae	*Strix leptogrammica*	Brown wood-owl	LC	0.26	5	0.14	4	3	2

The abbreviations in the column of animals under threatened status are: *LC* least concern, *NT* near threatened, *VU* vulnerable, *EN* endangered, *CR* critically endangered on the IUCN Red Lists (IUCN, 2021). The symbols with animal names denote the results of Fisher's exact test; $^{\dagger}P < 0.1$, $^{*}P < 0.05$, $^{**}P < 0.01$, and $^{***}P < 0.001$

Fig. 7.1 Animal photographs captured by camera traps: (**a**) Black-naped monarch, (**b**) Brown wood owl, (**c**) Crested serpent-eagle, (**d**) Bornean crestless fireback, (**e**) Bornean orangutan, (**f**) Yellow-throated marten, (**g**) Sun bear, (**h**) Southern pig-tailed macaque, (**i**) Sunda clouded leopard, and (**j**) Malay civet

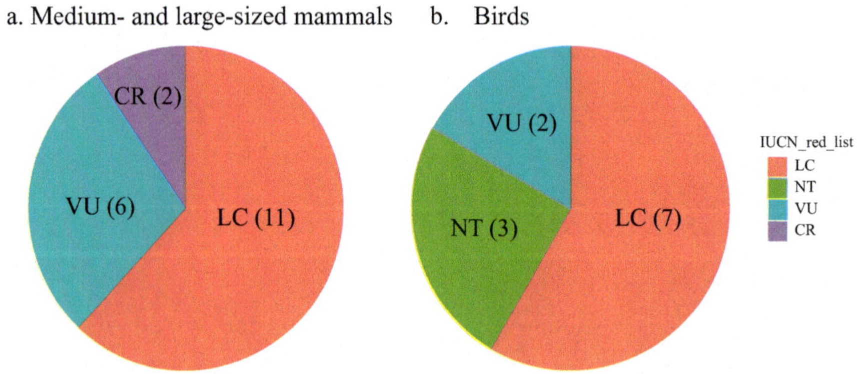

Fig. 7.2 Proportions of animals under threatened states for (**a**) medium- and large-sized mammals and (**b**) birds at the study site. The IUCN Red List categories and the number of species are given in each slice. The abbreviations are as follows: *LC* least concern, *NT* near threatened, *VU* vulnerable, *CR* critically endangered. The number in parentheses indicates the number of species

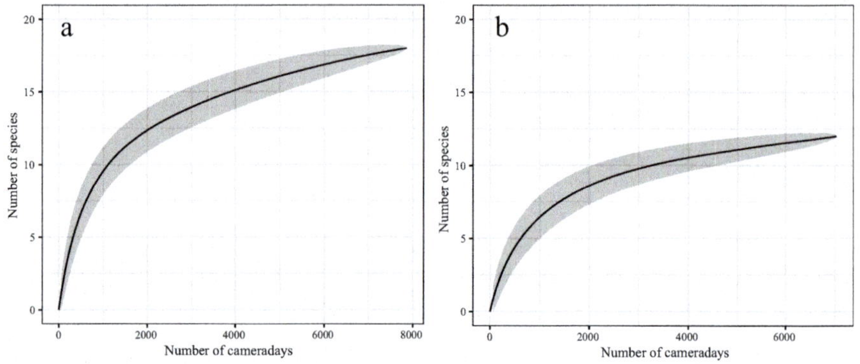

Fig. 7.3 Species accumulation curve of medium- and large-sized mammals recorded by camera traps (**a**) with bait and (**b**) without bait. The regressions did not reach the asymptotes

7.3.2 Characteristics of Mammal Fauna

Here, we compare the characteristics of the mammalian community of the peatlands and the Deramakot forest reserve, a lowland forest management area in Borneo. The number of recorded medium- and large-sized mammals in this study was 19. A survey conducted in Deramakot reported 34 medium- and large-sized mammals using a similar effort of approximately 15,000 camera days (Samejima et al. 2012). The significant difference between the mammal communities at Deramakot and the mammals recorded at this study site is that no porcupines were found in this study (Fig. 7.5), although three species of porcupine are distributed in Borneo and are all commonly found in lowland areas (Samejima et al. 2012; Jati et al. 2018). Additionally, Bornean yellow muntjac, a common species in the lowlands (Samejima

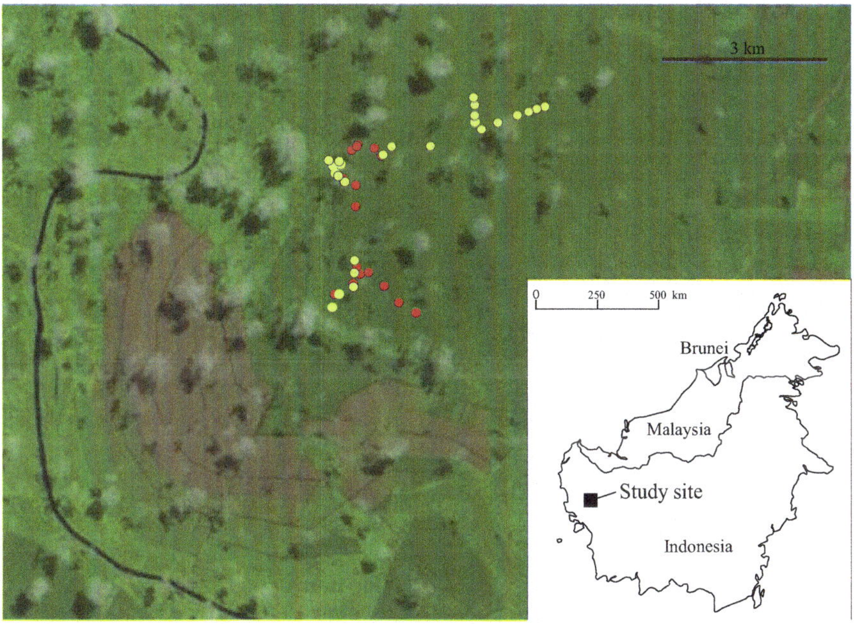

Fig. 7.4 The locations of the camera traps in the study site. The red points are the camera traps from August 2019–August 2020 with bait, and the yellow points are from September 2020–August 2021 without bait

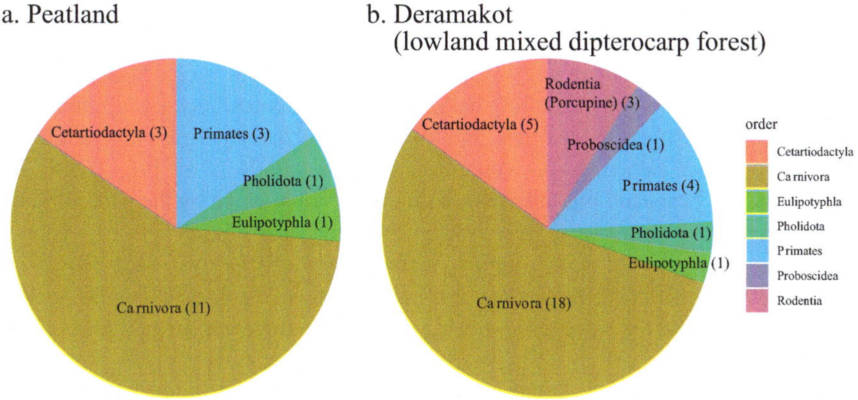

Fig. 7.5 Proportions of medium- and large-sized mammals in order level. (**a**) Peatland, (**b**) Deramakot forest reserve (after Samejima et al. 2012). The numbers in parentheses indicate the number of species

et al. 2012; Jati et al. 2018), was not recorded. The relative detection frequency of Cetartiodactyla is lower in this study site than in Deramakot (Samejima et al. 2012). In Deramakot, as many as 18 species of carnivores were recorded, whereas only

11 species were recorded in the peatland area (Fig. 7.5). This is the main reason for the difference in the number of mammal species recorded in the two areas. The carnivores recorded in Deramakot but not in the peatlands of this study site included many species with a low frequency of imaging, such as the mongoose, bay cat, and flat-headed cat. The number of camera locations and the distance between cameras were much larger in Deramakot, although the number of total camera days was roughly the same between the two sites. The success of recording mammals with camera traps can be affected by differences in experimental design, particularly for species that inhabit areas with low densities, move only short distances, or tend to use specific paths.

The species accumulation curves did not reach an asymptote, with and without baiting (Fig. 7.3). These results suggest that there are likely some species that were present but could not be recorded in this survey, and it is difficult to determine whether the number of carnivorous species in the peatland is lower than that in Deramakot using this survey. However, the lack of porcupines and Bornean yellow muntjac, which are commonly observed in the lowlands of Borneo, including Deramakot, is a characteristic of the mammal communities in the peatlands of this study site. The muddy environment of the peatland may limit their movement, because porcupines in Borneo and the Bornean yellow muntjac have low arboreal mobility.

7.3.3 Effects of Baits

The accumulation curve showed that more species could be recorded in fewer camera days with baits than without baits. It may be possible to record more species over a shorter duration by installing baits. Several omnivores, such as common palm civets and yellow-throated martens, and frugivores, such as squirrels and orangutans, tended to detect camera traps with baits more frequently ($P < 0.1$, Table 7.1). However, sun bears and pig-tailed macaques were not affected by the bait, although these species are frugivores. Research on baits revealed that camera trap capture rates in members of the weasel family increased when using fresh meat (Buyaskas et al. 2020). Using meat, fruits, and scents as bait effectively attracted civets, genets, and mongooses, while some prey species avoided passing near the bait (Mills et al. 2019).

Surveys using baits require periodic replacement. Additionally, frequent approaches of camera traps by humans to replace baits can reduce the frequency of capturing animals that are sensitive to the human presence (Buyaskas et al. 2020). It has also been noted that the effectiveness of baits varies with the number of days after the bait is placed (Mills et al. 2019), and it is expected to vary depending on the quality and quantity of food resources in the surrounding environment and the direction of the wind. Thus, studies using baits can pose difficulties for interpreting community data.

7.3.4 Finding New Habitats for Rare Species

This study recorded the Sunda stink badger (Fig. 7.6), which had reliable records only in northwestern Borneo in recent years (Samejima et al. 2016). This is an academically valuable record for predicting the distribution of Sunda stink badgers.

7.3.5 Conservative Value of This Study Site

The number of endangered medium- and large-sized mammals recorded in this survey was as high as 8 out of 19 (42%), and that of birds was 2 out of 12 (17%) (Fig. 7.2). The orangutan is a critically endangered species that is often detected in peatlands (Morrogh-Bernard et al. 2003), and it was also detected at this study site. These results suggest that this study site is an important area for conserving endangered species.

7.4 Conclusion

Camera traps provide researchers with an opportunity to gather detailed data on mammalian habitat use and behavior, which was previously inaccessible using traditional observation methods. It is especially effective in environments such as peatland, which is difficult to access. Camera trapping is a powerful tool for providing clear evidence of the existence of a targeted species in the study area. In

Fig. 7.6 A photograph of the Sunda stink badger recorded on May 14, 2020

the case study in West Kalimantan, the Sunda stink badger was found, which has not been known to inhabit the area. This is a good example of the effectiveness of camera traps. As camera trapping surveys are conducted more frequently in peatlands and greater insight is gained into the mammalian communities in these habitats, the neglected conservation value of peatlands will be re-evaluated. Additionally, video images enable researchers to obtain detailed behavioral ecology of mammals in peatland, where direct observation is difficult.

References

Buyaskas M, Evans BE, Mortelliti A (2020) Assessing the effectiveness of attractants to increase camera trap detections of North American mammals. Mamm Biol 100:91–100

Colwell RK, Chao A, Gotelli NJ, Lin SY, Mao CX, Chazdon RL, Longino JT (2012) Models and estimators linking individual-based and sample-based rarefaction, extrapolation and comparison of assemblages. J Plant Ecol 5:3–21

Jati AS, Samejima H, Fujiki S, Kurniawan Y, Aoyagi R, Kitayama K (2018) Effects of logging on wildlife communities in certified tropical rainforests in East Kalimantan, Indonesia. For Ecol Manag 427:124–134

Langner A, Miettinen J, Siegert F (2007) Land cover change 2002–2005 in Borneo and the role of fire derived from MODIS imagery. Glob Chang Biol 13:2329–2340

Mills D, Fattebert J, Hunter L, Slotow R (2019) Maximising camera trap data: using attractants to improve detection of elusive species in multi-species surveys. PLoS One 14:1–16

Morrogh-Bernard H, Husson S, Page SE, Rieley JO (2003) Population status of the Bornean orangutan (*Pongo pygmaeus*) in the Sebangau peat swamp forest, Central Kalimantan, Indonesia. Biol Conserv 110:141–152

Nakashima Y, Fukasawa K, Samejima H (2018) Estimating animal density without individual recognition using information derivable exclusively from camera traps. J Appl Ecol 55:735–744

Nowak K (2013) Mangrove and peat swamp forests: refuge habitats for primates and felids. Folia Primatol 83:361–376

Ofstad EG, Herfindal I, Solberg EJ, Sæther B-E (2016) Home ranges, habitat and body mass: simple correlates of home range size in ungulates. Proc R Soc B Biol Sci 283:20161234

Page SE, Hooijer A (2016) In the line of fire: the peatlands of Southeast Asia. Philos Trans R Soc B Biol Sci 371:20150176

Posa MRC, Wijedasa LS, Corlett RT (2011) Biodiversity and conservation of tropical peat swamp forests. Bioscience 61:49–57

Samejima H, Ong R, Lagan P, Kitayama K (2012) Camera-trapping rates of mammals and birds in a Bornean tropical rainforest under sustainable forest management. For Ecol Manag 270:248–256

Samejima H, Meijaard E, Duckworth JW, Yasuma S, Hearn AJ, Ross J, Mohamed A et al (2016) Predicted distribution of the Sunda stink-badger *Mydaus javanensis* (Mammalia: Carnivora: Mephitidae) on Borneo. Raffles Bull Zool 33:61–70

World Wildlife Fund (n.d.) Southeastern Asia: Indonesia, Malaysia, and Brunei. https://www.worldwildlife.org/ecoregions/im0104. Accessed 28 Nov 2021

Yamada I (2006) Peat swamp forests in Borneo and Sumatra—original state, development and disasters during the past 50 years with a proposal for future eco-resource management. Tropics 15:329–336

Yule CM (2010) Loss of biodiversity and ecosystem functioning in Indo-Malayan peat swamp forests. Biodivers Conserv 19:393–409

Part III
Eco-Evaluation Technology in Tropical Peatland

Chapter 8
SESAME System

Yohei Hamada, Yukihisa Shigenaga, and Hidenori Takahashi

Abstract The SESAME System is a comprehensive telemetry system specialized for field environments. Field data are obtained with a variety of sensors in a waterproof control device driven by solar energy with a rechargeable battery (without a commercial power supply); the data are periodically transferred through mobile phone networks to the cloud server and finally provided to users via a web application. Local wireless networks and/or a satellite-based data transfer service are also available out of the mobile phone service area. The SESAME System has been introduced not only in Japan but also in other countries in Southeast Asia and other regions, including in tropical peatland regions in Indonesia. Significant groundwater level drops during the dry season, in which wildfires have frequently occurred, were actually observed, suggesting the potential use of the SESAME System as an early warning system for wildfires in tropical peatland regions. Long-term field observations collected with the SESAME System succeeded in obtaining seasonal tree growth and land subsidence trends, which are strongly related to biomass storage in tropical peatland regions, namely, to aboveground rainforests and belowground peat soils, respectively. The SESAME System provides a good solution for the risk management of wildfire occurrences and can aid in long-term carbon balance evaluations in tropical peatlands.

Keywords Diameter at breast height (DBH) · Groundwater level · Ground truth · Internet of Things (IoT) · Land subsidence · SESAME System

---

Y. Hamada (✉) · Y. Shigenaga
Midori Engineering Laboratory Co., Ltd, Sapporo, Japan
e-mail: yham@midori-eng.co.jp

H. Takahashi
Hokkaido Institute of Hydro-Climate, Sapporo, Japan

M. Osaki et al. (eds.), *Tropical Peatland Eco-evaluation*,
https://doi.org/10.1007/978-981-99-6790-2_8

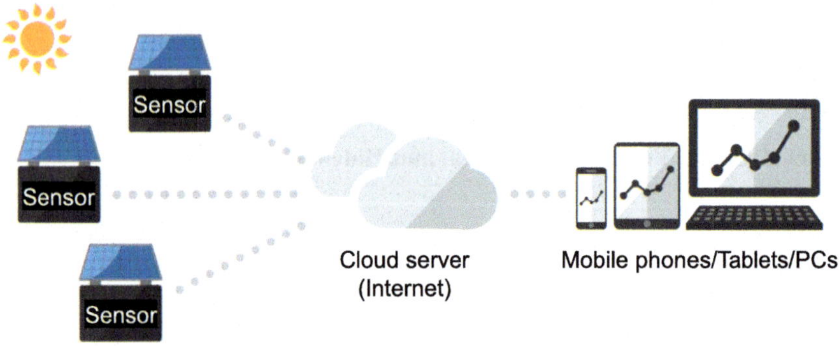

Fig. 8.1 Concept of the SESAME System

8.1 Introduction

The concept of the SESAME System is summarized in Fig. 8.1. The SESAME System is a one-stop telemetry service provided by Midori Engineering Laboratory Co., Ltd. (Sapporo, Japan, hereinafter MEL; "SESAME" is named after "*se*nsory data transmission *s*ervice *a*ssisted by *ME*L"). A variety of sensors, ranging from a simple combination of a water level sensor and a rain gauge to a high-sensitivity camera for day and night imagery, are connected to a control box driven by solar energy with a deep-cycle battery (a commercial power supply is also available).

Sensor data collected with the control box are periodically transmitted to a cloud server managed by MEL, usually via a mobile phone network available in the local area. For out-of-service areas, the SESAME System uses another type of wireless network, such as mobile satellite communication or LPWA technology. The sensor data stored in the cloud server are accessible with a sophisticated web application named SESAME-WEB. SESAME users can access their remote data from their own devices (desktop/laptop/tablet PCs and mobile phones), quickly check the data as a time-series plot, and download the original data for more detailed analyses.

8.2 Technology of Components

8.2.1 Sensors

A comprehensive list of sensors applicable to the SESAME System is shown in Table 8.1. Other sensor types can be connected, provided that (1) their output signal matches and (2) their power consumption needs can be covered by solar cells with rechargeable batteries (in cases where a commercial power supply is not available).

Table 8.1 Comprehensive list of sensors that can be equipped to the SESAME System

Item	Type	Methodology/ device name (manufacturer)	Output signal/ interface	Note[a]
Water level	Contact	Hydrostatic pressure	4–20 mA	
		Electrostatic capacitance		
	Noncontact	Ultrasonic wave reflection		
		Radio wave reflection		
Rainfall	Tipping bucket	Contact signal	0.2/ 0.5 mm per pulse count	
	Acoustic	WXT53x (Vaisala, Finland)	RS232C/ SDI-12	Rainfall/hailfall separated
Soil moisture	Dielectric	SM150T (Delta-T Devices, UK)	0–1 V	Soil temperature
		MP406 (ICT International, Australia)	4–20 mA	
		SLT500x (Murata Manufacturing, Japan)	RS232C/ RS485/ UART	Soil temperature, EC
	Tensiometer	HD-001 (Sensez, Japan)	4–20 mA	
Micrometeorology	Integrated	WXT53x (Vaisala, Finland)	RS232C/ SDI-12	Air temperature, relative humidity, rainfall, atmospheric pressure, wind direction, and speed
		SenSu-15xx (ITbook Technology, Japan)	RS485	Air temperature, relative humidity, rainfall, atmospheric pressure, wind direction and speed, illuminance, and UV
	Discrete	RHT-30-01 (Sensatec, Japan)	0–1 V (2 ch)	Air temperature and relative humidity
		Thermopile pyranometer	0–40/ 400 mV	Radiation intensity
Water quality	Integrated	U-53 (Horiba, Japan)	RS232C/ USB	pH, EC, DO, ORP, salinity, water temperature, turbidity, TDS, and water depth
	Discrete	WQ300 (Horiba, Japan)	RS485	pH/EC/DO/ORP with water temperature
			0–2 V	pH/EC with water temperature

(continued)

Table 8.1 (continued)

Item	Type	Methodology/ device name (manufacturer)	Output signal/ interface	Note[a]
		MM-4xDP (DKK-TOA, Japan)		
		DOS-20 (Optex, Japan)	4–20 mA	DO (luminescent)
		TC-100/500/3000 (Optex, Japan)	4–20 mA	Turbidity
Water flow	Indirect	FD-H series (Keyence, Japan)	4–20 mA	
	Submerged	MainProbe-V (Mainstream measurements, UK)	RS485	Water temperature
Photo image	Infrared	ELP-USBFHD0x (Ailipu Technology, China)	USB 2.0	
	Night vision	P1455-LE/P1375-E (Axis Communications, Sweden)	Ethernet	
Carbon storage				
Tree trunk growth	Dendrometer	MIJ-02 (Environmental Measurement Japan, Japan)	0–2 V with 2-V source	
Peat subsidence	Potentiometric	LP-FJ series (Midori Precisions, Japan)	0–2 V with 2-V source	
	Laser distance	IL-1000 (Keyence, Japan)	4–20 mA	

[a] Detailed items for the integrated sensors, choices of the items for the discrete sensors, and co-measured items for other sensors are shown

8.2.1.1 Water Level

Two main types of water level sensors are available: contact-type and noncontact-type sensors.

In contact-type sensors, the bottom of the sensing probe is fixed to a certain level below the water surface. The height of water above the fixed level is reflected in the physical and/or electrical properties of the probe, measuring a variable such as the water pressure or electrostatic capacitance. Water pressure sensors can be classified into the absolute pressure type and gauge pressure type. The water pressure data measured by absolute pressure sensors include the variation in the atmospheric pressure, which must be accounted for. Due to the small footprint of these probes, contact-type sensors are suitable for measuring water levels in narrow pipes.

In contrast to the contact-type sensors, noncontact sensors are fixed above the water surface. The distance between the probe and the water surface is determined by ultrasonic waves or radio waves. Water depth data can be obtained by subtracting this distance from the predetermined distance between the probe and the bottom of the water body. Strong winds and obstacles between the probe and the water surface may affect the measured values. Noncontact sensors are typically mounted on bridges and used to measure water levels in rivers and open channels.

8.2.1.2 Rainfall

Tipping-bucket-type rain gauge sensors emit a contact signal for every constant amount of rainfall (typically 0.5 mm) and can be connected to control boxes. After long-term measurements are obtained, dirt and mud may deposit on the wire screen above the funnel of the rain gauge. In tropical regions, ants often build nests inside rain gauges, inhibiting the movement of the tipping bucket. These deposits should be removed by regular maintenance.

Another type of rainfall sensor, called an acoustic rain gauge, converts the impact of falling rain drops to the amount of rainfall and is also available in the SESAME System (see Sect. 8.2.1.4). The advantages of this sensor are (1) less frequent maintenance is required to remove deposits and (2) the sensor collects rainfall and hailfall measurements separately according to the differential impacts.

Rainfall at the ground level would be underestimated if the forest canopy covered the rain gauge. Rain gauges must be horizontally fixed for the sake of the regular movement of the tipping bucket. Rainfall sometimes changes significantly over just a few kilometers, so a dense network of rain gauges may be needed to ensure correct evaluations.

8.2.1.3 Soil Moisture

Recently, dielectric-based soil moisture sensors have been widely used (time/frequency/amplitude domain reflectometry; TDR/FDR/ADR). The dielectric constant of soil water is much larger than those of soil mineral particles and soil air, so the dielectric constant of bulk soil depends mainly on the volumetric percentage of soil water. Some soil moisture sensors can measure not only soil moisture but also the temperature and electrical conductivity (EC) of the soil water.

Tensiometers, which measure the potential energy of soil water as the pressure head instead of the soil water content itself, can be used in the SESAME System. The pressure head can be converted to the soil water content according to their relationship expressed as a water retention curve, but this conversion is affected by the hysteresis in the relationship. Most tensiometers are equipped with a small water reservoir that should be filled with water, and lost water must be refilled by regular maintenance.

The use of soil moisture sensors in wetlands should be carefully considered. Most commercially available sensors are designed for well-drained surface soils, in which overdrying is more likely than overhumidity. Under nearly water-saturated conditions, the measured soil water content may be out of the official measurement ranges, and the data accuracy may drop. The water resistance of the probes should also be verified. In tropical peatlands, the groundwater level is generally more reliable than soil moisture as an indicator of drainage.

8.2.1.4 Micrometeorology

A simple weather station can be established using the SESAME System equipped with an integrated weather transmitter. Two weather transmitter models can be used: the WXT53x (Vaisala, Finland) and SenSu-15xx (ITbook Technology, Japan) model. The former is equipped with an ultrasonic anemometer and an acoustic rain gauge, while the latter is equipped with a three-cup anemometer, a wind vane, and a one-sided tipping-bucket rain gauge.

Air quality sensors can also be used in the SESAME System. In tropical peatlands, the use of carbon monoxide (CO) and suspended particulate matter (SPM) sensors may allow potential applications for the early detection of wildfires.

8.2.1.5 Water Quality

The water quality items that can be measured with the SESAME System include turbidity, pH, EC (or salinity), dissolved oxygen (DO), and oxidation–reduction potential (ORP). An array of 2–4 discrete sensors can be connected to the control box. The use of an integrated sensor (U53, Horiba) enables the measurement of up to 13 water quality items at once.

In general, pH sensors should be frequently calibrated. The calibration interval may be every day, every week, or every month at the longest, depending on the characteristics of the water and the required accuracy. DO in stagnant water should be measured with a luminescent DO sensor. Electrochemical-type DO sensors (known as galvanic and polarographic sensors) consume DO during the measurements, so if the water is stagnant, the DO near the sensor electrode would become lower than that in the surrounding bulk water. Both types of DO sensors are available in the SESAME System.

Related to the dynamics of greenhouse gas emissions from tropical peatlands, ORP and DO are important indicators. High DO concentrations promote the microbial aerobic decomposition of peat. Under very low ORP conditions, organic carbon decomposed by microorganisms is emitted as methane (CH_4) instead of carbon dioxide (CO_2), resulting in an increase in emissions on the basis of the global warming potential.

8.2.1.6 Photo Images

SESAME Camera is a general name for the camera subsystems used in the SESAME System. The basic operation of SESAME Cameras is to capture still images at predetermined time intervals repeatedly and upload them. Nighttime images can also be captured in infrared mode, which the system automatically switches according to the ambient brightness, or in night vision mode with high-grade cameras. Recently, a camera module suitable for security purposes in which moving objects are detected and automatically photographed and uploaded has also become available.

Although SESAME Cameras generally consume more power than other components of the SESAME System, most cameras can be driven by independent power sources consisting of solar cells and batteries (Sect. 8.2.2). The power supply voltage and housing temperature are transmitted with the captured image. These measured values can be used for remote maintenance of the camera system.

A sensor and a camera installed in the same place can be linked with the alarm output from the sensor (Sect. 8.2.2). When the camera receives an alarm signal from the sensor, the camera wakes up, captures an image, and uploads it even if it is not at the predetermined shooting time. These sensor data and images are also linked to each other in the web system, so the images can be referenced from the data plot (Sect. 8.2.5).

8.2.2 Control Box and Power Sources

The block diagram of a typical SESAME device is shown in Fig. 8.2. The control box is a 20-cm-wide and 15-cm-high plastic case in which an originally designed

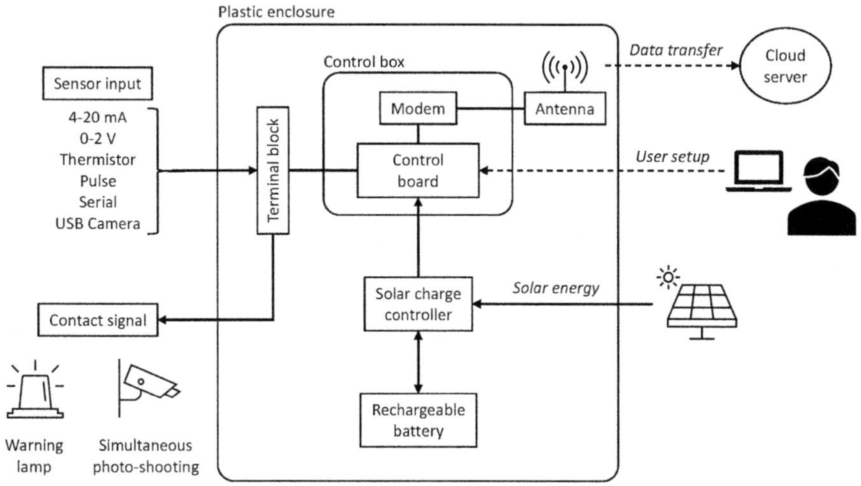

Fig. 8.2 A typical block diagram of the SESAME device

printed circuit board (control board) and a commercially available communication module (Sect. 8.2.3) are stored. The control box itself is then stored in a plastic waterproof enclosure with an antenna, a rechargeable battery, a solar charge controller, and terminal blocks.

The control board has the interfaces of a power input, sensor inputs, and contact outputs. It can also be connected with a PC via an RS232C-USB cable. Using a configuration program for the SESAME device, users can change the parameters, check the current sensor values, and run a data-sending test. The control boards have a built-in real-time clock (RTC) that is backed up with a coin cell battery and periodically synchronized with the GNSS-reported time or with the nearest base station to the mobile network.

The control board can receive a variety of sensor signals, such as current (4–20 mA), voltage, resistance (thermistor), and contact signals as well as serial data streams (USART, RS232C, RS485, and USB). The received data are temporarily kept in a built-in nonvolatile memory until being transferred to the cloud server and permanently stored in a storage media (SD card/USB thumb memory). The temporarily stored data that fail to be sent are resent at the next data transfer time.

As a power source of SESAME devices, an independent power supply consisting of solar cells and batteries is generally used. A solar panel can be directly connected to the control board in combination with a 7.2-V NiMH battery pack. An additional solar charge controller should be used when a 12-V lead-sealed battery is connected. A commercial AC 100-to-240-V power supply is also available through a 12-V AC-DC adapter. In this case, however, the system is susceptible to the influence of induced lightning.

When a sensor value exceeds a threshold, the SESAME device enters an alarm state. When the alarm state begins, the measurement and data transfer intervals are automatically changed. During this state, an open-collector-type contact signal is emitted at every measurement time. Using this function, a simple early warning system can be quickly established. For example, when the river water level monitored by a SESAME device exceeds the dangerous water level, a warning lamp and siren can be automatically triggered.

8.2.3 Wireless Network

In a SESAME device, a commercially available wireless communication module is attached on the control board. The module communicates with the control board by a serial data line (USART, RS232C or USB). Depending on the module, several types of communication protocols are available such as the TCP socket, FTP(S), and HTTP(S). An antenna for data transmission is directly connected to the module and is extended out of the control box. The antenna can be put in a plastic enclosure, whereas it must be pulled out of a metal cabinet.

According to the specific purpose and situation, an adequate communication module type can be selected. Inside of mobile phone service areas, SESAME devices

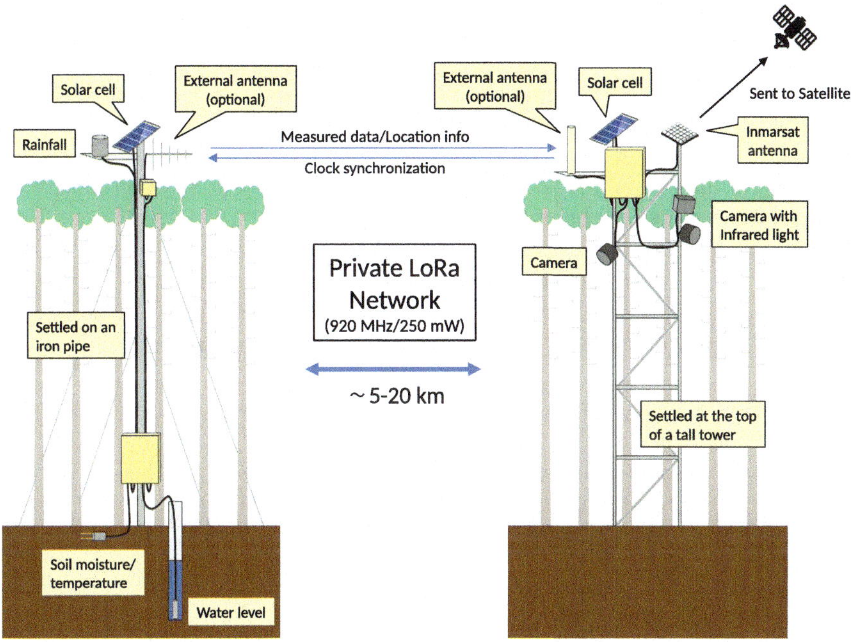

Fig. 8.3 Combination of a private LoRa network and a satellite-based data transmission service introduced in West Kalimantan, Indonesia

can use one of the 2.5G/3G/4G (LTE) networks provided by mobile carriers. Sensor data can be transmitted as ASCII text via 2.5G (GPRS/EDGE), but a 3G or higher-speed network is necessary for sending photographic images.

Out of mobile phone service areas, or in the case of many observation sites being located within a small area, a LoRa module can be used in the SESAME System (Fig. 8.3). LoRa (meaning "Long Range") is an LPWA technology. LPWA stands for Low-Power Wide-Area and is a generic term for wireless networking techniques designed for IoT services. The sensor data acquired with the SESAME devices installed at observation sites ("child" stations) are sent to a gateway ("parent" station) via the LoRa network. Since LoRa radio waves are largely attenuated by obstacles and the Earth, placing the transmitting and receiving antennas at high positions is recommended. If the location of the "parent" station is out of the mobile phone service area, a satellite-based data transmission service is available. In this case, the sensor data collected from each "child" station are compressed and unified into one file before being sent to the cloud server to reduce the transmission service fee via satellite.

8.2.4 Cloud Server

From the viewpoints of availability, the ease of scaling-in/scaling-out, and the integrity of the total system, all the data received and stored in servers in the SESAME System are migrated from on-premises to the cloud. A TCP-socket receiving program, HTTP and FTP servers, and other data-processing services are run on virtual server machines built in the cloud.

The SESAME System uses Amazon Web Service (AWS) as its main cloud. The data received and stored on the virtual server EC2 can be downloaded in CSV format as raw text data and provided in an easier-to-read format through a web application built on another virtual server (Sect. 8.2.5). Large data such as photographic images and short movies are permanently stored in S3, a simple storage service in AWS, and referenced as needed. Some countries require that primary data obtained domestically be stored on domestic servers. To comply with this regulation, SESAME System data acquired in Indonesia are received and stored in a virtual server built in BiznetGIO Cloud, a private cloud service provided in Indonesia.

8.2.5 Web Application

Users of the SESAME System can access their data in near-real time from anywhere in the world through a web application named SESAME-WEB. This application supports the latest versions of major web browsers (Chrome, Edge, Safari, etc.) that run on PCs, mobile phones, tablets, etc. Users log in to the web application using their individually issued user ID and password. Permissions are set for users, and users with administrator privileges can not only view data but also manage general users without administrator privileges (through registration, update, and deletion tasks), locational information on measurement points, and thresholds for issuing email alerts.

A typical dashboard screen of SESAME-WEB consists of seven panes (Fig. 8.4): a list of measurement points, a map in which the measurement points are plotted, a timeline showing recent time series of selected data, the latest values of the selected data, and time-series charts with different time spans (e.g., 1 day, 1 week, and 1 month). These panes are automatically rearranged according to the screen (window) size. In addition to the dashboard screen, SESAME-WEB has the following screens:

- Analysis screen: Users can select any measurement point(s) and period from the long-term stored data and plot the selected data on a graph or download it as a CSV format file.
- Alarm screen: Users can view lists of (1) the measurement points at which the current measured values exceed the alarm thresholds and (2) the past alarm history.

Fig. 8.4 Screenshot of the SESAME-WEB dashboard

- Maintenance screen: Users can take notes about their maintenance work, such as checking the measurement points and changing the web application settings.
- Monitoring screen: Users can check the latest measured values of each measurement point shown in order every 10 s.

Users with administrator privileges can also access several master screens within the databases.

SESAME-WEB supports multiple languages. As of December 2022, it supports five languages: English, Japanese, Indonesian, Thai, and Vietnamese (including partially supported languages). A default language for each user can be registered.

Images taken with SESAME Cameras are accessible from a specific website built for the camera systems. In addition, by clicking the time-series graphs of the sensors linked to the cameras, the images taken at the indicated time can be shown on SESAME-WEB.

8.3 Application in Tropical Peatlands

8.3.1 How to Install the SESAME System in Tropical Peatlands

The field situation of a typical SESAME device installed in a tropical peatland region is shown in Fig. 8.5. This device is equipped with a rain gauge and a pressure-type water level sensor to assess the risk of peat fires and the degree of aerobic peat decomposition. For this purpose, a soil moisture sensor is sometimes added.

After digging a guide hole, an iron pipe with a diameter of 5 cm is installed vertically on the ground. Peat soils are soft and usually do not need to be hammered.

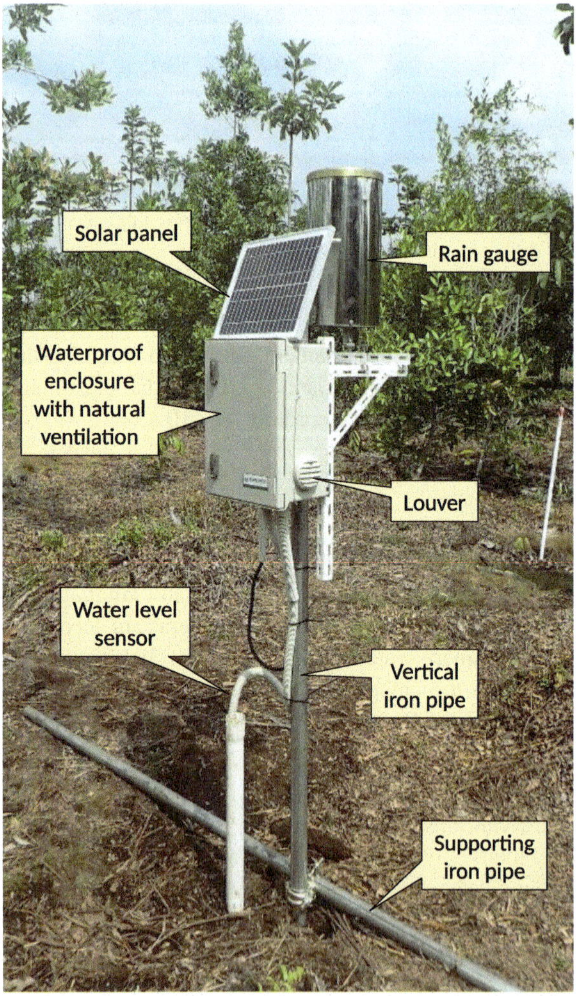

Fig. 8.5 Typical field settings of the SESAME device for groundwater monitoring in tropical peatlands

Rather, to prevent the pipe from sinking or tilting after installation, the vertical iron pipe should be supported by another iron pipe placed along the ground using an orthogonal clamp for stabilization. Since tropical peatlands are originally wetlands and the water level becomes high in the rainy season (higher than 1 m above the ground surface in some cases), the vertical iron pipe must be tall enough to keep the control box out of the water. The height of the pipe is also important for improving the signal strength of the mobile network by extending the waterproof antenna out of the control box and putting it at a high position along the pipe. Louvers on both sides of the enclosure promote natural ventilation to prevent hang-ups by heat. Ant repellent should be placed in the box and periodically sprayed around the base of the vertical iron pipe.

A perforated polyvinyl chloride (PVC) pipe is used for groundwater level monitoring. In well-drained peatlands, the groundwater level significantly drops toward the end of the dry season, so the PVC pipe should have a sufficient length. Before installation, the lower end of the PVC pipe is closed and sharpened. Similar to the vertical iron pipe, the PVC pipe is pushed into a predrilled guide hole. A water level sensor is then dropped to the bottom of the PVC pipe and fixed. The relative height of the top of the PVC pipe to the ground surface may vary due to the compression, decomposition, and burning of the peat layer (Sect. 8.3.3), so the relative height should be updated periodically to ensure the correct measurement of the depth of the water table from the ground surface. A rain gauge sensor can be mounted at the back of the control box. The mounting plate provided by MEL can level the rain gauge.

To ensure sufficient incident radiation enters the solar cells and allows the accurate measurement of rainfall, some of the vegetation surrounding the SESAME device may have to be removed. Even a small shadow cast on the corner of the solar panel can significantly reduce the power generation. In tropical regions, orienting the solar panel to face straight up is the most efficient way to generate electricity; however, in this position, the panel is subject to the accumulation of dirt that is difficult to clean off. When transmitting water level and rainfall data at one-hour intervals, a 3–5-W solar panel that stands upright works well. A solar panel that is mounted vertically should face east or west in the equatorial region; the battery would be charged mainly in the morning in the former case and in the afternoon in the latter case. If the panel faced north or south, the solar cells could not receive incident radiation for half the year, when the sun is in the opposite hemisphere.

One of the most frequently asked questions is "How many SESAME devices should be installed in a certain area?" The simplest answer to this question is "It depends on the purpose." Rainfall can vary widely over 2–3-km distances, especially in the case of transient and local thunderstorms common in tropical regions. The groundwater level is generally affected by the topographic relief, geology, and land use conditions. Under flat terrain and similar land use conditions, groundwater levels are likely not to vary extensively over relatively large areas. Conversely, the groundwater level can change rapidly within short distances near drainage canals. Around pumped-up wells, the water level drops depending on the pumping rate and the hydraulic conductivity (permeability of water) of the aquifer. It is therefore

Fig. 8.6 Antitheft trial of SESAME devices along an irrigation canal in Lampung, Indonesia. Left: field situation, right: how to maintain the system

important to determine the arrangement of measurement points according to the user's purpose.

Antitheft is one of the challenges faced when keeping long-term field observations. Some tropical peatland areas are used as arable land and forest plantations, while the other areas are classified as natural parks or are abandoned after wildfires. Due to the less attention paid to the sensors in the latter areas, ensuring antitheft of the instruments is essentially difficult. Figure 8.6 shows an example of antitheft measures in Lampung Province, Indonesia. In this case, tall poles made of reinforced concrete were built along an irrigation canal, and the control box, a solar panel, and a rain gauge were mounted at the top of these poles. This method is not suitable for peatlands because of the soft ground. Iron fences installed for antitheft are often themselves subject to theft. If there is private property near the observation site in which people actually live, it is recommended that the landowner be paid to allow the installation of the devices on their lands.

8.3.2 Groundwater Monitoring in BRG Projects

The Peatland Restoration Agency (Badan Restorasi Gambut, BRG) was established by the Indonesian government in January 2016. This agency was motivated by massive forest and land fires in 2015. The fires that occurred from June to November burned 2.6×10^6 hectares of land and resulted in thick smoke and haze. Peatlands occupied 37% of the burnt area (Peatland Restoration Agency of the Republic of Indonesia 2019). One of the factors that triggered these fires was the practice of draining, causing the peatlands to be more prone to fire, particularly during the dry season.

The BRG installed 40 and 102 sets of groundwater level monitoring devices in 2017 and 2018, respectively (Peatland Restoration Agency of the Republic of Indonesia 2019). As a part of these projects, 14 and 20 sets of SESAME devices were installed in 2017 and 2018, respectively. The distribution of these SESAME devices covered five provinces (Riau, Jambi, South Sumatra, West Kalimantan, and Central Kalimantan). All the data obtained with these monitoring devices are gathered to a BRG server and shown on SIPARAGA (sipalaga.brg.go.id), a web application developed and managed by the BRG.

Although at least 142 sets of monitoring devices were installed up until 2018 (some additional devices may be installed later), the number of monitoring posts shown on SIPARAGA was less than 70 as of November 2022. This means that more than half of the monitoring devices became nonactive before this time. The reasons for this may include device failures, theft, or burning. It is recommended that maintenance work be performed at least once a year, although many of the monitoring sites in tropical peatlands are difficult to access.

8.3.3 Potential Applications for Evaluating Carbon Storage

From the viewpoint of global climate change, tropical peatland ecosystems are very important as they form a vast terrestrial carbon reservoir. In contrast from the northern peatlands, in which moss and sedge are the main vegetation types, thick rainforests (mainly consisting of *Dipterocarpaceae* trees) cover tropical peatlands. Here, the authors introduce some examples of long-term field observations obtained using the SESAME Systems that may be applicable in evaluations of the carbon balance in tropical peat swamp forests.

8.3.3.1 Ground Surface Level Meter

In well-drained peatlands, surface peat is decomposed by intruding oxygen and subsequent aerobic microbial oxidation, resulting in CO_2 emissions and a decline in the land surface elevation. By measuring this land subsidence, it is possible to evaluate the loss of peat and the emission of carbon resulting from this aerobic decomposition.

The ground surface elevation in peatlands is affected not only by the peat decomposition but also by the surface load (including snow cover in the case of northern peatlands), drying-induced shrinkage, and swelling after groundwater level rise. These influences should be excluded by simultaneous measurements or estimations of the relevant terms. In addition, the reference point of the ground surface

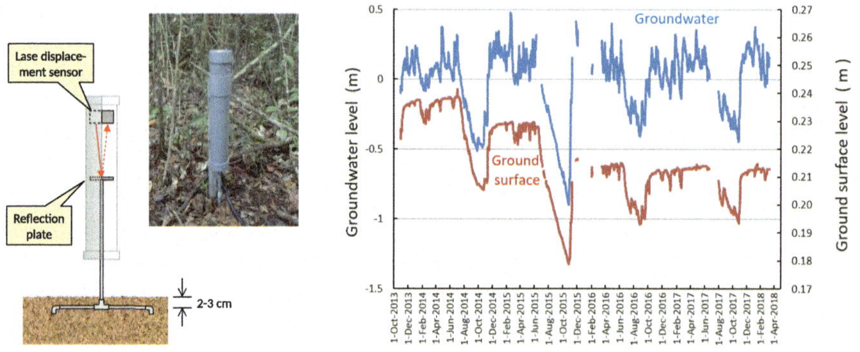

Fig. 8.7 The overview of a ground surface level meter used in the SESAME System (left) and the data obtained in a tropical peatland located in Central Kalimantan, Indonesia (right)

elevation must be fixed and not affected by the variation in the thickness of the peat layer. For this reason, an iron rod or some other hard object that provides a reference point must be driven down to the base layer below the peat layer.

An overview of a ground surface level (GSL) meter used in the SESAME System and the data obtained in a tropical peatland located in Central Kalimantan, Indonesia, are shown in Fig. 8.7. The distance between the reference point and the ground surface can be determined with a laser displacement sensor or a potentiometer. The observed data shown in Fig. 8.7 clearly show that the GSL was significantly affected by the groundwater level during the year, while this influence consistently dropped over the years of study.

8.3.3.2 Dendrometer

As mentioned above, not only below-ground carbon reservoirs such as peat soils but also above-ground carbon reservoirs such as peat swamp forests are important in tropical peatlands. When evaluating the carbon storage of a tree, the diameter at breast height (DBH) is one of the relevant parameters that can be calculated from the circumference at breast height (CBH).

Sensors designed for measuring CBH are commonly called dendrometers. Changes in CBH are detected as changes in the length of a wire wrapped around a tree at breast height. If the wire is in direct contact with the bark of the tree, friction with the bark may inhibit the smooth movement of the wire. To avoid such a situation, the wire should be passed through beads at regular intervals. The length of the wire extended from the dendrometer is limited, so it is necessary to visit the observation site and adjust the wire setting before the limit is reached. A wire that cannot extend any further not only results in incorrect CBH measurements but also clamps the tree trunk and inhibits its growth. Using the SESAME System, a user can know when to adjust the wire without going to the site.

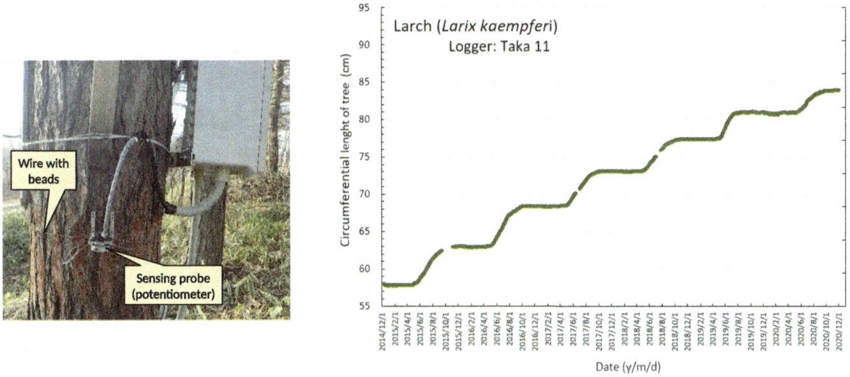

Fig. 8.8 The field settings of a dendrometer with the SESAME System (left) and long-term variation in the circumference at breast height of a larch stand in Hokkaido, Japan (right; Takahashi et al. 2018)

The field setting of a dendrometer with the SESAME System and the long-term variation in the CBH of a larch stand in Hokkaido, Japan, are shown in Fig. 8.8. The seasonal change in the CBH is clearly shown, reflecting growth in summer and stagnation in winter. In tropical regions, in contrast, there may be no seasonal changes due to the constant temperature, or seasonal changes may occur due to factors other than temperature, such as rainy-season growth and dry-season stagnation.

8.4 Conclusions

As a potential measure for acquiring field data in remote areas in near-real time, the SESAME System has been introduced not only in Japan but also in other countries in Southeast Asia and other regions. Especially in tropical peatlands, this system can be used in early warning systems for wildfires as well as in the frameworks of carbon-emissions assessments.

In ongoing projects as of December 2022, the SESAME System has been equipped with bilateral communications, not only for remote monitoring but also for remote control. In the near future, the use of artificial intelligence and edge-computing technologies will expand the applications of the SESAME System.

References

Peatland Restoration Agency of the Republic of Indonesia (2019) Three years of peatland restoration in Indonesia report, 38 p

Takahashi H, Shigenaga Y, Hamada Y, Osaki M, Setiadi B (2018) Near-real time remote monitoring of ground-water/surface levels and warning to carbon loss from tropical peatlands in Indonesia. In: The proceedings of the International workshop on forest ecological resources security for next generation: development and routine utilization of forest ecological resources and their domestication, pp 80–90

Chapter 9
Sensing, Monitoring, and Data Collection System (sPOTEKA-NET)

Kosuke Oku, Takashi Ogai, Kansuke Kotake, Asep Andi Yusup, and Riona Shimazaki

Abstract The correct understanding of environmental conditions, such as weather and groundwater level, is important for tropical peatland forest management. The IHI Corporation developed a real-time monitoring system equipped with a "high density remote weather observation system" named POTEKA (*P*oint *TE*nki *KA*nsoku in Japanese, where "Tenki" means weather, and "Kansoku" means observation). POTEKA has the capability to observe seven types of weather elements, including temperature, air pressure, humidity, sunshine, wind velocity and direction, rain detection, and precipitation. Power charged by solar panels and mobile phone network usage allows us to place POTEKA anywhere. The observed weather data, to be transferred to the cloud server, are accessible anytime and anywhere through the web. This cloud-based monitoring allows us to conduct remote quality checks and statistical analyses and to issue alerts when the monitored data, including the POTEKA status, exceed the limits previously specified. In addition, sPOTEKA provides the capabilities of groundwater level measurement and satellite network communication. Continuous monitoring of the groundwater level is essential for peatland forest management to prevent wildfire and to keep plantation trees healthy. As sPOTEKA consists of satellite network communication, it is possible to set sPOTEKA in remote areas. The base station of sPOTEKA collects data from monitor stations and sends it to the cloud server (sPOTEKA-NET) through the satellite network. The weather radio network is used to communicate between the base station and monitor stations.

Keywords Weather station · Groundwater level · Forest monitoring · Realtime · Remote · Robust system

---

K. Oku · T. Ogai (✉) · K. Kotake
IHI Corporation, Tokyo, Japan
e-mail: oku7570@ihi-g.com; ogai0919@ihi-g.com

A. A. Yusup
PT. Wana Subur Lestari/PT. Mayangkara Tanaman Industri, Jakarta, Indonesia

R. Shimazaki
Sumitomo Forestry Co., Ltd., Tokyo, Japan

M. Osaki et al. (eds.), *Tropical Peatland Eco-evaluation*,
https://doi.org/10.1007/978-981-99-6790-2_9

9.1 Introduction

Peatlands mainly consist of plant detritus that has accumulated due to incomplete decomposition under close to water-saturated conditions. Peatlands and their forests in the tropics, mostly in Southeast Asia, Africa, Central and South America, cover at least 820,000 km^2 globally (Xu et al. 2018), and total tropical peat carbon stocks are estimated to be 89 Gt (Dargie et al. 2017). These carbon stocks reach approximately 9 years of anthropogenic emissions based on the average emissions over 2011–2020 (Friedlingstein et al. 2022). Furthermore, forests are attracting attention not only as carbon sinks but also for their value in conserving biodiversity. In particular, tropical forests are known to be habitats for an extremely diverse range of organisms (Dirzo and Raven 2003). However, forest fires in tropical peatlands due to inappropriate land management have caused problems. In fact, during the El Niño dry season of 2015, it is estimated that the CO_2 emissions were approximately 0.89 Gt, and severe deterioration of air quality and health problems occurred due to the dense haze (Lohberger et al. 2017).

A correct understanding of environmental conditions, such as weather and groundwater level, is important for tropical peatland forest management. In particular, groundwater levels should be controlled properly to prevent wildfires in tropical peatlands (Hayasaka et al. 2016). Real-time monitoring with analytics is required to detect abnormal conditions in the field and monitoring system. Therefore, sensing technology and analysis technology for remotely obtaining, monitoring, and analyzing weather conditions and groundwater levels in real time play an important role in tropical peatland forest management.

Furthermore, in recent years, integrated tropical peatland forest management has been proposed that introduces a wide-area monitoring system that utilizes unmanned aerial vehicles (UAVs) and satellites (Tsuji et al. 2021). In this management system, the importance of data measured on the ground (ground truth data) has been mentioned as comparative evaluation information. Therefore, an accurate sensing of the surrounding environment on the ground plays an important role in future forest management using UAVs and satellites.

"sPOTEKA with network" (sPOTEKA-NET) provides the accurate sensing, monitoring, and data analysis method required for forest management in tropical peatlands (Iwashita et al. 2019; Yamashita et al. 2018).

9.2 Concept of sPOTEKA with Network (sPOTEKA-NET)

The development of "sPOTEKA" began in 2021 with the start of the collaboration between Sumitomo Forestry Co., Ltd. and the IHI Corporation. sPOTEKA with network (sPOTEKA-NET), a solution for forest management in tropical peatlands, consists of two main elements. The first is a sensing function for appropriately measuring and accurately understanding the surrounding environment, which is necessary for forest management in tropical peatlands. sPOTEKA's sensing function is developed based on a proven system, which has the advantage of being compact and being able to be installed anywhere. The monitoring station is responsible for this function. The second is a remote monitoring system based on cloud server technology and data analysis by building a network. The monitoring stations in the field transmit measured data to a base station via wireless communication. Then, the base station sends it to the cloud server (sPOTEKA-NET) via satellite communication. By using satellite communications, it is possible to communicate from remote areas where mobile communications are difficult to provide. Furthermore, by automatically alerting when the groundwater level, which is directly linked to fire risk, falls below a controlled value, the system advances the management of tropical peatland forests. Figure 9.1 shows the outline of sPOTEKA-NET, and Fig. 9.2 shows a mock-up of sPOTEKA and its role.

9.3 Sensing Function

The sPOTEKA-NET is based on "the high density remote weather observation system" named POTEKA (*P*oint *TE*nki *KA*nsoku in Japanese, "Tenki" means weather, and "Kansoku" means observation), and it is equipped with a sensing function. Table 9.1 shows sPOTEKA's sensor specification.

9.3.1 POTEKA System Technology

The sensor used in the POTEKA system is a compact meteorological sensor called the weather station developed by Meisei Electric Co., Ltd. based on the concept of being "compact, lightweight and easy to install." It measures seven meteorological elements (temperature, air pressure, humidity, sunshine, rain detector, wind velocity, and direction) with weather stations. Furthermore, the POTEKA system can add other observation elements, such as precipitation and lightning sensors. The sensors for temperature, air pressure, humidity, wind velocity, and precipitation passed the Japan Meteorological Agency (JMA) variable verification test. Its appearance is shown in Fig. 9.3.

The POTEKA sensors start working just after being powered on and run on a battery charged by a solar panel. Therefore, the system does not require an external, commercial, or any other power source. Thus, there are almost no restrictions on the installation location. It can be installed, for example, on the roof of a school or on a utility pole. Actual installation examples are shown in Fig. 9.4.

POTEKA's sensing technology has already been used worldwide, including by public institutions in Japan. Its uses are wide-ranging. For example, POTEKA was used for countermeasures against heat stroke at construction sites and for wind condition surveys for the structural design of buildings. As of September 2022,

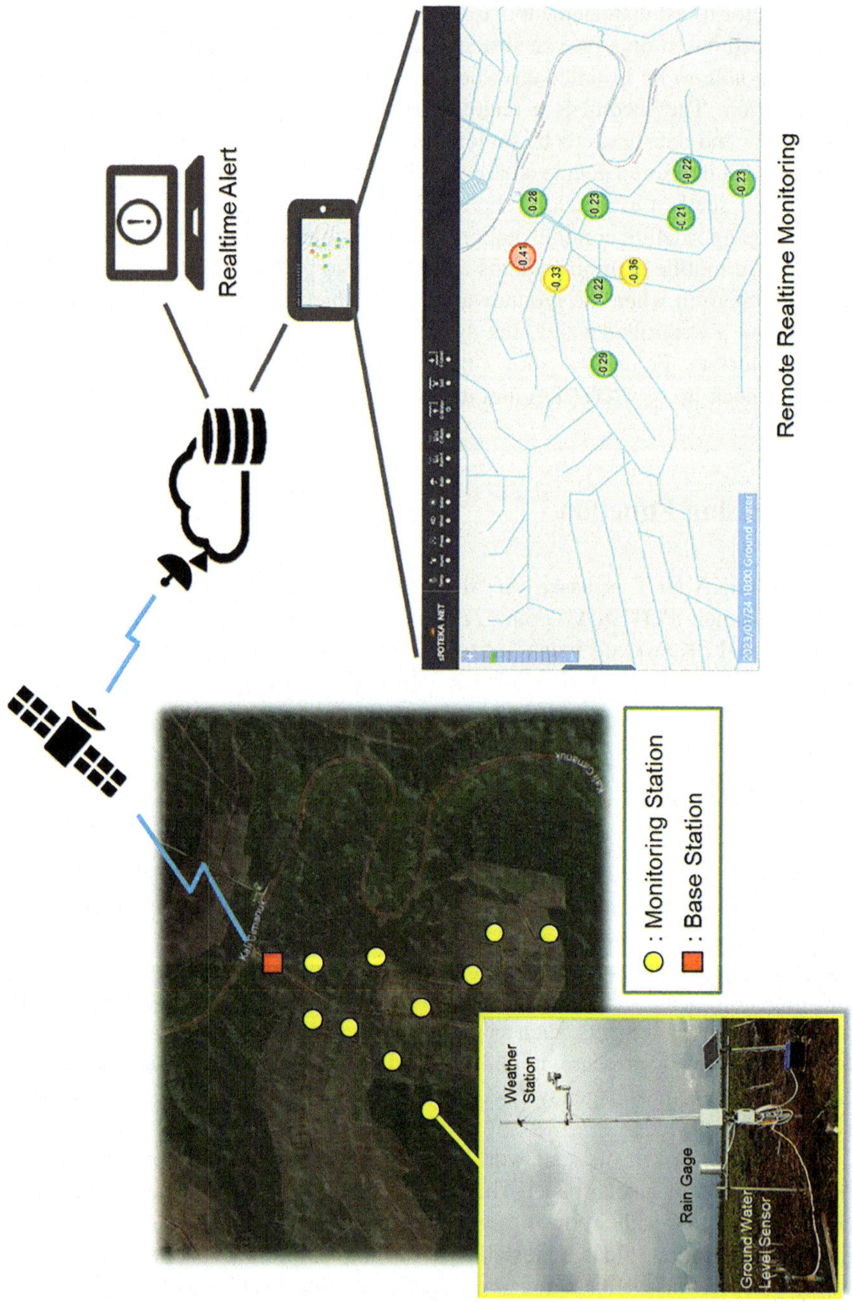

Fig. 9.1 Outline of sPOTEKA with Networrk (sPOTEKA-NET)

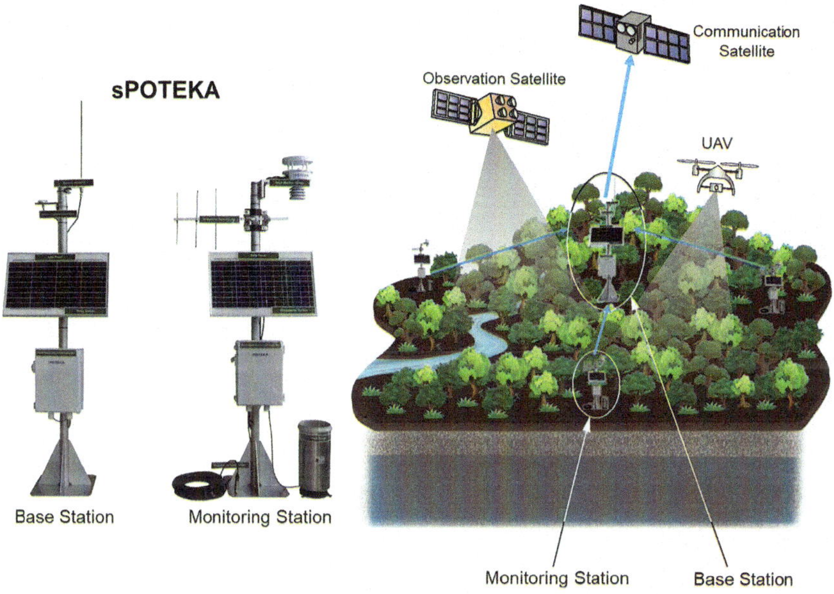

Fig. 9.2 Mock-up of sPOTEKA and its role

Table 9.1 Sensor specification of sPOTEKA

Element	Sensor type	Range	Accuracy	Remark
Temperature	Platinum resistor	−50.0 ~ +50.0 °C	±0.3 °C	JMA certification
Humidity	Capacitive	0.0 ~ 100.0% RH	±5.0%	JMA certification
Pressure	Capacitive	870.0 ~ 1050.0 hPa	±0.7 hPa	JMA certification
Wind direction	Ultrasound	1 ~ 360°	±10°	–
Wind speed	Ultrasound	0.0 ~ 30.0 m/s	±0.3 m/s (0 ~ 10 m/s) ±5% (Reading) (10 ~ 30 m/s)	JMA certification
Sunshine	Photodiode	0 ~ 1400 W/m^2	±10% (Reading)	–
Rain detector	Capacitive	0/1	–	–
Precipitation	Tipping bucket	0 ~ 200 mm/h	±0.5 mm/h (0 ~ 20 mm/h) ±3% (Reading) (20 ~ 200 mm/h)	–
Groundwater level	Pressure	0 ~ 10 m	±0.2% (Full scale)	–

JMA Japan Meteorological Agency

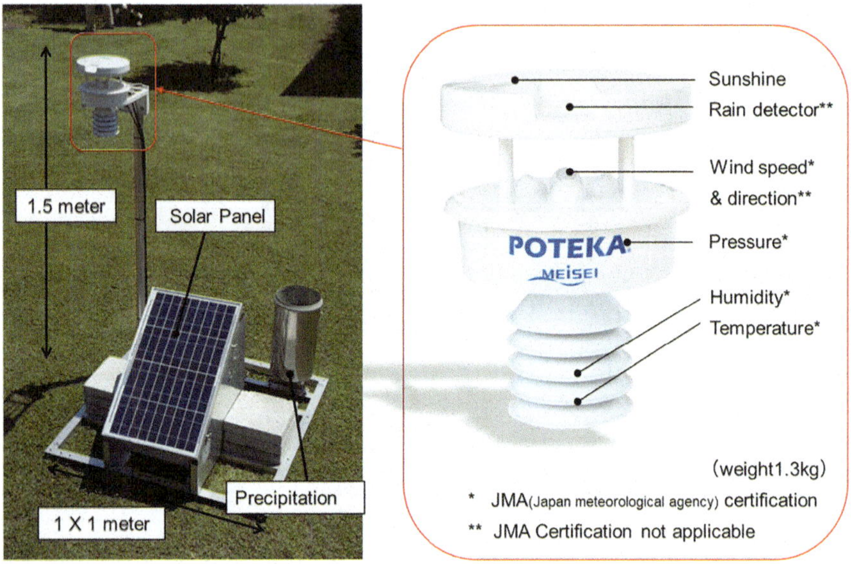

Fig. 9.3 POTEKA Compact Meteorological Sensor (Weather Station) and POTEKA Weather Observation Equipment

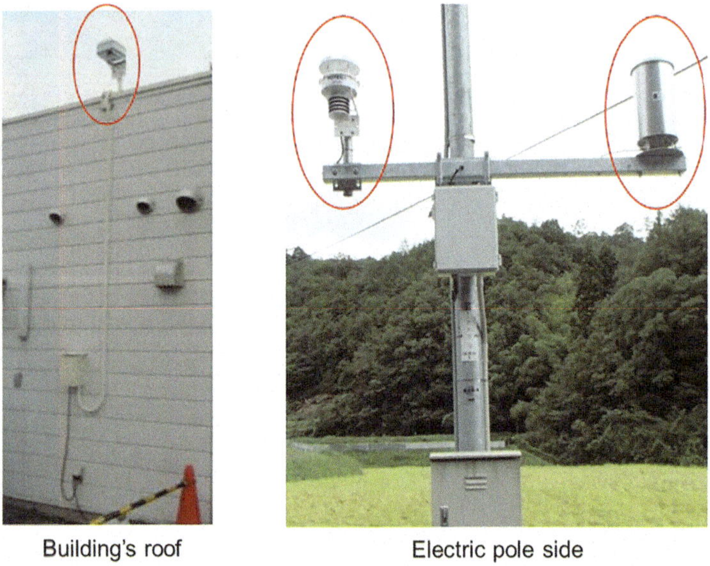

Fig. 9.4 Examples of installation locations

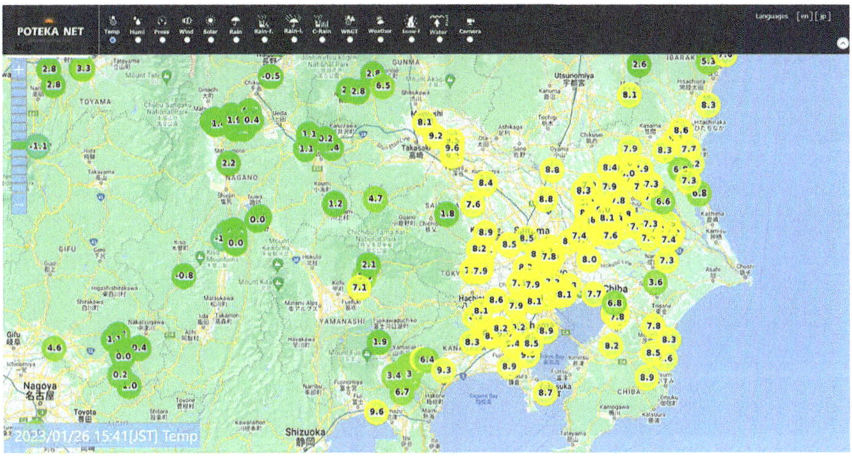

"POTEKA NET" homepage (http://www.potekanet.com/)
(This display is limited to publicly available data.)

Fig. 9.5 The POTEKA Service in Japan

approximately 1200 units have been produced for local governments, private companies, and universities. Figure 9.5 shows the POTKEA service in Japan.

9.3.2 sPOTEKA's Sensing Function

Sensing technology to measure weather conditions and groundwater levels is essential for proper forest management of vast tropical peatlands. Continuous measurement of the groundwater level is very important to prevent serious damage from peatland wildfires and to grow healthy forests in plantation areas. sPOTEKA has a groundwater level measurement function for the proper forest management of tropical peatlands. Figure 9.6 shows the sPOTEKA measurement function.

9.4 Monitoring Network and Data Analysis Function

Sensing solutions for tropical peatland forest management require remote monitoring in areas where mobile communications are difficult to provide. sPOTEKA-NET uses satellite communication and has a function that enables real-time data monitoring remotely by network communication. Figure 9.7 shows the monitoring system of sPOTEKA-NET.

The measuring equipment of sPOTEKA-NET consists of a base station (parent device) and a monitoring station (child device). First, the monitoring stations send

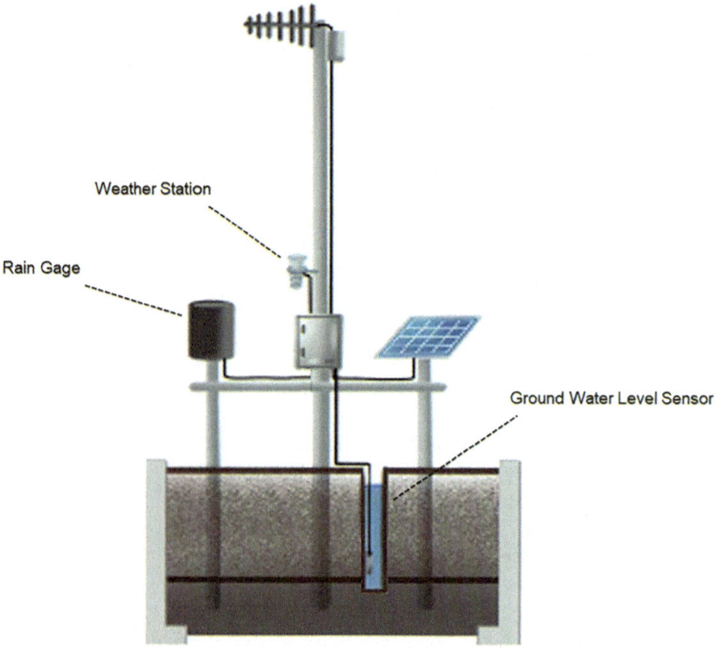

Fig. 9.6 sPOTEKA's sensing function

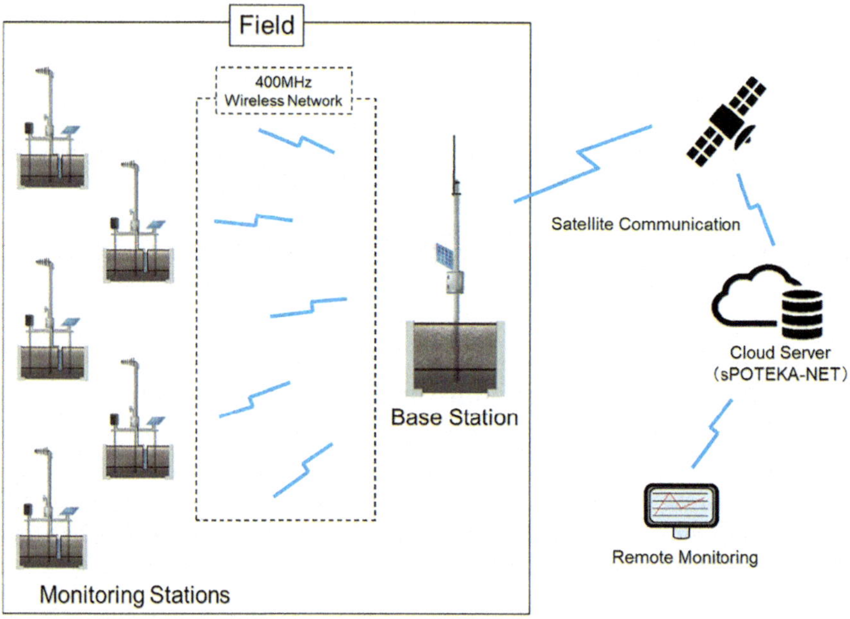

Fig. 9.7 sPOTEKA monitoring system

measured data to the base station by wireless communication. The base station then uses satellite communications to send data to the cloud server. This system enables remote monitoring even in remote areas where mobile communication is difficult to provide. In addition, the ability to remotely detect abnormal conditions through cloud data management and maintainability based on an after-sales service network has achieved a high operating rate for the system.

sPOTEKA-NET has been developed not only as a measuring device but also as a solution for forest management in tropical peatlands. For this reason, it has functions that provide useful information for forest management operations, such as automatically alerting when the groundwater level, which is directly linked to fire risk, falls below a controlled value.

In the future, it will be developed with the aim of developing a solution for more advanced forest management in tropical peatlands, such as a function that displays groundwater level predictions and fire risks calculated from data acquired by sPOTEKA-NET. Furthermore, we are planning to realize a function that predicts the anomaly of sPOTEKA in advance by analyzing the behavior of the obtained data and by monitoring the status of sPOTEKA. As a result, maintenance can be performed by condition-based maintenance, not by time-based maintenance, and the operating rate of the system can be greatly increased.

9.5 The Future of Tropical Peatland Management with sPOTEKA-NET

This section describes how sPOTEKA will advance tropical peatland forest management. Through sPOTEKA-NET, we aim to further improve water level control, which is important for forest management in tropical peatlands. When the water level drops below a certain value, the risk of fire increases. However, raising the water level too high has a negative impact on plant growth (Kato et al. 2021). Therefore, it is necessary to control the water level within an appropriate range for proper forestry operations and management. In the future, we plan to integrate data obtained from earth observation satellites and UAVs and use sPOTEKA-NET to display the distribution of groundwater levels in real time and predict the change in groundwater levels, aiming to be an information provision tool. Space utilization technology, ground truth environment measurement technology, and mid- to long-term forecasting technology will support effective forest management. Figure 9.8 shows the image of future forest management.

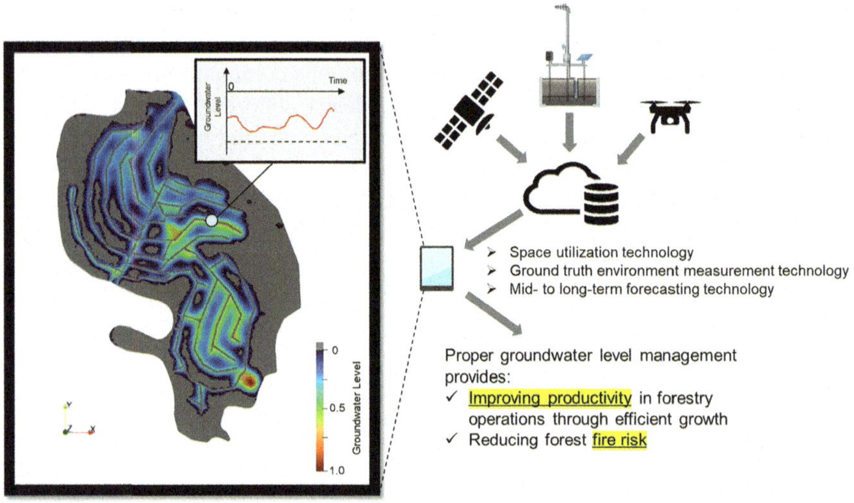

Fig. 9.8 The image of future forest management

9.6 Achievement of sPOTEKA-NET

As of September 2022, a prototype of sPOTEKA-NET installed on Kalimantan Island in Indonesia is in operation.

9.6.1 sPOTEKA-NET in Indonesia

To demonstrate the functionality of sPOTEKA-NET, we carried out measurement tests using a prototype on the island of Kalimantan, Indonesia. Figure 9.9 shows a photograph of sPOTEKA-NET installed in Indonesia. Figure 9.10 shows a display example of data measured by sPOTEKA-NET. The results demonstrated that the sensing function and the remote monitoring function worked properly.

9.7 Conclusion

Sensing technology that appropriately measures and accurately grasps weather conditions and groundwater levels is essential for proper forest management in vast tropical peatlands. Real-time monitoring with analytics is required to detect abnormal conditions in the field and monitoring system. The sensing function, including the groundwater level measurement, is based on established technology. Furthermore, although tropical peatlands are remote areas where mobile

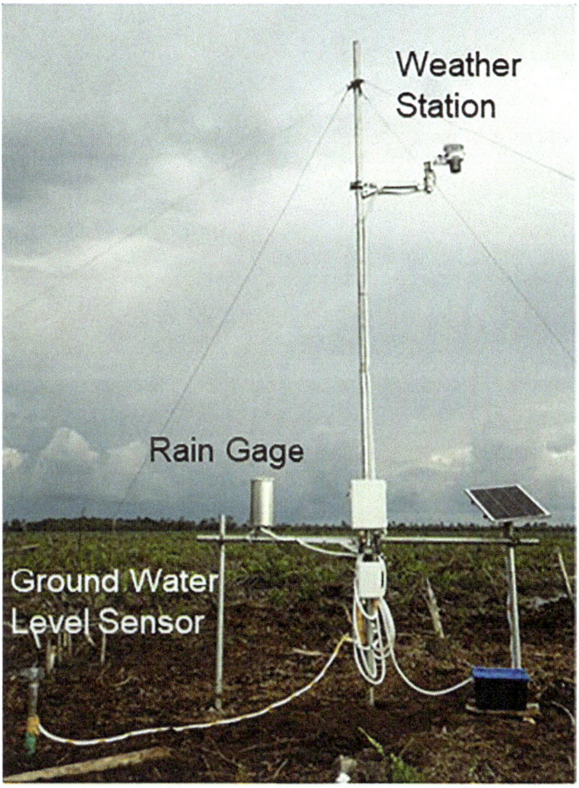

Fig. 9.9 Prototype of sPOTEKA at the Field Observatory of Sumitomo Forestry Co., Ltd., West Kalimantan, Indonesia

communication is difficult to provide, sPOTEKA-NET has communication functions using the cloud server and satellite network communication. As a solution, sPOTEKA-NET is attracting attention from worldwide, including at COP26. Figure 9.11 shows a photograph of the exhibition.

The appropriate management of groundwater levels by sPOTEKA not only improves the economic efficiency of forestry projects and reduces fire risks but also reduces CO_2 emissions (Hirano et al. 2009) and can be used to store objective data on water level management. Thus, sPOTEKA-NET is a powerful tool for the eco-management of tropical peatland, and it can provide functions such as groundwater level predictions, fire risks, and GHG emissions from peatlands.

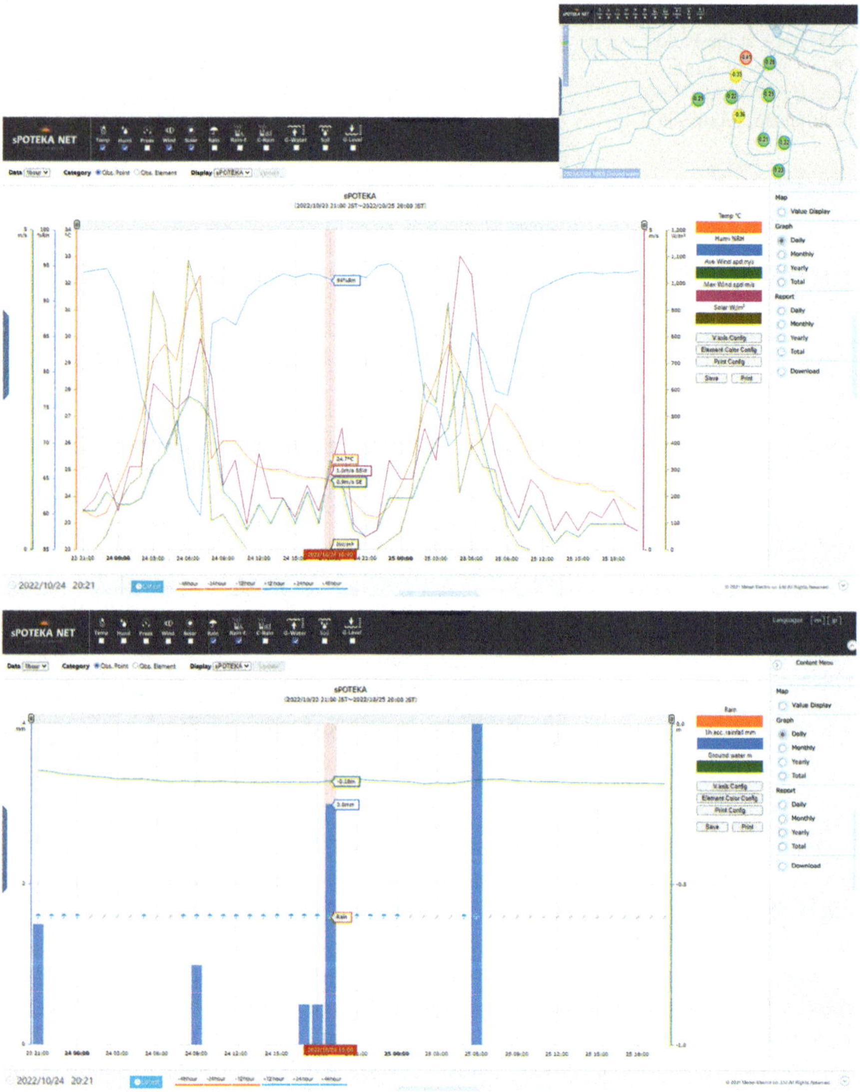

Fig. 9.10 Prototype of sPOTEKA Monitoring Data at the Field Observatory of Sumitomo Forestry Co., Ltd., West Kalimantan, Indonesia (the top chart shows temperature, humidity, sunshine, wind velocity and bottom chart shows precipitation, rain detector, groundwater level)

Fig. 9.11 Mock-up of sPOTEKA Exhibition. The left photo is in the Japan Pavilion, and the right photo is in the Peatland Pavilion

References

Dargie GC, Lewis SL, Lawson IT, Mitchard ETA, Page SE, Bocko YE, Ifo SA (2017) Age, extent and carbon storage of the Central Congo Basin peatland complex. Nature 542:86–90. https://doi.org/10.1038/nature21048

Dirzo R, Raven PH (2003) Global state of biodiversity and loss. Annu Rev Environ Resour 28:137–167. https://doi.org/10.1146/annurev.energy.28.050302.105532

Friedlingstein P, Jones MW, Sullivan MO, Andrew RM, Bakker DCE, Hauck J, Quéré CL, Peters GP, Peters W, Pongratz J, Sitch S, Canadell JG, Ciais P, Jackson RB, Alin SR, Anthoni P, Bates NR, Becker M, Bellouin N, Bopp L, Chau TTT, Chevallier F, Chini LP, Cronin M, Currie KI, Decharme B, Djeutchouang LM, Dou X, Evans W, Feely RA, Feng L, Gasser T, Gilfillan D, Gkritzalis T, Grassi G, Gregor L, Gruber N, Gürses Ö, Harris I, Houghton RA, Hurtt GC, Iida Y, Ilyin T, Luijkx IT, Jain A, Jones SD, Kato E, Kennedy D, Goldewijk KK, Knauer J, Korsbakken JI, Körtzinger A, Landschützer P, Lauvset SK, Lefèvre N, Lienert S, Liu J, Marland G, McGuire PC, Melton JR, Munro DR, Nabel JEMS, Nakaoka S, Niwa Y, Ono T, Pierrot D, Poulter B, Rehder G, Resplandy L, Robertson E, Rödenbeck C, Rosan TM, Schwinger J, Schwingshack C, Séférian R, Sutton AJ, Sweeney C, Tanhua T, Tans PP, Tian H, Tilbrook B, Tubiello F, Werf GR, Vuichard N, Wada C, Wanninkhof R, Watson AJ, Willis D, Wiltshire AJ, Yuan W, Yue C, Yue X, Zaehle S, Zeng J (2022) Global carbon budget 2021. Earth Syst Sci Data 14:1917–2005. https://doi.org/10.5194/essd-14-1917-2022

Hayasaka H, Takahashi H, Limin SH, Yulianti N, Usup A (2016) Peat fire occurrence. In: Osaki M, Tsuji N (eds) Tropical peatland ecosystems. Springer, Tokyo

Hirano T, Jauhiainen J, Inoue T, Takahashi H (2009) Controls on the carbon balance of tropical peatlands. Ecosystems 12:873–887. https://doi.org/10.1007/s10021-008-9209-1

Iwashita H, Morita T, Shibata K, Kobayashi F (2019) Downburst observations by a high density ground surface observation network (POTEKA). J Atmos Electr 38(1):23–35. https://doi.org/10.1541/jae.38.23

Kato T, Silsigia S, Yusup AA, Osaki M (2021) Large-scale practice on tropical peatland eco-management. In: Osaki M, Tsuji N, Foead N, Rieley J (eds) Tropical peatland eco-management. Springer, Singapore

Lohberger S, Stängel M, Atwood EC, Siegert F (2017) Spatial evaluation of Indonesia's 2015 fire-affected area and estimated carbon emissions using Sentinel-1. Glob Change Biol 24(2): 644–654. https://doi.org/10.1111/gcb.13841

Tsuji N, Kato T, Osaki M, Sulaiman A, Ajie GS, Kimura K, Hamada Y, Shigenaga Y, Hirose K, Silsigia S (2021) Evaluation of eco-management of tropical peatlands. In: Osaki M, Tsuji N,

Foead N, Rieley J (eds) Tropical peatland eco-management. Springer, Singapore. https://doi.org/10.1007/978-981-33-4654-3_5
Xu J, Morris PJ, Liu J, Holden J (2018) PEATMAP: refining estimates of global peatland distribution based on a meta-analysis. Catena 160:134–140. https://doi.org/10.1016/j.catena.2017.09.010
Yamashita K, Kubota H, Matsumoto J, Sato M, Momota E, Kojima S, Takahashi Y, Hamada J, Katahira Y (2018) Development of lightning observation system for short-term forecast of extreme weather events in the Philippine under ULAT project. In: 2018 34th International conference on lightning protection (ICLP), pp 1–4. https://doi.org/10.1109/ICLP.2018.8503378

Chapter 10
The Use of a Compact Fiber-Optic Spectrometer to Measure Atmospheric Carbon Dioxide in a Tropical Peatland

Masafumi Ohashi, Kenichi Tonokura, and Masahiro Kawasaki

Abstract A compact instrument for measuring the atmospheric column-averaged volume mixing ratio of carbon dioxide (XCO_2) in the near-infrared region was developed using an optical fiber Fabry–Perot interferometer. Direct sunlight that contains spectral information on the XCO_2 was collimated by a small fiber collimator installed on a portable sun tracker and then transmitted through an optical fiber into the data acquisition system. The daytime XCO_2 had one standard deviation (1σ) of 1.8 ppm. Here, a feasibility study is reported regarding field measurements obtained from a tropical peatland in Central Kalimantan, Indonesia.

Keywords Field measurement · Smoldering fire · Column-averaged · XCO_2 · CO · $PM_{2.5}$

10.1 Introduction

Peat is a type of charcoal formed from the accumulation of biological organic matter that has not yet decomposed. Since peatlands store one-third of the carbon contained in all soils worldwide, the stability of peatlands can have a significant impact on climate change. The tropical peat swamps and forests that were once common in Southeast Asia have been dehydrated and degraded due to drainage for large-scale plantings established in the 1990s. As a result, the organic matter in peat has continued to decompose, and a large amount of carbon dioxide (CO_2) is emitted into the air. In addition, dry peatlands are flammable, and peatland fires occur every

M. Ohashi
Department of Information Science and Biomedical Engineering, Kagoshima University, Kagoshima, Japan

K. Tonokura (✉)
Department of Environment Systems, Graduate School of Frontier Sciences, The University of Tokyo, Kashiwa, Japan
e-mail: tonokura@k.u-tokyo.ac.jp

M. Kawasaki
Arctic Research Center, Hokkaido University, Sapporo, Japan

M. Osaki et al. (eds.), *Tropical Peatland Eco-evaluation*,
https://doi.org/10.1007/978-981-99-6790-2_10

year in the dry season, damaging crops and causing serious health problems due to smoke pollution. Peat fires also emit large amounts of CO_2 emissions that contribute to global warming. Therefore, it is desirable to develop a method to measure CO_2 emissions from peat fires.

Due to their portability, which allows measurements in the field, small instruments are used to measure carbon emissions produced from dispersed and less predictable sources such as forest/peatland fires. The local CO_2 flux is measurable if CO_2 observation sites surround the targeted emission area, and if wind information is available. Every method has advantages and disadvantages. A ground-based in situ instrument samples a local air mass but not the complete atmospheric boundary layer. Space-based remote sensing, including satellite observations, can search for the boundary layer and survey a large area. Although clear sky conditions are preferable for stable observation, observations in the near-infrared (NIR) band are highly resilient to fluctuations in the intensity of solar radiation, which can be caused by occasional thin cloud coverage. The standard instrument used for such measurements is a Fourier transform spectrometer (FTS), which has been used by the Total Carbon Column Observing Network (TCCON) (Wunch et al. 2010). Since the FTS is a precise instrument that requires a well-prepared support infrastructure, it is not easy to use for field measurements. Recently, the mobile version of FTS has become commercially available but is still expensive (Hase et al. 2015). In contrast to the FTS, an alternative approach is the use of Fabry–Perot interferometry (Wilson et al. 2007; Kobayashi et al. 2010; Baily 2013). The instruments based on Fabry–Perot interferometry are small and robust and are easily installed for field measurements. Here, we report the results of a campaign in Central Kalimantan, Indonesia, that used a compact instrument called the Fiber-Etalon solar spectrometer for carbon dioxide (FES-C).

10.2 Instrumentation

10.2.1 Fiber Optics

The column spectrometer consists of an optical fiber-based spectrum analyzer with a resolution sufficient to resolve the CO_2 rotation lines in the NIR region. Direct sunlight is collected by a fiber collimator installed on a sun tracker and transmitted through a single-mode optical fiber into the spectrum analyzer to obtain the atmospheric CO_2 column densities, which are then converted to the column-averaged volume mixing ratios of atmospheric CO_2 (XCO_2). The FES-C instrument employs an optical fiber Fabry–Perot interferometer (FFPI) to measure the CO_2 column density through the absorption of solar light. Solar light passes through the atmosphere, where it undergoes appreciable absorption by atmospheric CO_2 in the NIR region at 1568–1576 nm. In this wavelength region, rotationally resolved CO_2 absorption appears. The output intensity of the FFPI reaches several nW and is sufficient for measurements with a commercially available NIR detector.

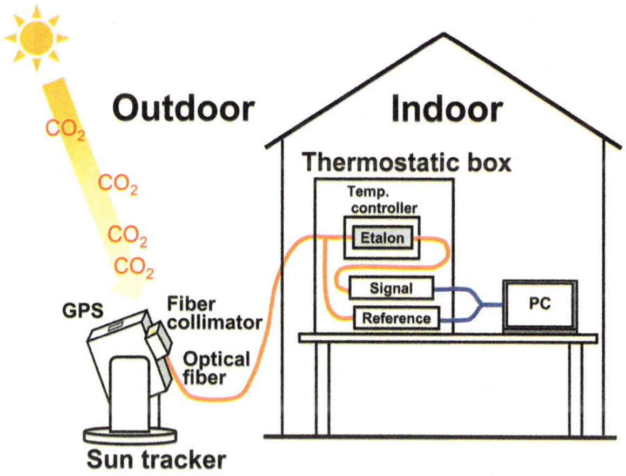

Fig. 10.1 Schematic of the sunlight collection and data acquisition system that are part of the CO_2 column spectrometer, FES-C

To measure the CO_2 column density, the Fabry–Perot interferometer utilizes quartz glass, which has a sufficient spectral resolution capable of resolving the rotational absorption lines. Wilson et al. (2007) first reported its application in the measurement of atmospheric CO_2. Their instrument was compact but seemed to require precise optical alignment. Additionally, a large solid etalon has a large heat capacity, and hence, it would be difficult to tune its temperature during a short measurement time as well as to handle it in the field under tropical weather conditions. Our tiny FFPI that has been used in the telecommunication industry had the same optical features as the solid Fabry–Perot interferometer (Kobayashi et al. 2010). FFPIs assembled using simple fiber optics do not require optical alignment and are compact, easy to set up, and highly resistant to shocks. Since the FFPI had a small heat capacity, the transmission wavelength produced by the FFPI quickly responded to temperature changes, which reduced the time interval for measurements.

A schematic of our FFPI is shown in Fig. 10.1. In brief, the outdoor components consist of a fiber collimator as a small solar telescope that is installed on a portable sun tracker. The indoor components consist of the FFPI (13 mm long with a ceramic shroud 1.25 mm in diameter), a solar intensity monitor, and a laptop computer. A long-wavelength pass filter is placed in front of an object lens of the solar telescope. The solar signal is directed to both the FFPI and the solar intensity monitor via a single-mode optical fiber cable attached to a beam separator. As shown in Fig. 10.1, one fiber is connected to the signal inlet of the FFPI, while the other is connected to that of the solar intensity monitor. The reference solar intensity is simultaneously monitored in the spectral range of 800–1800 nm with the FFPI signal to compensate for solar intensity fluctuations caused by thin-cloud coverage.

The transmission spectrum of the FFPI is shown in Fig. 10.2. The wavelength of the solar spectrum transmitted through the FFPI is controlled by varying its temperature because this variation induces a refractive index change in the core of the glass fiber or a wavelength shift within the transmission spectrum. The transmitted light is

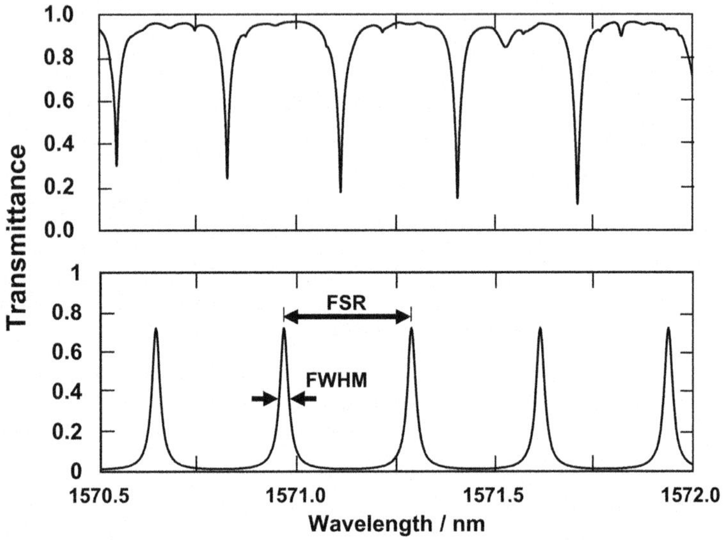

Fig. 10.2 Upper figure: Rotationally resolved spectrum of atmospheric CO_2. Lower figure: Transmission spectrum of the fiber Fabry-Perrot etalon optics. For the etalon, the free spectral range (0.324 nm, 1.25 cm^{-1}) and full width at half maximum (0.025 nm, 0.10 cm^{-1}) match the CO_2 rotational structure

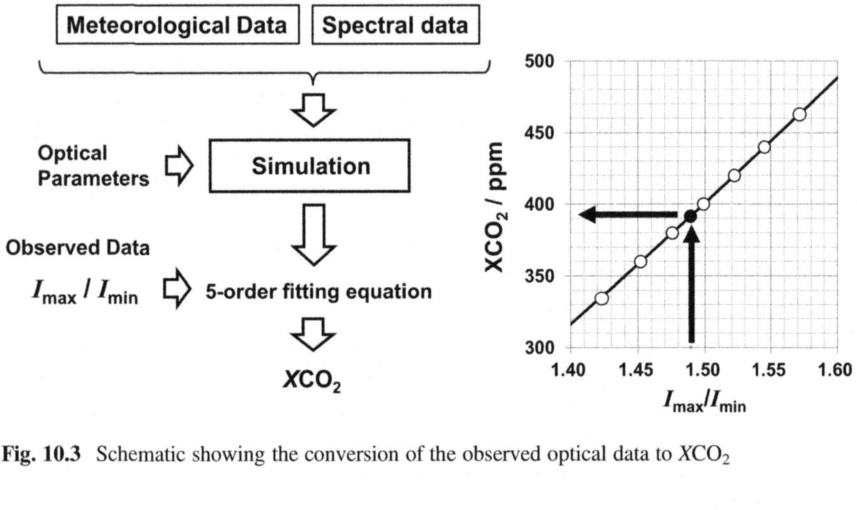

Fig. 10.3 Schematic showing the conversion of the observed optical data to XCO_2

thus aligned/unaligned with the CO_2 rotational lines that are centered at 1571 nm through a narrow band fiber filter.

By modulating the FFPI temperature at 40 s/cycle, the intensity ratio of the incident light to the transmitted light, I_{max}/I_{min}, as shown in Fig. 10.3, is deduced using the Beer–Lambert law that relates the absorption of light to the spectral properties of the CO_2 through which it is traveling. I_{min} is obtained when the rotational branches match the etalon spectrum. When they mismatch, I_{max} is measured. As will be described below, the optical transmission spectrum of FFPI optics is convoluted with the spectra simulated for various XCO_2. This allows XCO_2 to be estimated by simulating CO_2 spectra once one knows the optical characteristics of the FFPI and weather data along the solar path.

10.2.2 Column-Averaged Dry-Air Molar Mixing Ratios, XCO_2

For spectral analysis, the HITRAN2008 database was used (Rothman et al. 2009). Since the molecular absorption lines are pressure- and temperature dependent, the spectral simulation requires the meteorological parameters along the vertical profiles. From the Goddard Earth Sciences Data and Information Services Center at NASA, we incorporated the Modern-Era Retrospective analysis for Research and Applications weather data products, which consist of 1 min observations and have a horizontal resolution of 1 km, for the location of the station. To obtain the column density of atmospheric CO_2, a homogeneous vertical profile of XCO_2 from 0 to 48 km altitudes was assumed. To simulate a rotational spectrum for a certain value of XCO_2, first, the air column was divided into 28 sublayers. Second, the Beer–Lambert law was applied to determine the CO_2 absorption in each layer. By summing their absorbance for line-by-line calculation, the corresponding rotational line absorbance for the column density was obtained. To obtain XCO_2, the observed column density was divided by the total column density of dry air (Washenfelder et al. 2006). Air masses within the layers were calculated from the solar zenith angles (*SZA*) obtained from the GPS time and geographic data from the observation station. The subtracted water vapor masses were based on the reanalysis of relative humidity weather data for the ground and upper atmospheres.

There was a slight SZA dependency in the calculated XCO_2. For this dependency, we tested a correction formula implemented by Wunch et al. (2010). The correction influenced our results of the daily XCO_2 by less than 0.1%. With the use of calibration curves thus obtained, the observed column densities were converted to XCO_2, as shown in Fig. 10.3.

10.2.3 Validation of Instrument Sensitivity and Verification of Instrument Reliability

The XCO_2 data obtained with FES-C were compared with those of a colocated TCCON FTS between 12 March 12 and July 1, 2014 at the National Institute for Environmental Studies (NIES) in Tsukuba, Japan (36.05°N, 140.12°E).

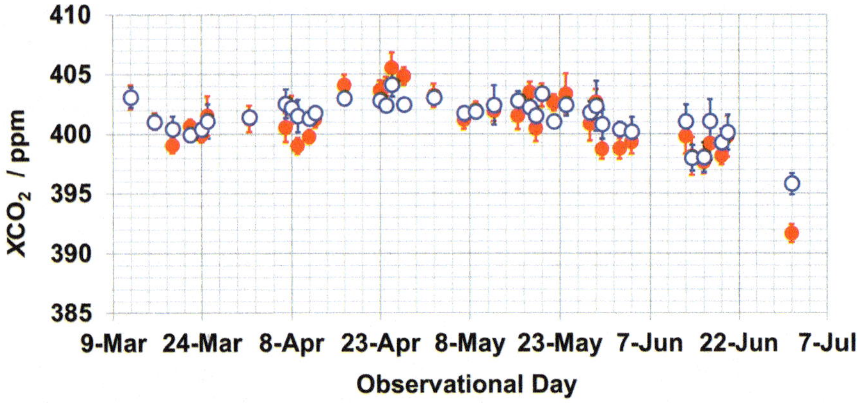

Fig. 10.4 Averaged data for solar time 10–14 h: (open circle) FTS at NIES; (filled circle) scaled FES-C, at Tsukuba, Japan, from March to July 2014. The vertical bars represent one sigma (Iriana et al. 2018)

Daily-averaged data for 40 days are plotted in Fig. 10.4. Referring to the TCCON FTS data, the FES-C scale factor is obtained as follows:

$$XCO_2(\text{scaled}) = XCO_2(\text{retrieved})/0.996$$

The one standard deviation (1σ) of the scale factors is 0.0046, which corresponds to 1.8 ppm, which includes the deviation (0.4 ppm) of the FTS data (Wunch et al. 2010).

Figure 10.5 (upper) shows the resiliency of the system to fluctuations in solar intensity due to occasional light cloud coverage, during which XCO_2 is resistant to solar intensity changes down to 20% of the full intensity. Intercomparison with the FTS from the NIES validated the performance of FES-C. Figure 10.5 (lower) shows an example of diurnal changes in XCO_2 when a CO_2-rich airmass reached the Tsukuba area from the Tokyo metropolitan area. Both XCO_2 signal intensities of FES-C and FTS hump at 13:00–15:00 JST.

10.3 Weather Radar Monitoring of Fire Plumes During the Dry Season

Figure 10.6 (left) is an image from the AHI/HIMAWARI-8 and the MODIS/Terra and Aqua/NASA satellites above a fire-prone equatorial peatland region, Palangka Raya, Central Kalimantan, Indonesia, at 09Z(16LT) on October 15, 2015. The image shows thick yellowish puffs of smoke that originated from forest/peatland fires. The weather radar measurement is shown in Fig. 10.6 (right) and shows the formation of the planetary boundary layer (PBL) at a height of several km, which the tropical

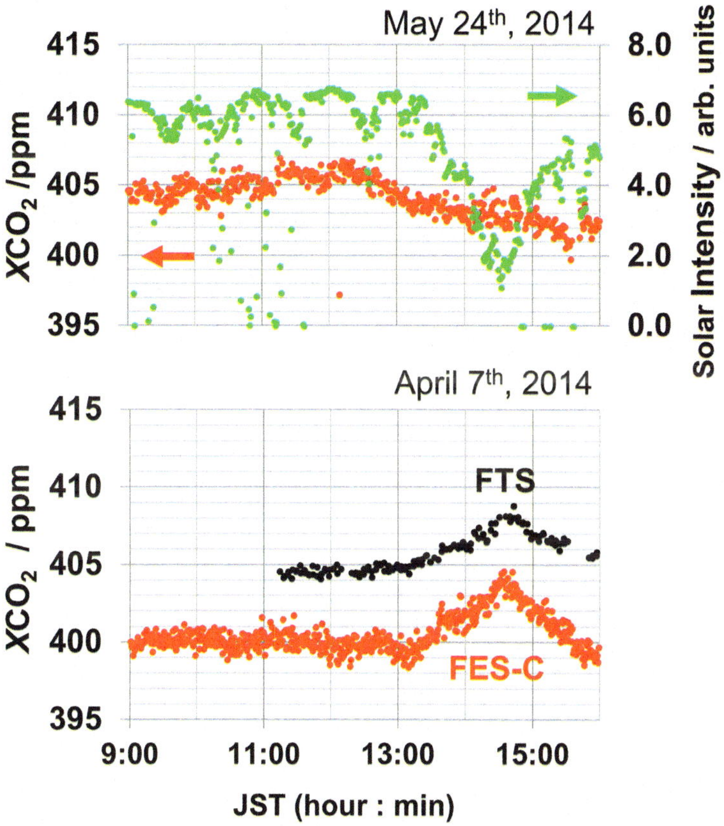

Fig. 10.5 Upper figure: The stability of the signal to fluctuations in the solar intensity. (Red) XCO_2 in ppm, (Green) near-infrared solar intensity on May 24, 2014, at Tsukuba, Japan. Lower figure: Intercomparison of XCO_2 by FES-C (red) and FTS (black). The FTS data are shifted upward by 3 ppm for visible clarity. Data acquired on April 7, 2014 at Tsukuba, Japan

diurnal cycle (or heating of the ground surface) induces after sunrise during the dry season. Atmospheric stratification by the PBL is responsible for capping air masses contaminated with aerosol and water vapor produced from peatland fires by suppressing atmospheric convection. The fire smoke layer-top trails are also visible in the radar images (Fig. 10.6), which extend from the MODIS fire hotspot points marked by red dots toward the northern areas. Smoke trajectories starting from the hotspots are shown by the red arrows in Fig. 10.6 (left), in which three trajectories are projected with use of the forward trajectories for 02Z–08Z at 1000-m altitude using the HYSPLIT model from NOAA. Comparing these trajectories with the measured top trails of the smoke layer seen in the radar images in Fig. 10.6 (right) reveals that they match each other. Based on satellite observations, Zender et al. (2012) reported that smoke traveled with prevailing southeast surface winds at a typical speed of ~3 m/s during the dry season in Central Kalimantan and formed

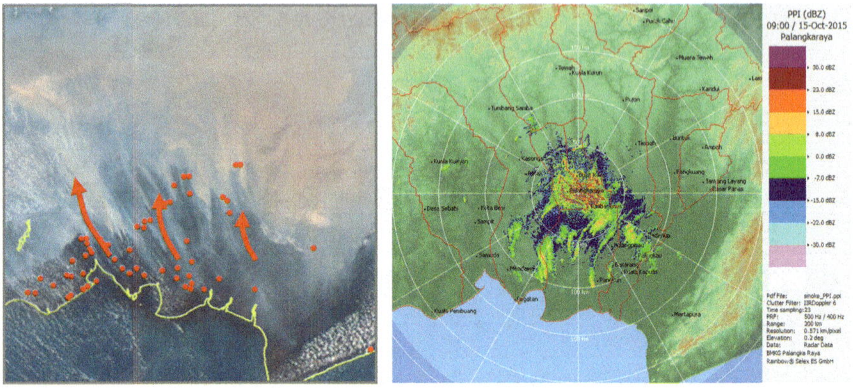

Fig. 10.6 Right figure: Radar image. Left figure: Smoke plumes are observed by the AHI/HIMAWARI-8/JMA satellite. The red dots are MODIS/NASA fire hotspots. The red arrows are forward trajectories at 1000-m altitude (Rahman et al. 2021)

ovoid plumes whose mean length, height, and cross-plume width were 41 km, 0.7 km, and 27% of the plume length, respectively. These reported sizes agree with the trail features observed in Fig. 10.6. These results suggest that airmasses contaminated with CO_2, CO, and PM entered into the northern region, Palangka Raya, for which the FES-C instrument measured the increments in XCO_2.

10.4 Field Measurements of Atmospheric CO_2, CO, and Fine Particulate Matter $PM_{2.5}$

There are two types of wildfires: surface flames and underground smoldering. Smoldering fires cause extensive air pollution that emits large amounts of carbon into the air (Rein and Huang 2021). In the dry season, large-scale wild/peatland fires occur during Aug–Oct at Palangka Raya. Surface fires burn trees, while underground fires burn peat. The fires increase atmospheric CO_2, carbon monoxide CO, and fine particulate matter $PM_{2.5}$ in the peatland area.

10.4.1 Measurements of Atmospheric CO_2 with FES-C

According to a BMKG report, the background value of XCO_2 in Indonesia was ~393 ppm at the end of 2014, which agrees well with our values presented in Fig. 10.7 for Nov–Dec 2014. The irregular change in XCO_2 during Aug–Oct was due to wild/peatland fires. Figure 10.7 shows that the fire season in the beginning-to-middle of September 2014 was characterized by high ΔXCO_2 and large hotspot counts as well as a low ground water level or low precipitation totals, where

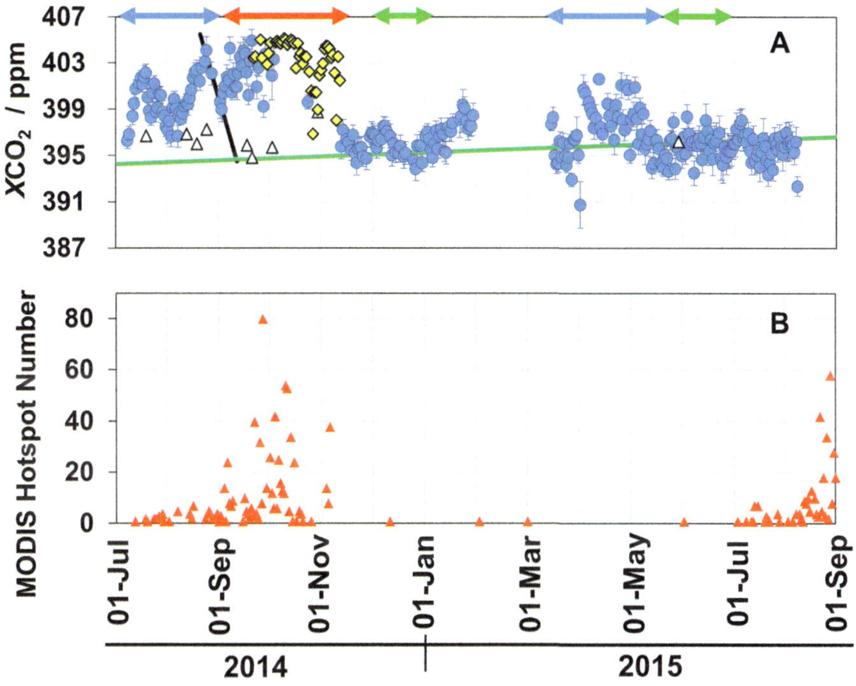

Fig. 10.7 Upper figure: Time series of XCO_2; (blue circle) FES-C average for UTC = 3–7 h. The error bar shows one standard deviation of the XCO_2 value, (white triangles) GOSAT data over the sea, (green line) period for background level, (yellow diamonds) ΔXCO_2 estimated from visibility data, (blue ↔) peat soil respiration period, (red ↔) fire period, (green ↔) background period. The black solid line is the estimated contribution of peat soil respiration to XCO_2. Lower figure: MODIS hotspot count per day in a 50-km circle centered at the observation site in Palangka Raya (Iriana et al. 2018)

increment ΔXCO_2 is defined as the difference between the observed XCO_2 in Palangka Raya and the background XCO_2 over the ocean. ΔXCO_2 increased up to 10.0 ppm (mean 6.0 ppm). The increase was caused mostly by peatland fires because the younger surface peat that was less decomposed was destroyed by fire, and the peat respiration activities were reduced. When sunlight did not reach the solar telescope of the FES-C under very heavy hazy conditions due to huge wildfires, the instrument could not measure ΔXCO_2. To estimate ΔXCO_2 during very hazy days, visibility data can be used in place (Field et al. 2009). Using the fire-origin ΔXCO_2 collected over 35 days via FES-C measurement, a linear relationship between ΔXCO_2 and visibility data was obtained (Iriana et al. 2018). Thus, estimated values are plotted in Fig. 10.7 with yellow triangles.

The CO_2 emissions from peat respiration or aerobic peat oxidation correlate strongly with lowering ground water levels. As marked by the two blue double arrows in the upper side of Fig. 10.7, ΔXCO_2 increased between July and August 2014 and mid-March and mid-May 2015, while no MODIS hotspots appeared

(Iriana et al. 2018). The increases in ΔXCO_2 are due to enhanced peat respiration by low ground water level or less precipitation and not by fire.

10.4.2 Measurements of CO and $PM_{2.5}$ with Small Sensors in the Dry Season

The Palangka Raya area was covered by thick haze in August and September 2019. The visibility was reduced to 1300 m at the Palangka Raya airport on September 29. Sensor (Model CO-B4, Alphasense) measurements indicated that the background level CO of in Palangka Raya was 0.4 ppm during the non-fire season and 3 ppm during the fire season. The concentrations of CO and $PM_{2.5}$, as measured with a small PM sensor (Shibata Science) in Palangka Raya, peaked in the second week of August and the third week of September 2019. On one extremely high-level day, CO increased up to 14 ppm and $PM_{2.5}$ up to 800 μg/m^3, as shown in Fig. 10.8. According to the public health guidelines, the US-AEGL (2013) (8 h) for CO is 27 ppm, and the WHO-AQGL (2021) (24 h) for $PM_{2.5}$ is 15 μg/m^3. Under these polluted conditions, the air quality standard index is unhealthy for humans and other living things. Figure 10.8 (right) shows a good correlation between CO and $PM_{2.5}$ because these pollutants were emitted from underground peatland fires. These results suggest that the enrichment in CO and $PM_{2.5}$ was caused by smoldering peat fires. The seasonal wind direction in the dry season was southeasterly as indicated by the red arrows in

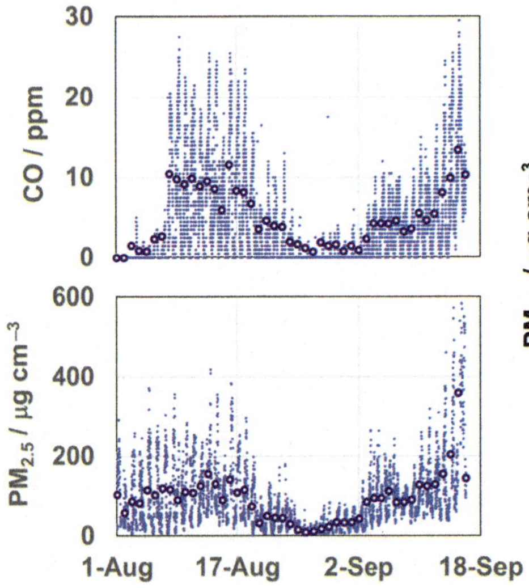

Fig. 10.8 Left figure: Temporal profiles of CO and $PM_{2.5}$ in Palangka Raya, 2019. Large diurnal variations in concentrations are due to nighttime trapping and daytime flow. Open circles represent daily-averaged values. Right figure: A linear correlation between CO and PM suggests these species are emitted from underground peat fires

Fig. 10.6. CO, $PM_{2.5}$ and CO_2 were advected under the PLB cap that extended from the peatland fire fields in the southern area to Palangka Raya in the northern area.

10.5 Conclusion

The performance of the convenient and robust spectrometer equipped with fiber optics in the tropical area of Central Kalimantan, Indonesia, was confirmed by field measurements in 2014–2015 during both dry and wet seasons. The increases in XCO_2 caused by wildfires were successfully observed by this stand-alone instrument without the need of support infrastructure. Peat degradation and peatland fires also emit methane (CH_4) as well as CO_2. The fiber optics instrument can be applied to the measurement of atmospheric CH_4. In the future, simultaneous measurements of CO_2 and CH_4 using this instrument will be collected for a more detailed understanding of greenhouse gas emissions in tropical peatlands.

Acknowledgments The authors thank BMKG (Indonesia), NASA (USA) and JMA (Japan) for providing us with the data. This work was supported by KAKENHI Grants (18KK0294, JP22K12406) of the Japan Society for the Promotion of Sciences and a project (No. 14200117) of Research Institute for Humanity and Nature.

References

Baily J (2013) A low-cost portable fibre-optic spectrometer for atmospheric absorption studies. Atmos Meas Tech Discuss 6:1067–1092. https://doi.org/10.5194/amtd-6-1067-2013

Field RD, van der Werf GR, Shen SP (2009) Human amplification of drought-induced biomass burning in Indonesia since 1960. Nat Geosci 2:185–188. https://doi.org/10.1038/ngeo443

Hase F, Frey M, Blumenstock T, Groß J, Kiel M, Kohlhepp R, Tsidu GM, Schäfer K, Sha MK, Orphal J (2015) Application of portable FTIR spectrometers for detecting greenhouse gas emissions of the major city Berlin. Atmos Meas Tech 8:3059–3068. https://doi.org/10.5194/amt-8-3059-2015

Iriana W, Tonokura K, Inoue G, Kawasaki M, Kozan O, Fujimoto K, Ohashi M, Rahman MA, Gunawan D (2018) Ground-based measurement of column-averaged mixing ratios of carbon dioxide in the peatland fire-prone area of Central Kalimantan in Indonesia. Sci Rep 8:8437. https://doi.org/10.1038/s41598-018-26477-3

Kobayashi N, Inoue G, Kawasaki M, Yoshioka H, Minomura M, Murata I, Nagahama T, Matsumi Y, Tanaka T, Morino I, Ibuki T (2010) Remotely operable compact instruments for measuring atmospheric CO_2 and CH_4 column densities at surface monitoring sites. Atmos Meas Tech 3:1103–1112. https://doi.org/10.5194/amt-3-1103-2010

Rahman MA, Nugroho DS, Yamanaka MD, Kawasaki M, Kozan O, Ohashi O, Hashiguchi H, Mori S (2021) Weather radar detection of planetary boundary layer and smoke layer top of peatland fire in Central Kalimantan, Indonesia. Sci Rep 11:367. https://doi.org/10.1038/s41598-020-79486-6

Rein G, Huang X (2021) Smouldering wildfires in peatlands, forests and the Arctic: challenges and perspectives. Curr Opin Environ Sci Health 24:100296. https://doi.org/10.1016/j.coesh.2021.100296

Rothman LS et al (2009) The HITRAN 2008 molecular spectroscopic database. J Quant Spectros Radiat Trans 110:533–572. https://doi.org/10.1016/j.jqsrt.2009.02.013

US-AEGL: National Ambient Air Quality Standards for Particulate Matter (2013). https://www.federalregister.gov/documents/2013/01/15/2012-30946/national-ambient-air-quality-standards-for-particulate-matter. Accessed 29 Nov 2022

Washenfelder RA et al (2006) Carbon dioxide column abundances at the Wisconsin Tall Tower site. J Geophys Res Atmos 111:1–11. https://doi.org/10.1029/2006jD007154

WHO-AQGL (global air quality guidelines) (2021) Particulate matter (PM2.5 and PM10), ozone, nitrogen dioxide, sulfur dioxide and carbon monoxide. World Health Organization, Geneva

Wilson EL, Georgieva EM, Heaps WS (2007) Development of a Fabry–Perot interferometer for ultra-precise measurements of column CO_2. Meas Sci Technol 18:1495–1502. https://doi.org/10.1088/0957-0233/18/5/040

Wunch D, Toon GC, Blavier J-FL, Washenfelder RA, Notholt J, Connor BJ, Griffith DWT, Sherlock V, Wennberg PO (2010) The total carbon column observing network. Philos Trans R Soc A 369:2087–2112. https://doi.org/10.1098/rsta.2010.0240

Zender CS, Krolewski AG, Tosca MG, Randerson JT (2012) Tropical biomass burning smoke plume size, shape, reflectance, and age based on 2001–2009 MISR imagery of Borneo. Atmos Chem Phys 12:3437–3454. https://doi.org/10.5194/acp-12-3437-2012

Chapter 11
Appraisal of LiDAR Measurements for Monitoring Tropical Peatlands

Hans-Dieter Viktor Boehm and Veraldo Liesenberg

Abstract Tropical peat swamp forests are important for their rich biodiversity and serve as a huge carbon pool. However, these endangered environments are decreasing due to deforestation, conversion into farmland, excessive draining, the use of shifting cultivation on a large scale, illegal logging, forest degradation, and palm oil plantations. Airborne laser scanning (ALS), also termed airborne light detection and ranging (LiDAR), is currently a good single sensor to investigate geophysical parameters in remote tropical rainforest areas (e.g., tree canopy height, which is strongly correlated with aboveground biomass).

This chapter aims to describe the LiDAR system and to present the applicability of LiDAR data to environmental work in tropical peatlands of Central Kalimantan, Indonesia. Several channels of the EX-Mega Rice Project (EMEP) and Sebangau National Park with peat swamp forest (PSF) are presented. LiDAR surveys were performed by helicopter surveys in 2007 and 2011, allowing us to perform change detection that highlighted vegetation dynamics and subsidence of peatlands. LiDAR shows promising opportunities for reducing emissions from deforestation and forest degradation (REDD) and environmental assessments.

Keywords Biomass · Digital surface model (DSM) · Digital terrain model (DTM) · Environmental verification · Geographic information system (GIS) · LiDAR · Peat Dome · Surface and terrain model · Topography

11.1 Introduction

A LiDAR system can rapidly transmit laser pulses at both daytime and nighttime. The pulse is, therefore, a reflection of both landscape and man-made features. The known speed of light and the measured time interval of the laser pulses from

H.-D. V. Boehm (✉)
Kalteng Consultants, Hoehenkirchen near München, Germany
e-mail: viktor@susiladharma.org

V. Liesenberg
Santa Catarina State University, Lages, Brazil

M. Osaki et al. (eds.), *Tropical Peatland Eco-evaluation*,
https://doi.org/10.1007/978-981-99-6790-2_11

transmission to return allow the determination of distances (Asner et al. 2012; Liesenberg et al. 2013; Asner & Mascaro, 2014). An aerial survey can be performed either by a fixed-wing airplane, helicopter, or drone with LiDAR equipment installed, which includes an inertial navigation system (INS) with gyros for measuring the platform vibrations and a GPS. The result is an innovative tool for studying the ground surface and all objects on it. Possible applications include hydrology, topography, archelogy, measurements of peat dome, extraction of tree height, estimation of **Above-Ground Biomass (AGB)**, etc. (Takahashi and Yonetani 1997; Liesenberg et al. 2013; Engelhart et al. 2012; Hajnsek et al. 2009; Kronseder et al. 2012; Schlund et al. 2016; Schlund and Boehm 2021; Osaki and Tsuji 2016). A LiDAR survey provides data beyond that of a conventional survey with traditional cameras and even field measurements (which are time-demanding and costly).

In summary, digital elevation models can be derived from LiDAR point clouds and used for a wide variety of tasks (Boehm & Frank, 2008). Another derived product is named the canopy height model (CHM) and can be used to extract objects such as buildings and even individual trees (Boehm et al., 2013). This product also allows us to derive vertical height metrics that may serve as input data in mathematical models to retrieve volume (Boehm et al., 2018). On the other hand, the digital terrain model (DTM), which is one of the main products derived from LiDAR, includes the design of infrastructure such as roads and water canals (Boehm et al. 2010).

The accuracy of the surface generated depends on several factors defined during LiDAR data acquisition, such as frequency and point density (i.e., 1–10 m^2). Generally, it lies between 15 cm in elevation (z), which seems to be optimal for agricultural planning, peatland research, firer analysis (Boehm et al. 2011a), river catchment surveys, flood mapping, and mining surveys (Boehm et al. 2011, 2012; Sweda et al. 2012; Osaki and Tsuji 2016). Interestingly, eroded areas can be accurately surveyed, even through a thick, dense grass layer.

The motivation of the abovementioned research involving LiDAR surveys was to obtain a better understanding of the tropical peatlands in Central Kalimantan to analyze the topography, tree heights, biomass, different peat dome profiles, carbon budget, and subsidence of peatlands (Silvius 1989; Jaenicke et al. 2008; Boehm et al. 2012; Hooijer et al. 2012; Sweda et al. 2012; Liesenberg et al. 2013) with this technology. Here, we present a review of the applications of helicopter surveys conducted with LiDAR over three tropical peatland areas during two flight campaigns in 2007 and 2011. First, the concept of generating DEMs is provided. Second, the LiDAR system is integrated into a helicopter. Then, LiDAR acquisition and preprocessing steps, including the results of the surveys, are presented. Finally, further perspectives and open research questions are presented as final remarks.

11.2 Generation of Digital Elevation Models

The eye-safe laser in the system has a very narrow beam called small-footprint LiDAR, which means that it can penetrate gaps in vegetation to reach the ground surface through tree gaps or other openings that are typically not visible from traditional stereo imagery. This penetration allows for an accurate determination of ground levels in vegetated areas. The laser also reflects above-ground objects, giving positions and heights of power lines and power poles, high and low vegetation, buildings, bridges, and fences, among others, building up a three-dimensional (3D) picture of the survey area. In addition, multiwave technology can also provide differentiated information about single trees or a forest canopy. We used the first and last laser pulses for tree analysis, but the multiwave information was stored (Boehm et al. 2011).

Due to the instrumentation in the LiDAR system that provides accurate position and attitude information for the laser points and an installed camera, very few ground control points are needed, which means that accurate surveys are possible in inaccessible areas. These include quarries, wetlands, floodplain measurements, etc. (Liesenberg et al. 2013). The LiDAR data can be processed and then made available with a geo-referenced position for visualization in a geographical information system (GIS). The LiDAR data overcome the common digital elevation model (DEM) by refining it into a **digital surface model (DSM)** with buildings and trees and a **DTM** with the topography of the landscape. The canopy height, also known as the normalized DSM (CHM or nDSM), can be obtained as follows: **CHM = DSM − DTM**.

11.3 LiDAR-System Integrated on Helicopters

Using airborne LiDAR, the topography of Central Kalimantan peatlands was measured to obtain both the DSM and DTM (Figs. 11.1, 11.2, 11.3, and 11.4). The flight was approximately 500 m above ground, and the footprint was 500 m.

With this information, hydrology models of peatland and the retrieval of biomass of Peat Swamp Forest (PSF) can be analyzed (Liesenberg et al. 2013). The data resolution depends on the type of LiDAR scanner and the flight height of the aircraft (helicopter, fixed-wing airplane, or drone).

Sometimes such a LiDAR-sensor system can also be combined with a high-resolution ortho-photo RGB camera and a hyperspectral scanner. For example, we used a high-resolution Hasselblad camera with a Riegl Laser-Scanner (RIEGL Laser Measurement Systems GmbH, Horn, Austria).

Figure 11.1 shows the principle of a laser scanner to archive good information from peatlands. The processing with filtering of the achieved data is very important to obtain a correct high-resolution 3D-peatland-GIS. Currently, many companies offer laser scanner acquisition services under variable prices according to survey

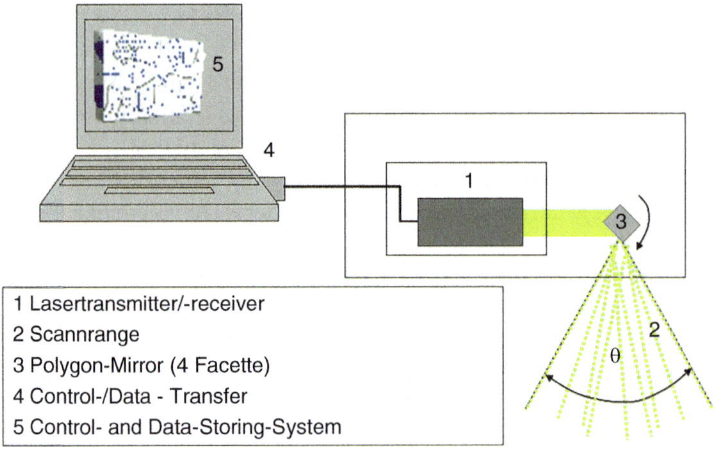

Fig. 11.1 Principle of a Laser Scanner—LiDAR System

Fig. 11.2 A LiDAR system integrated under the bottom of a Bell helicopter, 2007

characteristics. The results described here were possible using a helicopter and adapting the LiDAR system using an attached platform in the landing skids (Fig. 11.2) or using the platform on the back side of the engine (Fig. 11.3).

Additionally, an internal navigation and flight management system are necessary for a LiDAR scanner system. A LiDAR scanner system can be combined with a GPS antenna L1/L2 band, a DGPS ground station, and several recorders to obtain high-resolution accurate time and location information.

Fig. 11.3 A LiDAR system integrated in the back on a stabilized plate of a BK117 helicopter, 2011

11.4 LiDAR Acquisition and Preprocessing

The LiDAR transects in Kalimantan (Indonesia) were surveyed twice by Kalteng Consultants and Milan Geoservice GmbH (Fig. 11.4) from August 5–7, 2007, and August 4–5, 2011, using a Riegl LMS-Q560 Laser Scanner System. It was attached to a Bell 206 helicopter for the first flight trial and a BK117 helicopter for the second flight trial. The nominal height was on average 530 m for both surveys. A differential global positioning system (DGPS) reference station was mounted at the airport of Palangkaraya City, considering an elevation of 25.0 m or 82 ft. The position and orientation of the LiDAR system on the helicopter were measured by an inertial navigation system (INS) and a differential GPS located on the tail boom, respectively, on the cockpit roof at 256 Hz. A complete description of the LiDAR survey can be found in Boehm et al. (2010, 2011a, b). The technical details of the LiDAR systems are further detailed in Table 11.1.

Ground backscattering passing through dense PSF amounted from 1% to 3% of the total laser beams. The processed laser beams were classified into ground surface and aboveground classes using a terrain-adaptive bare earth algorithm for both dates. The algorithm is integrated with Cloud Peak software (LASEdit) and IDL software used by Milan, provides an unsupervised classification of the cloud points, and adapts it to a hypothetical bare earth condition. First, the triangular irregular network (TIN) was constructed based on a Delaunay triangulation. Then, a square grid of pixels was extracted for each TIN using linear interpolation for both ground and aboveground layers on both acquisition dates. Next, the classified laser representing the ground surface was converted into the DTM and DSM, both with a spatial resolution of 1 m. The difference between the DSM and DTM provided us with the CHM.

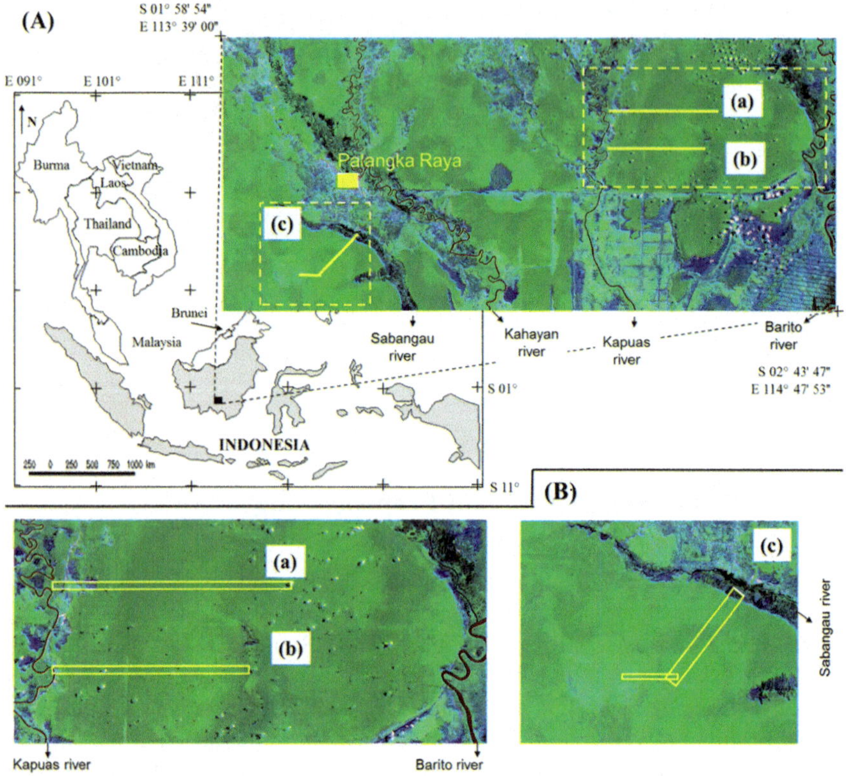

Fig. 11.4 Location of the study area in Central Kalimantan, Indonesia (A) and location of the three selected test sites in more detail (B). The yellow rectangle shows the location of Palangkaraya City. The solid yellow lines indicate the airborne LiDAR transects in **A**. The lower part of the image in **B** indicates the following: Mawas km228 south of the equator (a), Mawas km238 (b), and Sebangau (c). Each test site is shown in more detail in the Landsat-7/ETM+ images acquired on August 5, 2007 (3R4G5B)

Table 11.1 Specifications of the airborne LMS-Q560 LiDAR (Riegl) system and its data products

Parameter	Interval
Scan angle (field of view)	±30°
Swath width (m)	~500 m
Scan frequency (kHz)	66–100
Vertical laser beam accuracy (m)	≤0.15 m
Horizontal laser beam accuracy (m)	≤0.5 m (both *x*- and *y*-directions)
Laser beam (mrad)	0.5 (footprint up to 30 cm)
Laser wavelength (μm)	1.55 (near-infrared)
Point density (points/m^2)	1.4* and 3.5**
Ground resolution (pixel size)	0.5 m for both the DTM and DSM

Note: for the flight measurements taken in *2007 and **2011

11.5 LiDAR Surveys in Central Kalimantan

The three selected test sites on Sebangau National Park and in the Mawas area are relatively flat. According to Shepherd et al. (1997) and Page et al. (1999), the peat thickness of the Sebangau test site varies from 0 to 12 m. Estimates conducted by Jaenicke et al. (2008) using GIS techniques showed that the peat thickness varies from 0 to 18 m at Mawas sites. The climate of the entire study area, following the Köppen climate nomenclature, is humid tropical rainforest (Af).

Figure 11.5 shows the LiDAR transect in more detail with the raw count lines at meter intervals showing the typical flat characteristics of the PSF landscape. The resulting height profiles are shown in Fig. 11.6.

Fig. 11.5 LiDAR-DTM of Sebangau Nat. Laboratory transect with 1 m contour lines; transect begins at the river with 15 m altitude, and at the end, it is at a 25 m altitude superimposed on a Landsat image. The LiDAR transect width is 500 m. On the left lower corner are visible burnt scars from 1997. The violet in the image is the Sebangau catchment, black is the river, and green is peat swamp forest, which contains some roads used for logging

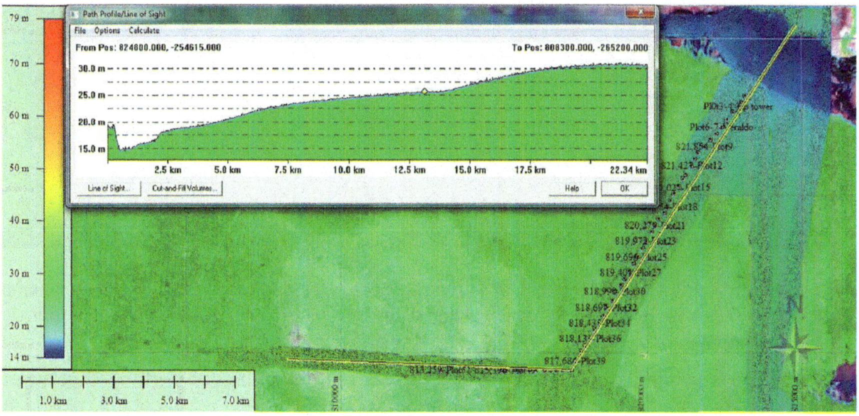

Fig. 11.6 LiDAR-DTM of Sebangau Nat. Laboratory transect with ground PSF plots of 50 m × 50 m measurements made in Aug. 2011. The image shows the peat surface profile

Fig. 11.7 DBH and tree height field measurements (Fig. 11.12) in the Sebangau Cimtrop peat swamp forest transect for LiDAR verification of the tree height collected in August 2011

Fig. 11.8 Ortho-Photo of Sebangau Cimtrop peat swamp forest transect. Left: PSF with a camp; right: Japanese tower for measuring atmospheric data

Field measurements are also performed over certain transects with special attention to forest height and diameter at breast height (DBH; Fig. 11.7) measured close to the Japanese flux tower (Fig. 11.8). Such measurements were collected from several field sample plots along the main transect shown in Fig. 11.9. The resulted

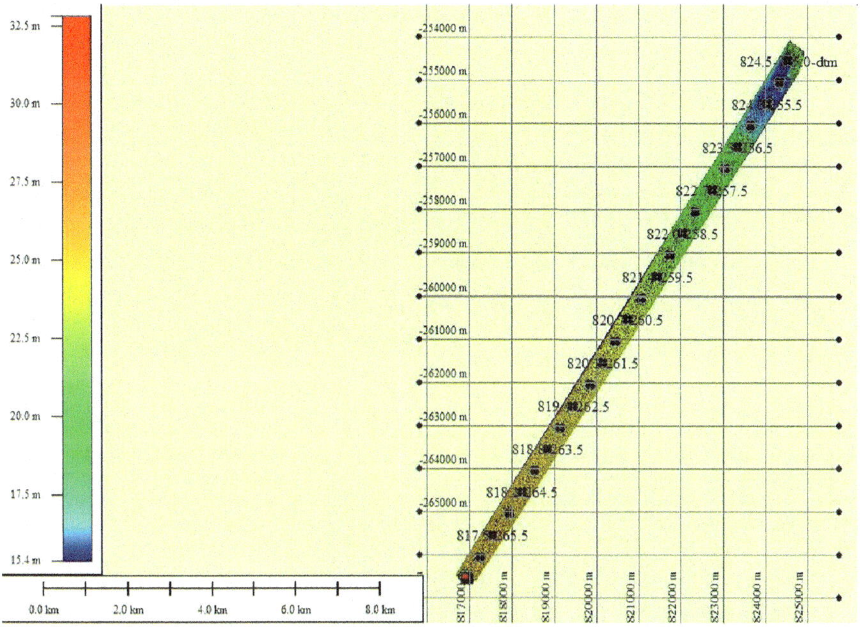

Fig. 11.9 Sebangau LiDAR-DTM Track with peat dome approx. 26 m altitude after 12 km LiDAR-DTM with 25 1-ha sample plots (100 m × 100 m) each extend 500 m to the south; Remark: along the transect, the steps are approx. 600 m long © 100 m × 100 m

measurements were useful to validate height metrics extracted from the LiDAR measurements (Fig. 11.10) and also very useful to estimate the ABG (Boehm et al. 2012) and leaf area index (LAI).

Anor LiDAR application was selected from the EMRP area in the so-called Blocks A + E.

Average tree height without the peat surface. A strong relationship between tree height and peat slope exists. At km 822.3, we have the tallest average trees at **17.1 m**. Much water and good soil nutrition are available here. At km 819.4, we found the lowest average tree height value of **12.6 m**. The tree heights increase as they were closer to the dome, which may be caused by low-intensity logging. No railway transect was found in the last three LiDAR measurements. The steepest peat surface is at km 822 with 0.7–0.8 m for a 600 m path length, which is approx. **0.13% max. slope**. The average tree height at 12 km is **14.6 m**. The Avg. + Max. tree height, up to 37.3 m at the slopes of the peat dome, are at km 822.

The ground field measurements allow us to assess the correlation of the DBH to tree height. Figure 11.12 shows the average tree heights for trees at 20 cm and the average tree height of all measured trees, which are represented by black stars and blue diamonds, respectively. In comparison, the LiDAR average data for the 52 plots are shown as line-columns. LiDAR monitoring does not observe the smallest trees; therefore, for comparison reasons, we used this approach only with trees at 20 cm in

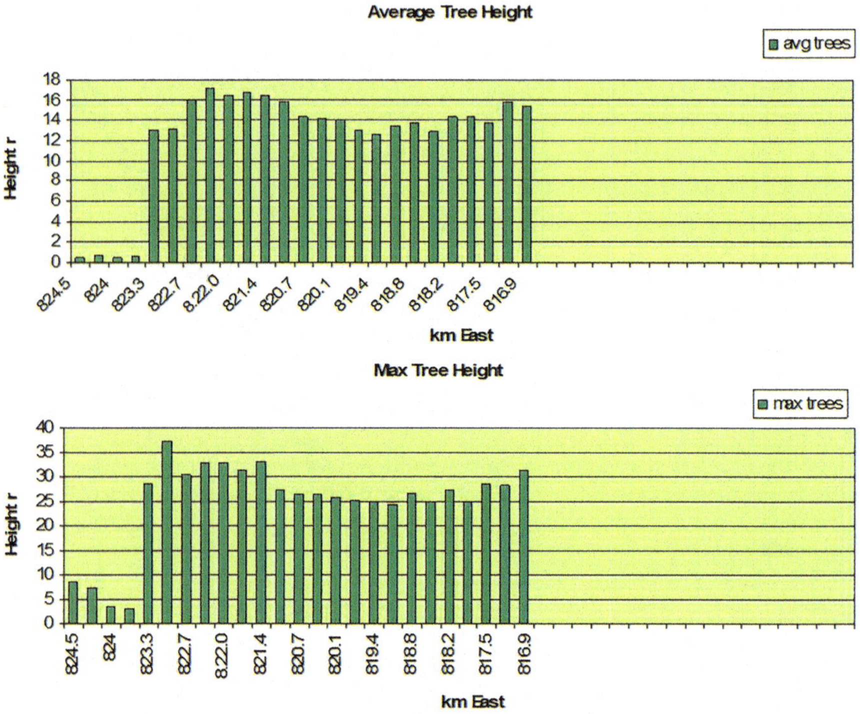

Fig. 11.10 Average and maximum tree height for the Sebangau Cimtrop transect

the collected field data. The respective LiDAR-CHM data are shown in Fig. 11.11 for the 52 plots. We analyzed the maximum, average-dominant (using a filtering method with a window of 10 × 10), and average tree heights for the 52 plots, including the burn scar from 1997 at the end of the transect.

There is a variation in tree heights between the different PSF types defined by (Shepherd et al. 1997): riverine forest (relatively small trees at the beginning), mixed pole forest (tall forest near the highest peat slope), low pole forest (in the middle), tall pole forest (near the end), and the burn scar from 1997 (at the end) (Fig. 11.12).

Figure 11.13 shows the aboveground biomass (AGB) methodology to combine the measurements from the ground survey with those of the 2011 LiDAR survey. The CHM was calculated by subtracting the DTM from the DSM, and Gaussian filtering was applied to smooth the CHM with less noise and more realistic trees. We used a local maxima method to find the treetops (Hyyppä et al. 2001) and searched in a window of 5 by 5 or 3 by 3 pixels for the highest point and then determined the DBH. We found little information about allometric equations suitable for PSF in the literature. Chave et al. (2005) reported a correlation between AGB and DBH for humid tropical forests.

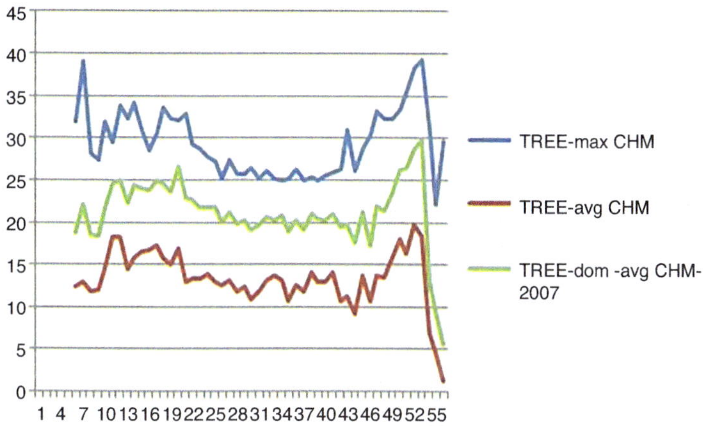

Fig. 11.11 LiDAR measurements of PSF max, avg-dominant, and avg tree height (CHM) in Sebangau transect 2007 with 52 plots up to the burnt scar, showing the tree heights in *y*-axis and the samples in *x*-axis

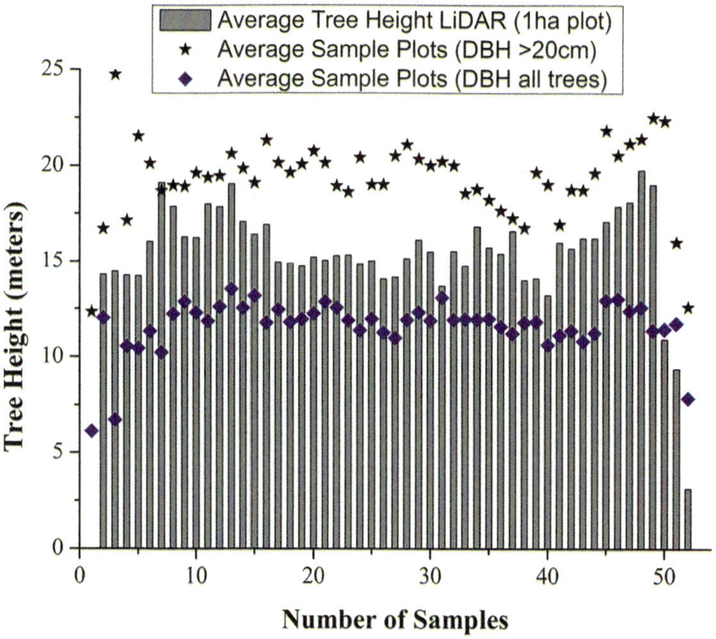

Fig. 11.12 Research area: Sebangau CIMTROP transect with 52 1 ha plots; field measurements and LiDAR-CHM

Figure 11.14 shows the bitemporal DTM and CHM for a sample plot located at the burned scar at the Sebangau transect (cf. Figs. 11.4 and 11.15). Forest regrowth at this site depends both on the characteristics of the ecosystem and specifically on

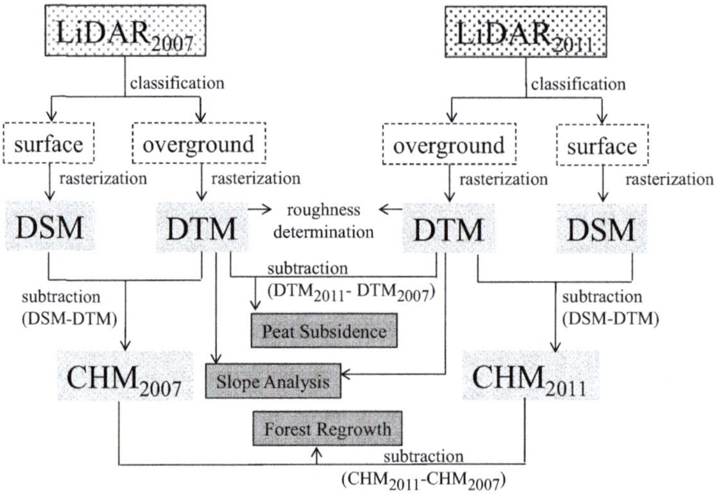

Fig. 11.13 Schematic diagram of the LiDAR data processing flow applied to this study

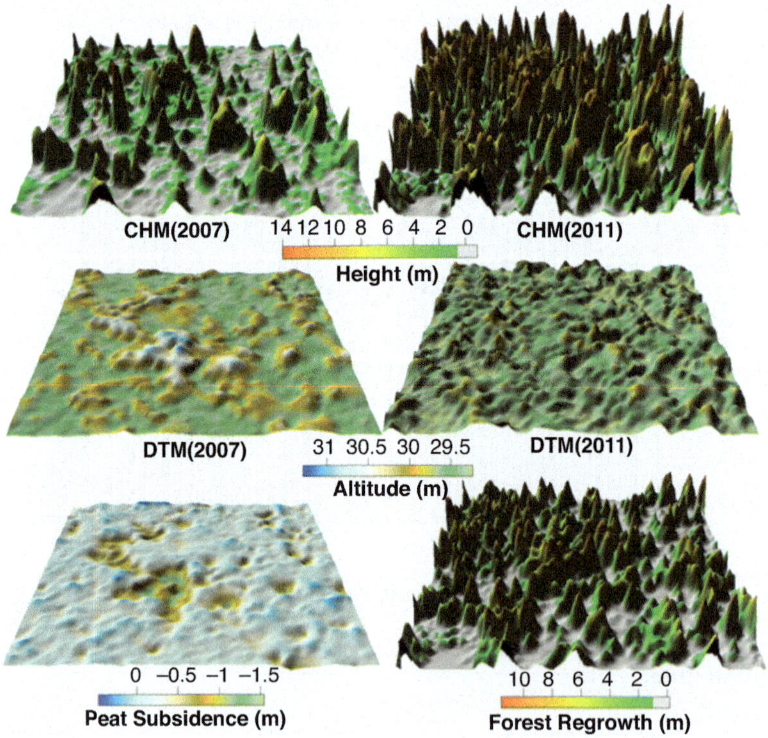

Fig. 11.14 DTM and CHM of a sample plot for 2007 and 2011 located at a burned scar area inside Sebangau National Park. Subsidence 2007 and 2011, left; regrowth 2007 and 2011, right

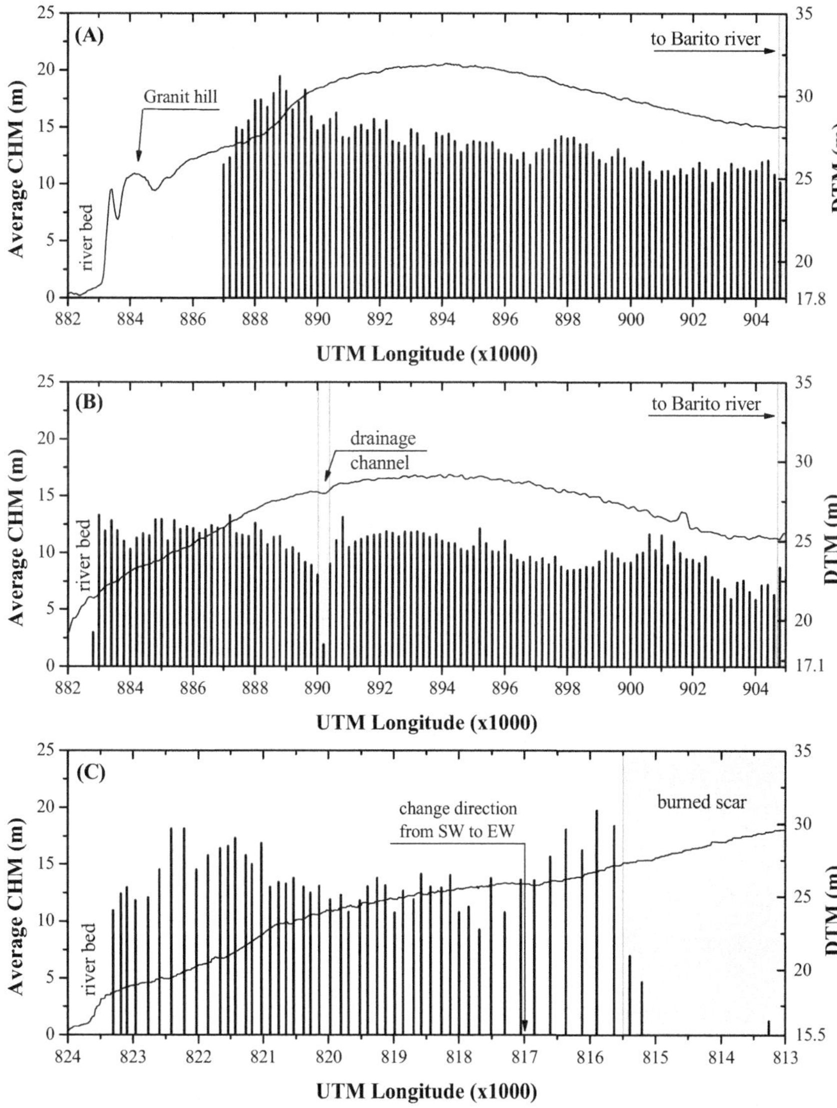

Fig. 11.15 LiDAR-derived digital terrain model (DTM) profile and LiDAR-derived canopy height model (CHM; average tree height) for Mawas km228 (**a**), Mawas km238 (**b**), and Sebangau (**c**). Refer to Fig. 11.4a for the test site description. The results are based on LiDAR measurements acquired on August 5–7, 2007. Each vertical bar is a 1-ha sample plot

the fire intensity and its recurrence (Hoscilo et al. 2011; Liesenberg et al. 2013). Few isolated trees and several small bushes, most likely sedges and fern bogs that are found in such environments, can be interpreted from the CHM in 2007, indicating widespread tree dieback by fires during the El Niño event in 1997 (Page et al. 2002;

Usup et al. 2004; Hoscilo et al. 2011; Boehm et al. 2012; Liesenberg et al. 2013). In the Landsat image acquired on August 20, 2001, this burned site showed high red and low NIR values due to the large proportions of nonphotosynthetic vegetation. Fire recurrences were also reported in Kalimantan in 2002, 2004, 2006 and 2009 (Hirano et al. 2007; Miettinen & Liew, 2010; Hirano et al. 2012). However, they were not evident at the Sebangau test site by analyzing the Landsat images. The red response tends to decrease from August 20, 2001, to August 5, 2007, and then to June 13, 2011, while NIR increases as more photosynthetic vegetation is present (Liesenberg et al. 2010).

Another application was the analysis of the Ex-MRP in Blocks A + E called the Mawas area. We flown several transects over this peat swamp forest, called km228 and km238 (Liesenberg et al. 2013). The results of the measurement are shown below with the averaged tree heights and the peat dome.

The average tree height varied widely over the selected transects (Fig. 11.14). This might be due to different intensities of past log intervention to the forest that may explain part of the heterogeneity in the roughness and the variations in tree height among the selected transects (Fig. 11.14). While Mawas km228 faced less intervention by forest logging, Sebangau and Mawas km238 experienced strong intervention by selective logging through different concession companies until 1997 (Boehm and Siegert 2004). In such forest logging practices, large trees are harvested, which directly implies large gaps in the canopy in addition to damage to neighboring trees that further causes a drastic loss in the amount of organic matter inputs. Thus, harvesting practices also include the construction of small railways and channels to transport logs out of the forest. These factors lead to significant changes in peat floor heterogeneity and ecohydrological function. The forest disturbances are still visible on the LiDAR-derived products (i.e., CHM and DTM) as well as based on visual interpretation (Fig. 11.4) and the spectral analysis of the Landsat scenes. Figure 11.16 shows six peat profiles measured with LiDAR using the DTMs in the Mawas area and in Block A of the Ex-MRP.

11.6 Final Remarks

A LiDAR survey was performed by helicopter over PSF and opened peat land in Central Kalimantan, Indonesia, during two time periods in 2007 and 2011. Parallel ground field measurements were performed in August 2011 to collect PSF parameters such as DBH, tree height, and LAI values. We evaluated carbon changes in this tropical lowland ombrogenous PSF: an average forest regrowth of 1.9 m was measured over an interval of 4 years. Additionally, we estimated the AGB for 2011 of the study area Sebangau CIMTROP transect. We found an AGB value of over 300 Mg ha in a forest area with the highest peat slope of 1.7 m km. The peat profile with the peat dome between two rivers can be measured by a LiDAR system very easily. We found different shapes of peat surfaces and domes: symmetrical shapes, asymmetrical forms, and double peat domes.

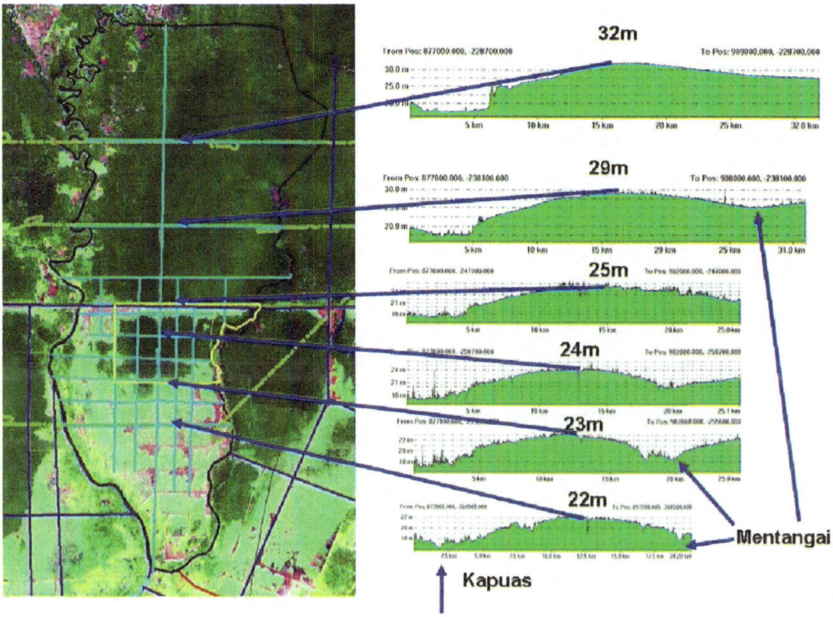

Fig. 11.16 Peat Dome analysis with LiDAR-DTM 2007 in Block E-Mawas + Block A of Ex-MRP (Mega Rice Project) between rivers Kapuas (left) and Mentangai (right) superimposed on a Landsat image from July 2000. In this case, the difference from the river level at 15 m up to the peat dome at 32 m is 12 m. Remark: The reference point with the DGPS is located at the Palangkaraya airport at 25.0 m altitude

With its drainage and subsidence, the opened peat land in this part of Central Kalimantan is now becoming a carbon source, which is supported by fires during the dry period that occur approximately every 4–5 years.

A LiDAR system is a very good tool for REDD verification, showing this article's few examples of surveys in Central Kalimantan. The development of such a sensor package has been very quick and offers a cheaper budget for a drone flight over a small area. Depending on the peat size being monitored and the survey location, a good system can be tailored.

References

Asner GP, Mascaro J (2014) Mapping tropical forest carbon: calibrating plot estimates to a simple LiDAR metric. Remote Sens Environ 140:614–624. https://doi.org/10.1016/j.rse.2013.09.023

Asner GP, Mascaro J, Muller-Landau HC, Vieilledent G, Vaudry R, Rasamoelina M, Hall J, van Breugel M (2012) A universal airborne LiDAR approach for tropical forest carbon mapping. Oecologia 168(4):1147–1160

Boehm H-DV, Frank J (2008) Technical Report, Airborne laser scanning measurements for CKPP to achieve high-resolution digital elevation models of tropical peatlands, PSF, in EX-MRP of Central Kalimantan. https://doi.org/10.13140/RG.2.1.2079.1206

Boehm H-DV, Siegert F (2004) The impact of logging on land use change in Central Kalimantan, Indonesia. Int Peat J 12:3–10

Boehm H-DV, Liesenberg V, Frank J (2010) Relating tree height variations to peat dome slope in Central Kalimantan, Indonesia using small-footprint airborne LiDAR data. In: Proc. 10th Int. Conf. LiDAR applications for assessing forest ecosystems, SilviLaser-Freiburg, pp 216–228

Boehm H-DV, Liesenberg V, Frank J, Limin S (2011a) Multi-temporal Helicopter LiDAR- and RGB-Survey in August 2007 and 2011 over Central Kalimantan's Peatland. In: Proceedings of 3rd International workshop on wild fire and carbon management in peat-forest in Indonesia, 22–24 September 2011, Palangkaraya, Indonesia

Boehm H-DV, Liesenberg V, Frank J, Limin S (2011b) Characterizing peat swamp forest environments with airborne LiDAR data in Central Kalimantan, Indonesia. In: Proc. 11th Int. Conf. LiDAR applications for assessing Forest ecosystems, SilviLaser-Hobart, pp 1–11

Boehm H-DV, Liesenberg V, Sweda T, Tsuzuki H, Limin S (2012) Multi-temporal airborne LiDAR-survey in 2007 and 2011 over tropical peat swamp forest environments in Central Kalimantan, Indonesia. In: Proceedings of IPC, June 3–8, Stockholm, Sweden

Boehm H-DV, Liesenberg V, Limin S (2013) Multi-temporal airborne LiDAR-survey and field measurements of tropical peat swamp forest to monitor changes. IEEE J Sel Top Appl Earth Observ Remote Sens 6(3):1524–1530

Boehm H-DV, Schlund M, Liesenberg V, Kuntz S (2018) Biomass situation of Mawas Region in Central Kalimantan between 2007 and 2015 using LiDAR- and TerraSAR-X data. In: Proceedings of IPC 2018, Aug 3–8, Kuching, Sarawak, Malaysia

Chave J, Andalo C, Brown S, Cairns MA, Chambers JQ, Eamus D, Folster H, Fromard F, Higuchi N, Kira T, Lescure JP, Nelson BW, Ogawa H, Pulg H, Riera B, Yamakura T (2005) Tree algometry and improved estimation of carbon stocks and balance in tropical forests. Oecologica 145:87–99

Engelhart S, Keuck V, Siegert F (2012) Modeling aboveground biomass in tropical forests using multi-frequency SAR data—a comparison on methods. IEEE J Sel Top Appl Earth Observ Remote Sens 5:298–306

Hajnsek I, Kugler F, Lee SK, Papathanassiou KP (2009) Tropical-forest-parameter estimation by means of Pol-InSAR: the INDREX-II campaign. IEEE Trans Geosci Remote Sens 47(2): 481–493

Hirano T, Segah H, Harada T, Limin S, June T, Hirata R, Osaki M (2007) Carbon dioxide balance of a tropical peat swamp forest in Kalimantan, Indonesia. Glob Change Biol 13(2):412–425

Hirano T, Segah H, Kusin K, Limon S, Takahashi H, Osaki M (2012) Effects of disturbances on the carbon balance of tropical peat swamp forests. Glob Chang Biol 18:3410–3422. https://doi.org/10.1111/j.1365-2486.2012.02793.x

Hooijer A, Page S, Jauhiainen J, Lee WA, Lu XX, Idris A, Anshari G (2012) Subsidence and carbon loss in drained tropical peatlands. Biogeosciences 9(3):1053–1071

Hoscilo AC, Page SE, Tansey KJ, Rieley JO (2011) Effect of repeated fires on land-cover change on peatland in southern Central Kalimantan, Indonesia, from 1973 to 2005. Int J Wildland Fire 20(4):578–588

Hyyppä J, Kelle O, Lehikoinen M, Inkinen M (2001) A segmentation-based method to retrieve stem volume estimates from 3-D tree height models produced by laser scanners. IEEE Trans Geosci Remote Sens 39(5):969–975

Jaenicke J, Rieley JO, Mott C, Kimman P, Siegert F (2008) Determination of the amount of carbon stored in Indonesia peatlands. Geoderma 147(3–4):151–158

Kronseder K, Ballhorn U, Boehm V, Siegert F (2012) Above ground biomass across forest types at different degradation levels in Central Kalimantan using LiDAR data. Int J Appl Earth Obs Geoinf 18:37–48

Liesenberg V, Boehm H-DV, Gloaguen R (2010) Spectral variability and discrimination assessment in a tropical peat swamp landscape using CHRIS/PROBA data. GIScience Remote Sens 47(4):541–564

Liesenberg V, Boehm H-DV, Joosten H, Limin S (2013) Spatial and temporal variation of above ground biomass in tropical dome-shaped peatlands measured by Airborne LiDAR. In: Proceeding of International symposium on wild fire and carbon management in peat-forest in Indonesia, 24–26 September 2013, Palangka Raya, Indonesia. ISSN 2338-9532

Miettinen J, Liew SC (2010) Status of peatland degradation and development in Sumatra and Kalimantan. Ambio 39(5–6):394–401

Osaki M, Tsuji N (2016) Tropical peatland ecosystems. Springer, Tokyo, p 651

Page SE, Rieley JO, Shotyk ØW, Weiss D (1999) Interdependence of peat and vegetation in a tropical peat swamp forest. Philos Trans R Soc Lond B Biol Sci 354(1391):1885–1897

Page S, Siegert F, Rieley JO, Boehm H-DV, Jaya A, Limin S (2002) The amount of carbon released from peat and forest fires in Indonesia during 1997. Nature 420(6911):61–65

Schlund M, Boehm H-DV (2021) Assessment of linear relationships between TanDEM-X coherence and canopy height as well as aboveground biomass in tropical forests. Int J Remote Sens 42(9):3409–3429

Schlund M, Poncet FV, Kuntz S, Boehm H-DV, Hoekman D, Schmullius C (2016) TanDEM-X elevation model data for canopy height and aboveground biomass retrieval in a tropical peat swamp forest. Int J Remote Sens 37(21):5021–5044. https://doi.org/10.1080/01431161.2016.1226001

Shepherd PA, Rieley JO, Page SE (1997) The relationship forest vegetation and peat characteristics in the upper catchment of Sungai Sebangau, Central Kalimantan. In: Rieley JO, Page SE (eds) Biodiversity and sustainability of tropical peatlands. Samara Publications, Cardigan, UK, pp 191–210

Silvius MJ (1989) Wetland in Indonesia. In: Scott DA (ed) A directory of Asian wetlands. IUCN the World Conservation

Sweda T, Tsuzuki H, Maeda Y, Boehm H-DV, Suwido L (2012) Above- and below-ground C budget of degraded tropical peatland revealed by multi-temporal airborne laser altimetry. In: Proceedings of IPC, June 3–8, Stockholm, Sweden

Takahashi H, Yonetani Y (1997) Studies on microclimate and hydrology of peat swamp forest in Central Kalimantan, Indonesia. In: Rieley JO, Page SE (eds) Biodiversity and sustainability of tropical peatlands. Samara Publications, Cardigan, UK, pp 179–187

Usup A, Hashimoto Y, Takahashi H, Hayasaka H (2004) Combustion and thermal characteristics of peat fire in tropical peatland in Central Kalimantan, Indonesia. Tropics 14:1–19

Chapter 12
Topography Monitoring by Unmanned Aerial Systems (UASs)

Kotaro Iizuka, Kato Tsuyoshi, Niken Andika Putri, Sisva Silsigia, and Asep Andi Yusup

Abstract The integration of unmanned aerial systems (UASs) has made great strides and impacts in the field of remote sensing in recent years, providing unprecedented capabilities for scientific discovery and strategic planning. The flexibility of UASs enables researchers to perform in-depth analysis of local issues and obtain broader perspectives by utilizing various onboard sensors. In this chapter, we explore the application of UASs in the peat environment of West Kalimantan, Indonesia, where acacia plantation practices are widespread. We begin with a comprehensive introduction to UASs and their scientific applications, with a focus on photogrammetry. We then delve into two major research topics: (1) the fraction of vegetation cover (FVC), which is determined using a multispectral camera and provides insight into the land condition, and (2) the thermal characteristics of the peat soil, which are observed using a thermal camera onboard UASs. The thermal trends provide valuable information and predictions on acacia tree growth and growth patterns. Overall, this chapter highlights the immense potential of UASs in scientific research and the critical role of UASs in environmental monitoring and management.

Keywords Remote sensing · Geographical information system (GIS) · Unmanned aerial vehicle (UAV) · Forest cover · Thermal infrared

---

K. Iizuka (✉)
Center for Spatial Information Science, The University of Tokyo, Kashiwa-city, Chiba, Japan
e-mail: kiizuka@csis.u-tokyo.ac.jp

K. Tsuyoshi · N. A. Putri
Sumitomo Forestry Co., LTD., Tokyo, Japan

S. Silsigia · A. A. Yusup
PT Wana Subur Lestari, Pontianak, Indonesia

M. Osaki et al. (eds.), *Tropical Peatland Eco-evaluation*,
https://doi.org/10.1007/978-981-99-6790-2_12

12.1 Introduction

In ancient Greek mythology, the famous story of Icarus describes his use of a bird's feathers and wax to create wings to escape from the labyrinth into the sky. The warning from his father not to fly too high was ignored by Icarus, resulting in the wings' wax melting from the sun's heat, leading to a tragedy where he met his end. Historically, many people have craved the sky, challenges have been considered, and experiments have unfolded. Hot air balloons were developed, the first aircraft flight was accomplished by the Wright brothers, technology advanced to launch multiple satellites into space, and today, individuals can own a drone. It is such an exciting era.

Remote sensing technology and its application have been indispensable as a sensing tool for diagnosing the global environment. Remote sensing, in general terms, refers to technologies that can observe objects, areas, or phenomena without directly contacting them. These can be radiosondes for observing meteorological data, aircraft for broad-scale observation and collection of images of land surfaces, satellites such as optical and radar sensors for broader Earth observations, and local-scale but flexible operations using drones or, to be more specific, unmanned aerial systems (UASs). Depending on the target, resolution, area, temporal scale, etc., each will consider the platform and data for their work. Conventionally, remote sensing had common sense reflected in satellite remote sensing, where many users and researchers used its platform along with the development of the geographical information system (GIS), a platform that allows handling and analyzing geographic information (such as maps). With the emerging trend of the UASs being introduced at an individual consumer level, the platform is now popular among researchers, who use it to operate and implement their plan for data acquisition.

According to Japanese aviation law, UAVs (unmanned aerial vehicles) are defined as "airplanes, rotorcraft, gliders, and airships that are structurally inaccessible to passengers and can be flown by remote control or autopilot, and have a weight over 100 g." The term UAS refers to everything that composes the whole "system," including the vehicle, inertial measurement unit (IMU), digital compass, global navigation satellite system (GNSS), and multiple payload sensors (e.g., optical camera, multispectral, LiDAR); additionally, the operating system, ground stations, networking, etc., are all aspects that construct the system. Therefore, when using UAS terms, UAVs (all drones, or UAVs) are a component of the UAS. Hence, hereafter, we will use the term UAS when comprehensively describing this system.

The advantages of utilizing UASs can be separated into three general categories: (1) the high-definition data that can be collected via ultrafine-resolution sensors and videos at a low altitude, (2) the flexibility of the UAS that can be operated at any geographical location and at any time, even in cloudy or rainy conditions, and (3) the capacity to replace sensors with different types other than optical cameras (Fig. 12.1).

UASs have been developing in a way that can collect not only optical images but also multispectral information of the electromagnetic spectra, laser scanning, and

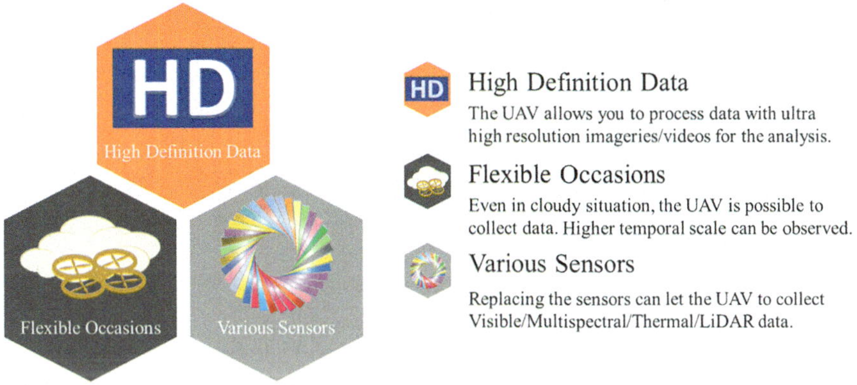

Fig. 12.1 Advantages of UAS as a tool

more. Recently, UASs have been shown to be able to fly even in rainy conditions, where the IP rating (the standard used to classify a product's resistance against dust and water) of the vehicle is IP45 for the new DJI Matrice 30 series (DJI, Shenzhen, China), and that of Yamaha YMR-II (Yamaha, Shizuoka, Japan) is IP55. The single positioning method of the UAV embedded with the GNSS was conventionally untrustworthy due to its low geolocation accuracy (especially for the *Z*-axis). However, currently, it is no longer rare to find a vehicle with a system of real-time kinematics (RTK) in improving the relative or the absolute positioning (improves to the cm scale).

12.2 Photogrammetry Technology

Structure-from-motion (SfM) and photogrammetric approaches are used to calculate the position of the camera shoot and reconstruct the feature points of the scene. Then, multi-view stereo (MVS) is processed to reconstruct a dense point cloud of the target to develop a complete point cloud of the structure based on the selected stereo pairs.

Photogrammetry is the science of making measurements from photographs, and it is especially used for surveying and mapping purposes. This method has a long history in various fields, such as architecture, engineering, archaeology, and geomorphology. Especially in geomorphology and geography, the technique is essential in obtaining terrain information and the contours of the reliefs. The method works by using two images of the same location or objects that are slightly shifted. The slightly different geometry in the two images, looking at one with the left eye and the other with the right eye, will form a stereoscopic vision. Typically, we use equipment such as a mirror stereoscope, as shown in Fig. 12.2, and from the stereo view, contours can be drawn. This principle is now extended to UASs to enable terrestrial photogrammetry with the use of the SfM-MVS technique.

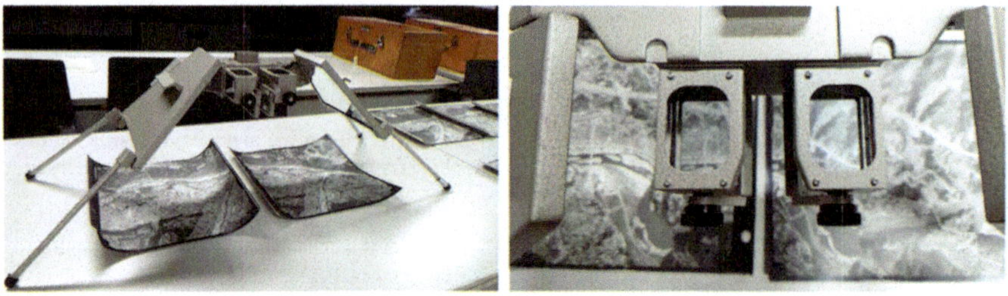

Fig. 12.2 The mirror stereoscope equipment for photogrammetry

SfM is the process of estimating the 3D structure of an object/scene from a set of more than two images (can be more like 100 s or even 1000 s). The computer will configure and align each image by finding a pair that corresponds to a similar geometry, matching the features in multiple images. This step produces a sparse point cloud of x, y, and z coordinates. At this point, we can start to identify the structure of the scene, but the next process is generally progressed for dense reconstruction, called MVS. After creating the dense 3D scene of the area, it will be possible to create either a 3D mesh or a digital surface model (DSM) of the area in a 2D map. The final product that we desire is an orthoimage (Fig. 12.3). This orthoimage consists of two different processes for generation. One is the orthorectification process, which geometrically corrects the distortion of the imagery, where the objects in the image are distorted (inclined away from the center) due to camera tilt, distance, and lens distortions. The corrected data provide an orthographic view, where the whole scene is fixed as if it is all viewed from a nadir angle. Finally, each of the images is stitched together to create one large image of the scene to be used, such as a map. This process is called mosaicking. Hence, the name of the final product is an orthoimage or an orthomosaic. Good flight designs and photo shoots are necessary to collect enough information to allow the computer to recreate the scene. This can depend on the resolution of the images, the size of the objects, the types of the objects (artificial objects, natural objects), and even how the data are collected (overlapping percentages between images, at nadir or oblique images, etc.), which is one challenge for UAS observations.

12.2.1 Photogrammetry to Estimate Tree Heights and Peatland Topography

The sustainable management of Indonesian peatlands has been a challenge. Large-scale land clearing for plantations and illegal logging activities are both major causes of peatland fires (Cattau et al. 2016; Sumarga 2017) that produce smoke haze pollution and carbon emissions in Indonesia (Stockwell et al. 2016). Government, nongovernment, and private sectors have undertaken high-level efforts to reduce peatland deforestation and fires and restore degraded peatlands. In large countries

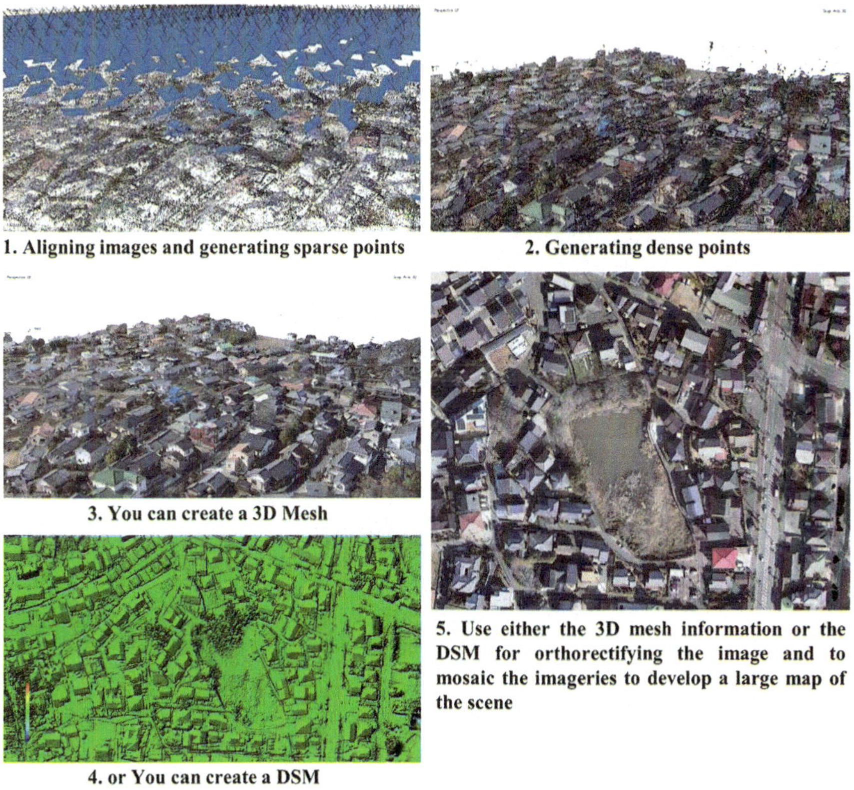

Fig. 12.3 Overall process and samples of each step in the SfM-MVS method and orthomosaicking

such as Indonesia, monitoring the natural environment is a hard-working task since not every location can be monitored and secured sustainably. For continuous broad-scale observations, conventional approaches utilize satellites to diagnose the condition of the Earth's surface (e.g., land, forests, ocean). However, the tropics have frequent cloud cover that limits the viewing of the Earth's surfaces, blocking observations. As mentioned, UAS technologies can overcome these situations, and this section explains what kind of information we can obtain when monitoring the peat area.

The experimental site is located in West Kalimantan, and the site is managed by an industrial plantation company. The area was planted in January 2017, with *Acacia crassicarpa* as the main commercial species. The objective here is to implement UAS observation of the plantation area to see if it can successfully generate the site map, extract the tree structure, and monitor the thermal condition of the peat soil. Each tree could be effectively interpreted through aerial photos, and these photos provided a clear view of the bare soil; therefore, the area was appropriate for this study. The experimental area was 34 hectares, and we focused on a smaller area at the site. The observed area was approximately 400 m from west to east and 200 m

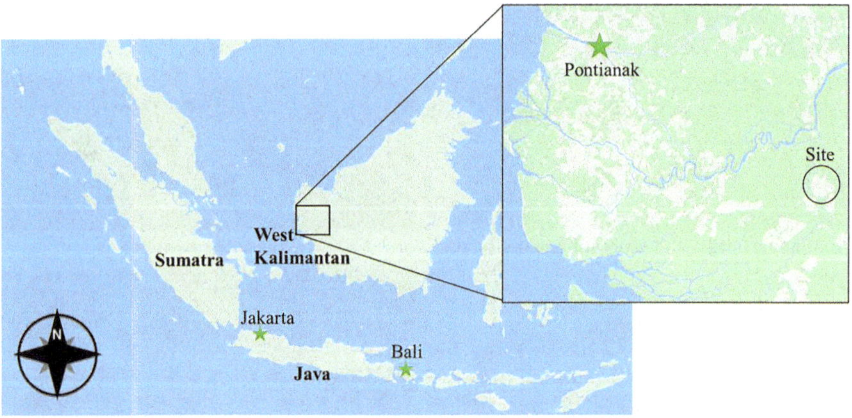

Fig. 12.4 Location of the plantation site located at West Kalimantan. (From Iizuka et al. 2018)

Fig. 12.5 UAS Mavic Pro (left) and Inspire 1 Pro (right)

from north to south. The annual precipitation in the area is approximately 3140 mm, and the air temperature can range from 23.2 to 36.6 °C depending on the season. The elevation of the study area is approximately 8 m above sea level. Figure 12.4 shows the site location (Iizuka et al. 2018).

The UAS observation was implemented using a DJI Mavic Pro multicopter (DJI, Shenzhen, China) for collecting optical data (Fig. 12.5). This type of UAS is very handy when operating in extreme locations due to the small and lightweight product, so it can easily operate in any location. One issue that needs to be considered is that these inexpensive types of UASs have slight limitations in camera sensors and their shutter types. The Mavic Pro has a CMOS sensor size (1/2.3 in.) and a rolling shutter system. This shutter system generates deformation in the captured imagery depending on the shooting design. Compared to types such as Phantom 4 Pro with a 1-in. sensor size and a global shutter mechanism, the quality of the images declines. Nevertheless, it can still operate, and the first challenge here was to develop a method to recreate the 3D model of the area successfully. For thermal data, DJI Inspire 1 Pro (DJI, Shenzhen, China) was used with the Zenmuse-XT (FLIR Systems, Wilsonville, USA) thermal camera attached and operated to collect for the test site (Fig. 12.5). To ensure accurate data collection, careful consideration should be given to selecting an appropriate thermal camera. For example, while a typical thermal camera can provide relative temperature differences within an observed scene, certain products, such as Zenmuse-XT or XT2, can record absolute temperature data for each pixel. This feature is especially important when absolute temperature data are required for analysis.

12.2.2 Photogrammetry to Estimate Tree Growth Retrieval

The SfM process, which generates point clouds, has a wide range of applications beyond the creation of DSMs for orthorectification. Further processing of the point cloud can provide valuable information, such as the heights of objects within the surveyed area. How does it work? The original point clouds generated from images create the top surface layer of the objects and the scene. This means that if the object is a house, you will obtain the point cloud information of the roofs. If it is trees, then you obtain the canopy, and if it is the bare surface, you will obtain the terrain or the area's topography. The digital information reflecting the top surface layer of all these objects and areas is named the DSM. In addition to DSM, there is a digital terrain model (DTM). This is where the digital information reflects all the terrain information and excludes the natural or built features on top of such a surface. Therefore, if we have these two pieces of information, the DSM and DTM, simply taking the difference between those heights can generate additional information about the feature's height relative to the ground. This is named the canopy height model (CHM). Figure 12.6 illustrates the definition of this digital information.

By generating the CHM, we can obtain tree height information, which is an essential benchmark for understanding tree growth. There are two ways of doing this. One manually selects the lowest points within a specific ground range to be categorized as ground points. Manual selection can take time; therefore, there are alternatives, such as the second option, which is the automated procedure for ground point classification. The two main algorithms may be preferred for automation. One is the improved progressive TIN (triangulated irregular network) densification filtering algorithm by Zhao et al. (2016), and the second is the cloth simulation filter by Zhang et al. (2016). We will not go into the details of these algorithms, but they are the two algorithms typically used for obtaining the ground points. If the point cloud is obtained by light detection and ranging (LiDAR) technology, it is robust in extracting the CHM. However, the optical data have challenges in delineating the true ground points for generating a suitable DTM. For example, LiDAR can penetrate through the canopy so that the underlying ground surface can be observed. However, the optical camera cannot, therefore, estimate the ground surface from the surrounding visible ground layers. This is usually a trial-and-error method of the extraction process.

Figure 12.7 shows the actual DSM, DTM, and CHM for one of the compartments in the plantation site. The elevation becomes flat, and only the tree height relative to

Fig. 12.6 Digital surface model (DSM), digital terrain model (DTM), and canopy height model (CHM) illustration

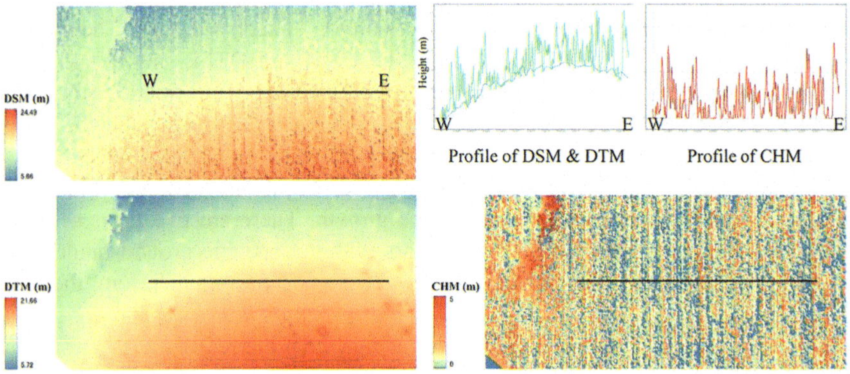

Fig. 12.7 Results of the different digital models (DSM, DTM, and CHM) and their profiles along the scene

the ground is computed. Unfortunately, there are some errors in these data, e.g., on the west side of the site, a blurred area is found, which is caused by the errors in the reconstruction of the 3D model. This can occur from misses that occurred during the photo shoot, where the lens fails to focus correctly, producing a blurred image. It is best to fix the camera lens to infinite focus to avoid such issues. Another problem is the failed reconstruction of geometry caused by lens distortions (i.e., the doming effect). We will not go into detail here; however, related works can be found in Iizuka et al. (2022).

12.3 Thermal Image Sensing

The Inspire 1 Pro and Zenmuse-XT were used to collect thermal images of the same location. The great thing about utilizing UASs, as described, is the flexibility of the flight designs. Since the flight can be implemented at any time of the day, we have been challenged in observing the temporal characteristics of the thermal conditions of the compartment, especially by making multiple observations within the same day. The purpose of this was to identify whether moisture conditions on the surface could be detected through multitemporal thermal observations. The idea behind this was based on two perspectives: one is about the water table affecting peat oxidation. Under dry conditions, it is said that the microbial oxidation rate increases, and it is necessary to keep the water table at a certain level for suppression. We wanted to know if something was observable relating to this aspect. Another is that the growth conditions of acacia trees are affected by the wetness of the soil. Acacia is vulnerable to flooding conditions obstructing the growth of trees. Therefore, we wanted to find if such a trend could be found from the CHM and the thermal data. The thermal images available for UASs have a very low resolution compared to optical cameras. Additionally, processing thermal data through SfM is difficult due to the lack of features that can be extracted; therefore, it is important to ensure high overlapping of the images (recommend >90%).

12.3.1 *Thermal Mosaic Multitemporal Images*

We operated the UAS a total of four times: three times on the first day, from 07:00, 14:00, and 19:00, and 07:00 on the following day, local time. Figure 12.8 shows the successful reconstruction of the thermal mosaics, showing a heterogeneous distribution of the thermal trends through space and time.

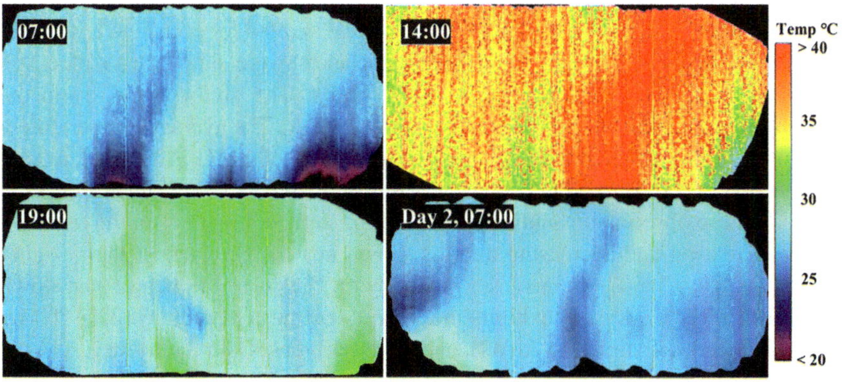

Fig. 12.8 Thermal mosaic multitemporal images at different times of the day

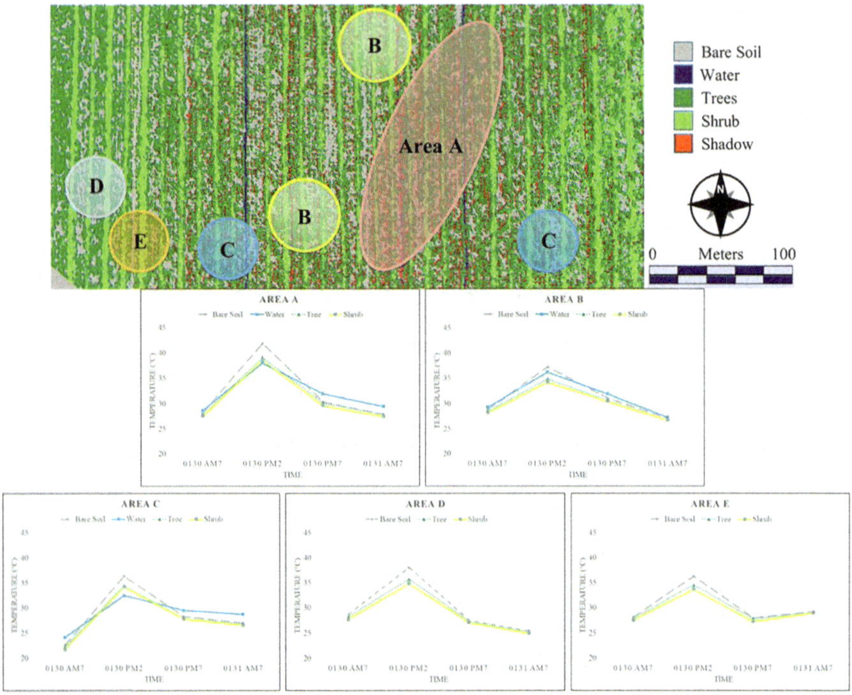

Fig. 12.9 Areas A–E were extracted based on the visual interpretation of areas from the PCA components that displayed different among space. The bottom graph shows the temporal characteristics of temperature trends in each area for different land cover types. (Modified from the original figure of Iizuka et al. (2018))

We further analyzed the four images with principal component analysis (PCA) to compress the images and output the general patterns of the thermal trend through space. From the components generated, we selected five different areas within the site to see what is shown in the temporal trend of the thermal condition. Figure 12.9 displays a graph illustrating the mean temperature extracted for each land cover class at different times of the day. Four general classes were output: (A) bare soil, (B) water body, (C) acacia tree, and (D) shrubs. The study found distinctive temperature trends for each land cover type in Area A. The average temperature at 07:00 was 28 °C, which gradually increased to its peak of 40 °C. Afterward, as solar radiation declined, the cooling process was initiated, and the temperature dropped to 30.5 °C. These observations indicate high heating and cooling rates in the region. A slowly increasing and decreasing trend is shown in Area B, where it starts slightly warmer at 28.7 °C at 07:00, increases gradually to 35.8 °C at 14:00, and then slowly decreases to 31 °C. Area C exhibits the lowest starting temperature (22.7 °C) but gradually increases, similar to Area B, and then it decreases, similar to Area A. Areas D and E have some different random trends, but they are similar to other areas. Perhaps the biggest difference is observed for Area E, where the temperature increases on the following day at 07:00. By analyzing the thermal behavior of different land cover types, it was observed that the bare soil responded quickly to changes in temperature, whereas the water bodies exhibited slower heating and cooling rates compared to the bare soil. In particular, the water bodies tended to have the warmest temperatures in the morning and at night. This can be attributed to their high capacity to absorb heat energy, which causes radiation absorption and release to take longer, resulting in warmer temperatures in the morning and at night and cooler temperatures at midday.

12.3.2 Tree Growth Estimation by Thermal Sensing

Using the information derived from the CHM, the average tree height for each area was computed. Figure 12.10 shows the CHM image and the boxplot of the tree and shrub heights for Areas A–E. The box is the interquartile range, the solid line represents the median, the bar graph represents the maximum and minimum, and the dots on the top are the outliers. At the top of the boxplot, the results of the Tukey HSD test are provided as letters to indicate the differences between each area. Overall, Area D displayed the highest average tree and shrub heights, and Area B exhibited the lowest heights. Areas C and E had similar tree heights, but there were few high values of the shrub height in Area E. We would like to focus on Area B, where the growth is shown to be the lowest among all areas. The thermal trend of Area B showed low increasing and decreasing temperature trends. This kind of trend is similar to how it is shown in the water class in other areas. As explained, water has a higher capacity for absorbing heat energy. Therefore, it is likely that the area exhibiting such temperature trends is characterized by a higher moisture content (Al-Kayssi et al. 1990). This may be due to a high water table or microtopography that causes the area to be slightly flooded. This method could be used to gain insight into the water content of the area (Lakshmi et al. 2003).

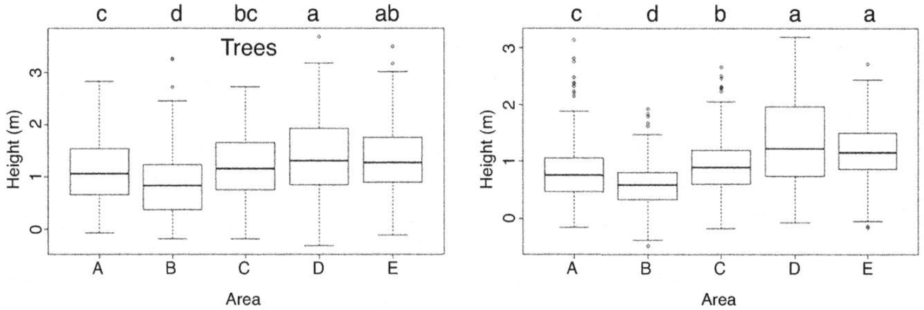

Fig. 12.10 Canopy height model (CHM) of the study site and difference of the height growth among the areas focused on the different thermal trends. The letters on the top of the graph indicate the areas are significantly different from each other (computed by Tukey HSD method ($p < 0.05$)). (Figure from Iizuka et al. 2018)

12.3.3 Thermal Mapping

Thermal mapping has successfully worked through the SfM process, and it is one of the challenges for UAS remote sensing. The broad-scale observation of thermal conditions spatially has large potential and possibilities, and the SfM method can enhance the work to accomplish the mapping of the area. The approach can contribute to various fields, such as urban studies relating to heat islands (Iizuka and Akiyama 2020). Understanding the thermal trends across the site can reveal the daily cycle of thermal trends. In this work, each land cover's thermal characteristics are shown, and the patterns of temporal data indicate how the temperature warms and cools, and the heterogeneity among spaces is revealed. This approach may be used as a method for mitigating the impacts of forest fires by monitoring the temperature of the local sites. Additionally, for carbon emissions, if the peat oxidation process is continuously increasing with the temperature trends shown by Jauhiainen et al. (2014) and Hilasvuori et al. (2013), the majority of the area is warmer than 25 °C, and in the afternoon, it can even peak at 50 °C. If so, then the progress of decomposition accelerates more than expected, and there is a possibility of underestimating the possible emission rates of peat areas. These kinds of discussions could be carried out if new findings can be brought up from UAS remote sensing focusing on the local scale at an ultrafine resolution.

12.4 Fractional Vegetation Cover (FVC)

Quantitative assessments of the green vegetative cover of terrestrial environments are essential for understanding ecosystem dynamics. The functions of green environments (e.g., vegetation and forests) provide important benefits to ecosystems, such as controlling air quality through photosynthesis, generating an energy supply from woody biomass, preventing soil erosion, improving water quality, and balancing the heat fluxes of the Earth (Yao et al. 2012; Iizuka and Tateishi 2015; Masocha et al. 2017; Gomes et al. 2019; Liu et al. 2019). The worldwide terrestrial environment is currently showing rapid changes in particular regions from anthropogenic activities, causing land degradation that engulfs the natural environment (Taubenböck et al. 2012). Some areas show transitions away from green areas, resulting in substantial impacts on local to global ecosystems and sociocultural and economic impacts (Satterthwaite et al. 2010; Iizuka et al. 2017). To mitigate these impacts, known cooperative agencies and organizations are planning actions to recover green areas (Climate Focus 2015; Di Lallo et al. 2017). The consistent monitoring of the rapid environmental changes in green vegetative coverages is important for conservation and maintaining the sustainability of the natural environment (from Iizuka et al. 2019).

12.4.1 Detection of Land Cover Change

In the plantation site in Kalimantan, Indonesia, the rapid growth of acacia trees shows dramatic changes in the land conditions. This is not a 1- or 2-year scale; rather, it is more like a monthly scale (Fig. 12.11). Usually, satellite images fail in estimating green conditions due to the long revisit time and low data availability due to frequent cloud cover. Therefore, operations utilizing UASs from lower altitudes are necessary to immediate update the environmental conditions.

Since the conventional multicopter UAS lacks a spatial extent, the challenge here is to utilize a different type of UAS that can observe broader areas. In such a sense, a fixed-wing UAS is appropriate for our objective. The fixed-wing type of UAS is similar to an airplane but has a smaller-sized platform. Several types exist; however, we utilize the vertical take-off and landing (VTOL) fixed-wing UAS, a hybrid system integrated from both multicopter-type hovering and airplane-type gliding. The system has a longer flight duration and broader observation potential due to the benefit of aerodynamic lift, which increases efficiency during flight. We used the FireFly6 Pro (BirdsEyeView Aerobotics, New Hampshire, USA) VTOL with a multispectral sensor onboard to collect images that can be collected from specific electromagnetic spectra. The SlantRange 3p multispectral sensor (SlantRange, San Diego, CA, USA) can collect green, red, red-edge, and near-infrared bands (Fig. 12.12). The fraction of vegetation cover (FVC) will be analyzed using the data collected from such a UAS.

12.4.2 Data Collection, Processing, and Computing of Vegetation Indices

The fixed-wing UAS flew along the north-south direction of the compartment, collecting multiple images along the area. Forward and side overlap was set to 40%, and the flying altitude was 150 m, corresponding to an approximate 6 cm

Fig. 12.11 Example of temporal differences for fast-growing species. The left image shows the fast-growing Acacia trees in their early stages in August 2018, while the right image shows their rapid growth in October 2018. Even with small temporal differences, the situation of the land area can change dramatically (Iizuka et al. 2019)

Fig. 12.12 Operating the FireFly6 Pro VTOL and the SlantRange 3p multispectral sensor

ground sampling distance (GSD). The images are further calibrated with the software using the information collected from the ambient illumination sensor, which calibrates each image for solar illumination. The images are further processed with the SfM method to create a large area map for each spectral band. Three different vegetation indices (VIs) were computed from the obtained data. The normalized difference vegetation index (NDVI), green NDVI (GNDVI), and red-edge NDVI (ReNDVI) were used, where each index can be computed using the following formulas:

$$\mathrm{NDVI} = (\mathrm{IR} - \mathrm{R})/(\mathrm{IR} + \mathrm{R}) \tag{1}$$

$$\mathrm{GNDVI} = (\mathrm{IR} - \mathrm{G})/(\mathrm{IR} + \mathrm{G}) \tag{2}$$

$$\mathrm{ReNDVI} = (\mathrm{IR} - \mathrm{Re})/(\mathrm{IR} + \mathrm{Re}) \tag{3}$$

where IR is the infrared band, R is the red band, G is the green band, and Re is the red-edge band. The functions are similar among the VIs; however, the slightly different characteristics of the wavelength will show different sensitivities to the status of the land cover. In the case of the most common NDVI, the IR and R are used to characterize the activity (greenness) of the vegetation. This works because the leaves' chlorophyll absorbs the red and reflects the infrared spectra. The alternative use of the green band for the GNDVI shows a wider dynamic range and higher sensitivity to chlorophyll. The ReNDVI uses the red-edge band, which is between the red and near-infrared spectra, corresponding to vegetation status, especially water stress. Therefore, the three VIs here are examined in relation to the FVC and other classes to understand their sensitivity.

12.4.3 Estimating FVC and Comparing the Different Spatial Resolutions

The orthomosaic image generated for the study area is processed with image classification to transform the original information into a cluster of different land cover classes. The thematic map shows a total of four general classes of the landscape: acacia, bare soil, water bodies, and grass/shrubs. Four scaled gridded polygons, or fishnets, were generated using 5, 10, 30, and 60 m sizes. The percentage of each land cover type within each grid was computed, indicating the fraction of each cover type. The size of the grids coincides with other satellite remote sensing data, such as Rapideye with a 5 m resolution, Sentinel-2 for with a 10 m resolution, Landsat series with a 30 m resolution, and an additional size of 60 m to examine the sensitivity when the resolution is upscaled to a coarser one.

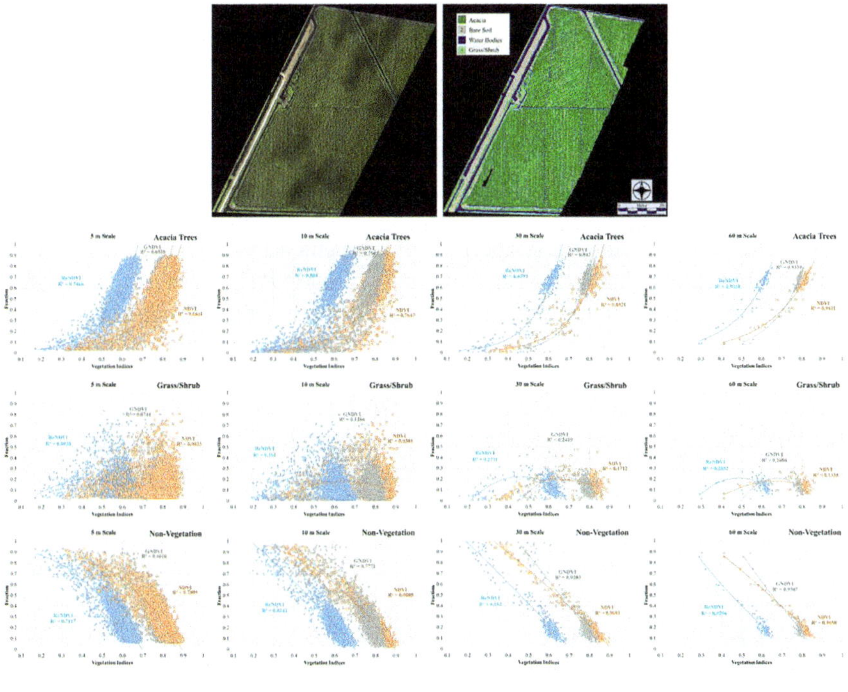

Fig. 12.13 Land cover map of the compartment and the relationship analysis between the different vegetation indices and the fraction of land cover types. (Modified from Iizuka et al. 2019)

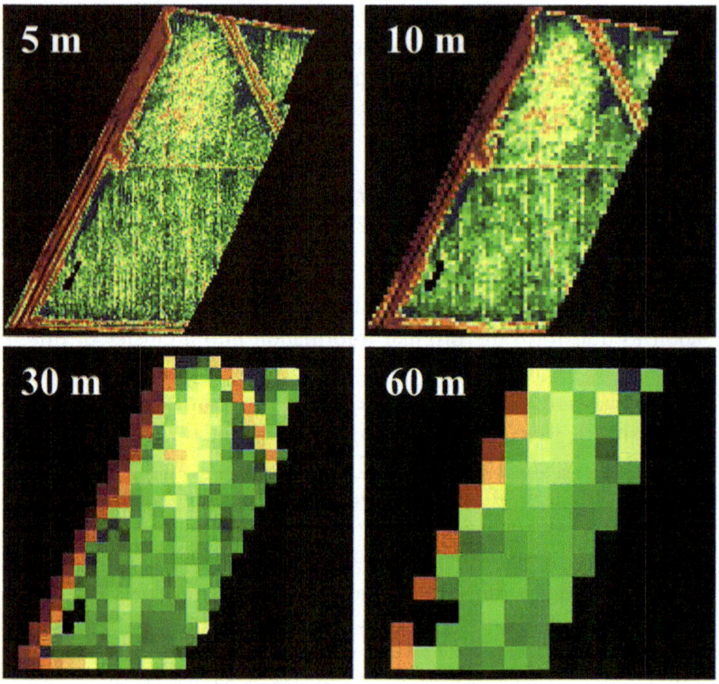

Fig. 12.14 Estimated FVC for each grid resolution from the model used in the relationship analysis. (Modified from Iizuka et al. 2019)

Figure 12.13 shows the land cover map of the site and the relationship between the fraction of land cover and each VI. Each VI was examined for three different cover types: acacia, grass/shrubs, and nonvegetation (i.e., bare soil + water bodies). Depending on the classes, the relation shows each characterized trend. Acacia shows a clear exponential relationship between the FVC and VIs. However, the grass/shrub shows a two-stage relation with a positive correlation in the lower VIs but a negative trend in the higher VIs (a second-order polynomial trend line is shown as a reference). The nonvegetation area presents a clear negative correlation to the fraction of nonvegetative cover (hereafter FnVC). It is a clear trend for the FVC and FnVC since the increases in the fraction of FVC result in decreases in the FnVC and vice versa. It is necessary to interpret grass/shrub trends correctly, and this can be ensured by carefully selecting where the information is being collected. If the grass/shrub class is extracted from the sandy platform pathway in the west of the area, the VI will increase as the fraction of grass/shrub increases. However, if this is seen within the plantation site, then a similar decreasing pattern will be shown with the increase in FVC.

The resolution differences characterize the sensitivity ranges of the FVC and VIs, where a finer resolution can detect the lower and upper ends of the fraction, and the coarser grids average out, reducing the variation in data, and both ends are excluded.

Figure 12.14 shows how it will be present when the models are used to compute the FVC at each grid resolution. It seems that a 5 m resolution is needed to understand the overall characteristics of the site. For example, the strip plantation is observable where the southern compartment is lined along the north-south direction, but the northern compartment is in a diagonal line. The situation of forest management can be monitored using this FVC output.

12.5 Conclusion

The emerging trend of using UAS remote sensing is increasing globally. The information we have discussed in this chapter introduces only a small portion of what this technology can do. UAS technology continues to advance. We can see today that synthetic aperture radar (SAR) is now experimenting with being utilized onboard a UAV, and electrical prospecting is already being used to explore the underground environment using the UAS from the air. LiDAR technology is much smaller and light weight, and it is common to find affordable-priced equipment that can be attached easily to an existing UAS. UAVs are also advancing. Hybrid types of UAVs using two or more energy sources to expand their flight duration are available on the market, and even solar-powered fixed-wing UASs are available—these can fly for almost half a day. UAVs that can hold a heavy payload of up to 30 kg and still have a flight duration of 55 min have recently been announced, and more are likely to be developed.

The airborne and spaceborne sensing field is continuously expanding, revealing its distinctive capability for Earth observations. We hope that the emergence of UAS technology presents novel opportunities for advancing this field, particularly in areas where other sensing tools may be insufficient. Furthermore, the integration of UAS technology with conventional sensing technologies is anticipated to advance future Earth observations and generate new insights.

References

Al-Kayssi AW, Al-Karaghouli AA, Hasson AM, Beker SA (1990) Influence of soil moisture content on soil temperature and heat storage under greenhouse conditions. J Agric Eng Res 45:241–252

Cattau ME, Harrison ME, Shinyo I, Tungau S, Uriarte M, DeFries R (2016) Sources of anthropogenic fire ignitions on the peat-swamp landscape in Kalimantan, Indonesia. Glob Environ Chang 39:205–219

Climate Focus (2015) Forests and land use in the Paris Agreement. The Paris Agreement Summary. http://www.climatefocus.com/publications/cop21-paris-2015-climate-focus-overall-summary-andclient-briefs. Accessed 22 Dec 2015

Di Lallo G, Mundhenk P, Zamora López SE, Marchetti M, Köhl M (2017) REDD+: quick assessment of deforestation risk based on available data. Forests 8:29

Gomes L, Simões SJC, Dalla Nora EL, de Sousa-Neto ER, Forti MC, Ometto JPHB (2019) Agricultural expansion in the Brazilian Cerrado: increased soil and nutrient losses and decreased agricultural productivity. Land 8:12

Hilasvuori E, Akujärvi A, Fritze H, Karhu K, Laiho R, Mäkiranta P, Oinonen M, Palonen V, Vanhala P, Liski J (2013) Temperature sensitivity of decomposition in a peat profile. Soil Biol Biochem 67:47–54

Iizuka K, Akiyama Y (2020) Assessing the micro-scale temperature-humidity index (thi) estimated from unmanned aerial systems and satellite data isprs Ann. Photogramm. Remote Sens. Spatial Inf. Sci. V-3-2020:745-750 https://doi.org/10.5194/isprs-annals-V-3-2020-745-2020

Iizuka K, Tateishi R (2015) Estimation of CO_2 sequestration by the forests in Japan by discriminating precise tree age category using remote sensing techniques. Remote Sens 7:15082–15113

Iizuka K, Johnson BA, Onishi A, Magcale-Macandog DB, Endo I, Bragais M (2017) Modeling future urban sprawl and landscape change in the Laguna de Bay Area, Philippines. Land 6:26

Iizuka K, Watanabe K, Kato T, Putri NA, Silsigia S, Kameoka T, Kozan O (2018) Visualizing the spatiotemporal trends of thermal characteristics in a peatland plantation forest in Indonesia: pilot test using unmanned aerial systems (UASs). Remote Sens 10(9):1345. https://doi.org/10.3390/rs10091345

Iizuka K, Kato T, Silsigia S, Soufiningrum AY, Kozan O (2019) Estimating and examining the sensitivity of different vegetation indices to fractions of vegetation cover at different scaling grids for early stage acacia plantation forests using a fixed-wing UAS. Remote Sens 11(15): 1816. https://doi.org/10.3390/rs11151816

Iizuka K, Ogura T, Akiyama Y, Yamauchi H, Hashimoto T, Yamada Y (2022) Improving the 3D model accuracy with a post-processing kinematic (PPK) method for UAS surveys. Geocarto Int 37(14):4234–4254. https://doi.org/10.1080/10106049.2021.1882004

Jauhiainen J, Kerojoki O, Silvennoinen H, Limin S, Vasander H (2014) Heterotrophic respiration in drained tropical peat is greatly affected by temperature—a passive ecosystem cooling experiment. Environ Res Lett 9:105013

Lakshmi V, Jackson TJ, Zehrfuhs D (2003) Soil moisture-temperature relationships: results from two field experiments. Hydrol Process 17:3041–3057

Liu Y, Xue Y, MacDonald G, Cox P, Zhang Z (2019) Global vegetation variability and its response to elevated CO_2, global warming, and climate variability—a study using the offline SSiB4/TRIFFID model and satellite data. Earth Syst Dynam 10:9–29

Masocha M, Murwira A, Magadza CHD, Hirji R, Dube T (2017) Remote sensing of surface water quality in relation to catchment condition in Zimbabwe. Phys Chem Earth A/B/C 100:13–18

Satterthwaite D, McGranahan G, Tacoli C (2010) Urbanization and its implications for food and farming. Philos Trans R Soc B 365:2809–2820

Stockwell CE, Jayarathne T, Cochrane MA, Ryan KC, Putra EI, Saharjo BH, Nurhayati AD, Albar I, Blake DR, Simpson IJ et al (2016) Field measurements of trace gases and aerosols emitted by peat fires in Central Kalimantan, Indonesia, during the 2015 El Niño. Atmos Chem Phys 16:11711–11732

Sumarga E (2017) Spatial indicators for human activities may explain the 2015 fire hotspot distribution in Central Kalimantan Indonesia. Trop Conserv Sci 10:10

Taubenböck H, Esch T, Felbier A, Wiesner M, Roth A, Dech S (2012) Monitoring urbanization in mega cities from space. Remote Sens Environ 117:162–176

Yao J, He XY, Li XY, Chen W, Tao DL (2012) Monitoring responses of forest to climate variations by MODIS NDVI: a case study of Hun River upstream, northeastern China. Eur J Res 131:705–716

Zhang W, Qi J, Wan P, Wang H, Xie D, Wang X, Yan G (2016) An easy-to-use airborne LiDAR data filtering method based on cloth simulation. Remote Sens 8(6):501. https://doi.org/10.3390/rs8060501

Zhao X, Guo Q, Su Y, Xue B (2016) Improved progressive TIN densification filtering algorithm for airborne LiDAR data in forested areas ISPRS. J Photogramm Remote Sens 117:79–91. https://doi.org/10.1016/j.isprsjprs.2016.03.016

Chapter 13
Multifootprint Observation Lidar and Imager (MOLI) Mission for Peatland Observations

Rei Mitsuhashi, Yoshito Sawada, Tadashi Imai, Youhei Okawa, Daisuke Sakaizawa, and Taishi Sumita

Abstract It is necessary to monitor aboveground forest biomass and vegetation phenology to understand global environmental changes and their driving factors, as these changes play an essential role in the carbon cycle over extended areas. Large-scale fires and human land development are vast sources of carbon emissions. In particular, peatland forests, a global carbon reservoir, require remote sensing observations over a wider area than field surveys can cover. However, it is challenging to estimate forest biomass accurately in areas with high forest biomass, such as the Amazon basin and Southeast Asia, using optical satellite imaging and synthetic aperture radar (SAR) data due to signal saturation. Lidar data from ICESat (Ice, Cloud, and land Elevation Satellite)/Geoscience Laser Altimeter System (GLAS), launched in 2003, have revealed that even high-biomass forests can be estimated with high accuracy and no signal saturation using space-based Lidar observations. Because of various problems with ICESat/GLAS observations, JAXA is planning the installation of a multifootprint observation Lidar and imager (MOLI) instrument on the Japanese experiment module (JEM) onboard the International Space Station (ISS). MOLI will be the first space Lidar mission observing Japan and has some unique features. MOLI estimates the slope of the ground with a multifootprint technique by detecting the signal time lag corresponding to each footprint because an inclined surface causes inaccurate canopy height estimates due to height differences. MOLI has a multiband imager to measure vegetation indices and determine lidar footprint geolocations. This chapter presents an overview of the MOLI mission and its objectives. We describe the effectiveness of forest biomass observations for this mission using simulated data expected to replicate those that MOLI will return. We also evaluate its effectiveness in precisely estimating the elevation of the forest ground, how data will be acquired to form a secondary dataset, and the ways that those data will be utilized.

R. Mitsuhashi (✉) · Y. Sawada · T. Imai · Y. Okawa · D. Sakaizawa · T. Sumita
Japan Aerospace Exploration Agency (JAXA), Tsukuba City, Ibaraki, Japan
e-mail: mitsuhashi.rei@jaxa.jp

M. Osaki et al. (eds.), *Tropical Peatland Eco-evaluation*,
https://doi.org/10.1007/978-981-99-6790-2_13

Keywords Earth observation · Aboveground biomass · Lidar · International Space Station (ISS) · Deep learning

13.1 Background of Biomass Observations by Spaceborne Lidar

13.1.1 Significance of Forest Biomass Observations

Increased atmospheric carbon dioxide has caused strange weather and climate change, directly affecting the global water cycle, ecosystems, biodiversity, and agricultural production. Extreme phenomena, such as supertyphoons, are occurring more frequently. These issues affect traditional social systems, and new measures are required for disaster prevention and mitigation. Several researchers have developed climate system models to project future climate change. However, these models are based on different scenarios (e.g., different assumptions about the extent to which greenhouse gas emissions can be reduced). The calculations involved in the simulation of these models can be massive. Previous projections by various sources and models have yielded quite different results. However, research is underway to verify and compare the results of different models. Such research will assess the uncertainty in the predicted results more accurately and thereby improve the models (Friend et al. 2014).

Improving the reliability of climate-change models rests firmly on verifying how well the models can reproduce current climate conditions using the measured concentrations of various gases and the land/ocean greenhouse gas budgets. International projects have recently attempted to assess the budgets of important greenhouse gases, such as carbon dioxide and methane, based on global observations (Ciais et al. 2014). This accumulated knowledge was published in the Global Carbon Budget Archive and included in the Fifth IPCC. The comprehensive findings were published in the Global Carbon Budget Archive, which consists of papers and documents that provide an essential scientific basis for the Fifth IPCC Assessment Report ("Global Carbon Budget Archive" n.d.; Friedlingstein et al. 2022).

Both top-down and bottom-up methods have been used to assess the global carbon cycle among the Earth's geosphere, hydrosphere, atmosphere, and biosphere. Top-down estimations calculate the amounts of carbon dioxide absorbed and emitted at the Earth's surface through inverse estimations using atmospheric carbon dioxide concentrations obtained from satellite observations and atmospheric transport models. The concentration of carbon dioxide fluctuates due to absorption and emission from anthropogenic sources and natural sources, such as oceans and vegetation. These processes can be quantitatively expressed in terms of the amount of carbon dioxide that moves to and from the Earth's surface each hour (flux). Thus, by expressing the interrelationship between the distribution of the atmospheric carbon dioxide concentration and fluxes at the land surface in a model and by incorporating a data-assimilation method (e.g., 4D-Var or LETKF) into the model and fusing it with observed data, sound absorption and emission estimations can be obtained retroactively (Kondo et al. 2015; Ciais et al. 2013; Joosten and Clarke 2002). Bottom-up estimation methods involve terrestrial ecosystem models that simulate the distribution of vegetation on the land surface and plant production activities to calculate the amount of carbon dioxide released by plants into the atmosphere. Global absorption and emission are estimated from a spatial scale containing observation points using a model incorporating processes such as plant photosynthesis, respiration, and organic matter decomposition. For their estimates, Yang et al. (2007) and Ueyama et al. (2013) optimized the blank areas between ground observation networks. Carbon dioxide absorption and emissions were measured using eddy correlation methods with the support vector regression used in machine learning (Page et al. 2008; Dargie et al. 2017). Both methods have made progress in collecting and maintaining observation data, and the uncertainties in carbon budget assessments based on observations can now be discussed meaningfully (Kondo et al. 2015). Of course, these methods need to be further refined to improve the resulting accuracy when setting reduction targets for greenhouse gas emissions and when improving the accuracy of climate-change projections that provide the basis for adopting new and improved measures.

However, a regional comparison of the results obtained with these two approaches shows that there are still differences between them, especially in the tropics (Ciais et al. 2013). These differences are due to factors related to tropical forests. Global observation data representing the forest biomass and its changes, which are critical in the carbon cycle ecosystem models used in bottom-up approaches, need to be more available and currently have significant estimation errors.

13.1.2 Forest Biomass in Peatlands

Peatland ecosystems play an essential role as carbon stores in global warming mitigation (Joosten and Clarke 2002), and their impact on greenhouse gas emissions has also been studied (Page et al. 2008). In particular, according to remote sensing data, tropical peat swamp forests have been found to contain approximately 20% (104.7 Pg) of the world's peatland carbon in field studies (Dargie et al. 2017). However, the contribution of peat swamp forests to the carbon stock is threatened by the large-scale expansion of drainage-based agriculture, such as oil palm and pulpwood production, on peatlands. This is supported by data from Landsat 7 and Landsat 8 collected between 1990 and 2015 in Peninsular Malaysia, Sumatra, and Borneo, which identified the rapid expansion of industrial plantations on peatlands (Miettinen et al. 2016). These forests are not fire-resistant but serve as fuel and ignition sources; peat fires in Sumatra, Kalimantan, and Papua in 2015 emitted 6.2 megatons of carbon dioxide (Setyawati and Suwarsono 2018).

13.1.3 Forest Biomass Observation by Satellite Remote Sensing

Satellite remote sensing methods are effective for obtaining global observation data on forest biomass and its changes. Low-resolution optical sensors, such as AVHRR and MODIS, provide high-frequency observations with a spatial resolution of several hundred meters and are used to map aboveground biomass on a continental scale. Typical methods for estimating aboveground biomass by regression models are based on band-specific reflectance and vegetation indices, among which the normalized vegetation index (NDVI) and simple ratio (SR) are often used (Dong et al. 2003). However, these methods have a disadvantage in that they are ineffective in tropical forests, for example, because spectral information and vegetation indices are saturated in forests with closed canopies and lose their correlation with the aboveground biomass (Baccini et al. 2008; Foody et al. 2003). The methods available for estimating forest biomass using shortwavelength infrared bands and vegetation indices acquired by optical sensors with a spatial resolution of several tens of meters, such as Landsat and ASTER (Avitabile et al. 2012), or by the tree canopy coverage or image texture (Sarker and Nichol 2011), have also been studied. Such sensors also have the disadvantage that the signal becomes saturated in high-biomass forests with densities above 150–200 t/ha. For example, Zhao et al. calculated the relationship between the reflectance of the shortwavelength infrared band and the aboveground biomass. They found that although there were differences among forest types, on average, saturation was reached at 156 t/ha (Zhao et al. 2016).

In addition to optical sensors, other methods available for observing forest biomass include radio wave observations using SAR (there is a strong positive correlation between the backscattering coefficient and forest biomass) (Le Toan et al. 1992). Microwaves at L-band wavelengths (15–30 cm) are dominated by scattering from large branches and trunks and can be used to estimate high-biomass forests in forest biomass observations and in research involving forest biomass estimations. ALOS/PALSAR and ALOS-2/PALSAR-2 have also been used to this end (Santoro and Cartus 2018).

The saturation point of L-band SAR is 100–150 t/ha (Shugart et al. 2010). Considering that the tropical rainforests with the highest biomass average 350 t/ha (Penman et al. 2003), there is a limit to covering all global biomass observations with L-band SAR satellites.

13.1.4 Principle of Observation of Spaceborne Lidar

Spaceborne Lidar can observe the forest structure within its observation footprint by emitting a laser beam from an altitude range of 400–600 km, above airborne Lidar scanning (ALS), and can record the TOF of the light scattered from between the tree canopy and the ground (Fig. 13.1). Unlike optical sensors and L-band SAR, this

Fig. 13.1 Mechanism of a waveform-recording Lidar

method directly observes the tree height, which is highly correlated with the forest biomass. This method can thus measure even high-biomass forests where optical sensors and SAR would face signal saturation if the laser beam reached the ground. There are two types of spaceborne Lidars: waveform-recording Lidars and photon-counting Lidars. Waveform-recording Lidars include ICESat/GLAS, which NASA operated from 2003 to 2010, and GEDI, which has been operating since 2019. ICESat-2/ATLAS, a photon-counting Lidar, has been in operation since 2018.

13.1.5 Features and Problems of Spaceborne Lidar

ICESat/GLAS has a footprint diameter of 60 m and observes the area directly below the satellite orbit at intervals of 172 m. The primary purpose of ICESat/GLAS has been to observe ice sheet dynamics. Many studies have shown that this sensor is also effective for estimating the forest canopy height and aboveground biomass. Methods for estimating aboveground biomass from GLAS data include estimating the canopy height from waveforms and converting it using allometric equations (Healey et al. 2012) and estimating aboveground biomass using waveform statistics. Another method involves making estimates based on empirical equations, such as regression models using waveform statistics (Ballhorn et al. 2011). In both cases, the footprint diameter of ICESat/GLAS is larger than several to several tens of centimeters of the aircraft-mounted Lidar, which has a diameter of 70 m. If the ground has a large slope, the ground surface elevation difference within the footprint decreases the measurement accuracy of the forest canopy height (Saatchi et al. 2011). Furthermore, the only sensor ICESat has is a Lidar (GLAS), so it cannot acquire image data, making it challenging to identify the observation point of the Lidar (Popescu et al. 2011).

Although GEDI, in operation since 2019, has a smaller footprint diameter (25 m) than ICESat/GLAS and the observed waveform is less affected by the topography, the geolocation error during the first 8 months of observation has been reported to be 23.8 m (1σ) (Beck et al. 2021). This effect has been shown to impact the tree crown significantly (Roy et al. 2021). There was an attempt to reduce this geolocation error by comparing Pearson's product factor correlation coefficients between actual and simulated data using the GEDI simulator (Hancock et al. 2019) and ALSs near GEDI observation sites (Lang et al. 2022). However, the correction range was limited to only the area where ALS data exist.

Unlike GEDI, ICESat-2/ATLAS uses the photon-counting method to calculate the ground surface height and tree height using the ATL08 product, which is equivalent to a histogram of the ATL03 product that records measured photon data. However, it has been noted that the accuracy of these ground surface height and tree height estimates deteriorates in tropical forests with high tree canopy coverage, mainly because it is difficult to distinguish between noise and signals from the canopy and ground (Neuenschwander and Magruder 2016).

13.2 Multifootprint Observation Lidar and Imager Mission

13.2.1 Mission Overview

JAXA has been conducting research and developing the Multifootprint Observation Lidar and Imager (MOLI) mission, a simultaneous Lidar/imager mission to be carried onboard the International Space Station (ISS) ("MOLI Multi-Footprint Observation LIDAR and Imager Mission" n.d.). The mission has three goals:

1. To demonstrate the high accuracy of 3D maps
2. To reduce uncertainty in terrestrial biomass by measuring the forest canopy height
3. To demonstrate spaceborne Lidar technology

The accuracy demonstration uses AW3D data (Tadono et al. 2014) measured by ALOS/PRISM to produce a global 3D map based on photogrammetry. However, this method cannot record the height of the ground in forests where the ground is not visible. Currently, the ground height is manually corrected using points around the forest where it can be seen. However, the use of these locations in forests as reference points has been demonstrated by measuring the ground height with Lidar and using those data to construct a highly accurate digital terrain model (DTM). Joint research with a private company is underway to realize this technology ("JAXA and NTT DATA Launch Joint Research on Space-Borne Laser Altimeter to Create Advanced 3D Map" n.d.).

As mentioned in the previous section, MOLI is expected to contribute to improving climate models and developing climate-change countermeasures by globally observing forest biomass, as such observations cannot be achieved with current satellite observations. This mission is expected to reduce the uncertainty in biomass data.

The technological demonstration of MOLI will be Japan's first spaceborne Earth-observing Lidar. Based on the results from MOLI, we aim to realize a spaceborne Lidar. Lidar data for Earth observation are used in meteorology, forestry, and polar research activities. Many issues remain regarding the in-orbit use of such instruments, such as ensuring the lifetime of the laser, demonstrating a noteworthy performance, and managing heat removal. A long-life operation ground test for the MOLI mission was performed in a vacuum chamber to refine the technology for operating a powerful laser in space (Sakaizawa et al. 2018).

13.2.2 Instrument Overview

The specifications of the MOLI mission are shown in Table 13.1. Unlike the JEM-EF (Japanese experiment module Kibo, exposed facility) standard payload, MOLI is mounted on i-SEEP (IVA-replaceable, small, exposed experiment platform) (Figs. 13.2 and 13.3). Although smaller in size than the standard payload, it was transported to the ISS in the cargo transfer bag of the pressurized logistics carrier onboard the transfer vehicle and mounted by the ISS crew through Kibo's airlock. Unlike the transport process by the unpressurized logistics carrier, the launch conditions, such as the vibration conditions, were mitigated, which was advantageous for the MOLI mission with its precision optical instruments. In an emergency or at the end of the mission, the MOLI will be returned to the pressurized JEM section. If the equipment is replaced or when the mission is over, the laser will be returned for ground staff to investigate any damage to optical components and obtain necessary data for future Lidar research.

Table 13.1 MOLI specifications

Item	Specifications
Mission instrument	• Lidar – Laser wavelength: 1064 nm – Number of beams: 2 beams – Beam power: 40 mJ – Pulse width < 7 ns – Footprint radius: Φ25 m • Imager – Band (TBD) 550–630 nm (green) 630–740 nm (red) 740–880 nm (NIR) – Spatial resolution: 5 m – Swath: 1 km
Size	1605 × 640 × 830 mm
Mass	Approximately 300 kg
Operation	>1 year
Operational orbit	ISS orbit (inclination: 51.6°) Nonsynchronous at 400 km

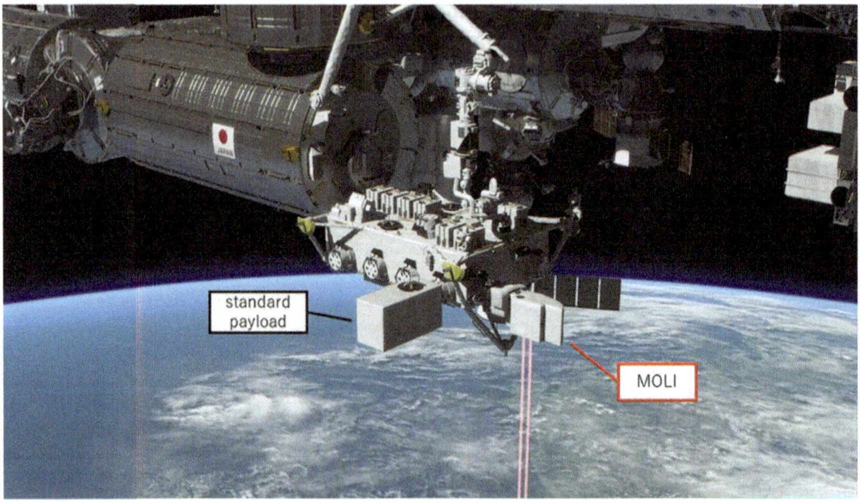

Fig. 13.2 MOLI attached to the ISS

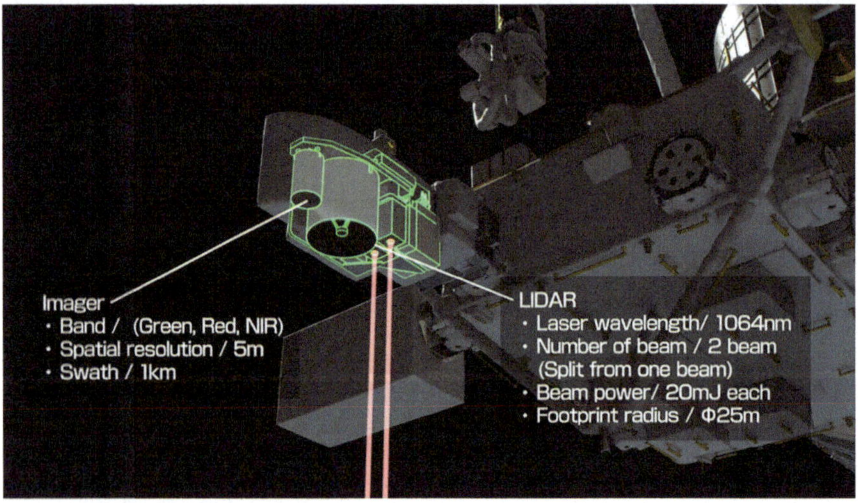

Fig. 13.3 MOLI observation sensor

13.2.3 MOLI Mission Features

The MOLI mission is characterized by multifootprint slope estimation and simultaneous imager and Lidar observations. In ICESat/GLAS, the effect of the footprint slope has been reported to reduce the accuracy of the tree height and AGB estimates (Hayashi et al. 2013). The effect of the slope on the accuracy of tree height estimations was also reported in GEDI (Fayad et al. 2021). Therefore, in the

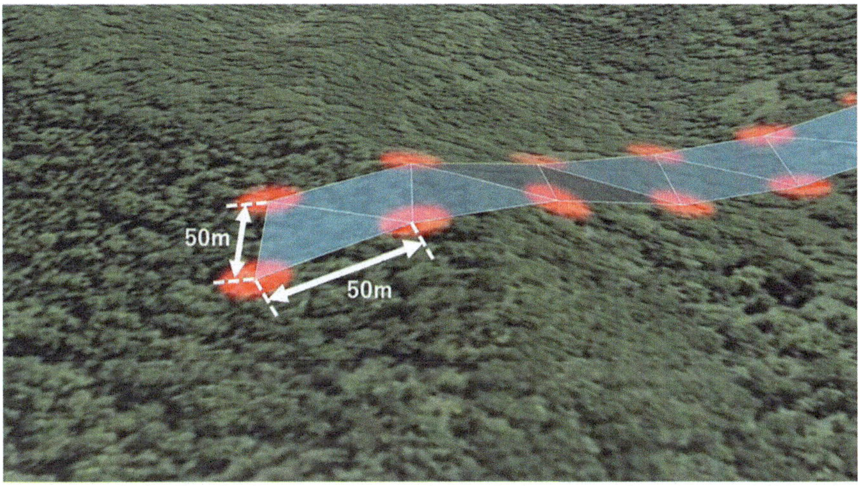

Fig. 13.4 Multifootprint Lidar observation

MOLI mission, to reduce the influence of the slope on the forest canopy height measurements, the footprint was made smaller (diameter of 25 m), two lasers were designed to make simultaneous observations in two areas with high irradiation frequencies (150 Hz; 50 m apart) in the direction of travel, and the ground slope was estimated from the elevation of each footprint (Fig. 13.4). This design enables highly accurate tree height estimations by correcting for the effect of the slope from the MOLI observation data, even if no reliable topographic information is available as an external reference source at the observation site.

Another feature is the simultaneous observations of the Lidar and imager, as shown in Fig. 13.3. Since the Lidar and imager are installed in the same optical system, the relative relationship between the imager and Lidar's lines of sight toward the Earth is constant except for thermal distortion. The Lidar observations are performed at a specific point within the observation range (Fig. 13.5). This enables the correction of observation points that are difficult to correct using Lidar alone, using the same parameters as the geometric correction involving the imager's ground control points (GCPs). In addition, the simultaneous observation by the Lidar and imager enables vegetation information, such as cloud-screening data of Lidar observation data and normalized difference vegetation index (NDVI) data of the Lidar observation points, to be measured. In a previous study, the tree height information observed by ICESat/GLAS was combined with a high-resolution satellite (RapidEye) to estimate tree heights in the Mekong River basin and the central Amazon (Sawada et al. 2015) and to construct a global tree height map using GEDI and Landsat-8 observation data (Potapov et al. 2021). In this way, Lidar alone provides only spatially discrete data, but combining data from areal imagers produces a composite analysis. Simultaneous observations also make it possible to eliminate the effects of land cover changes during both the Lidar and imager observations.

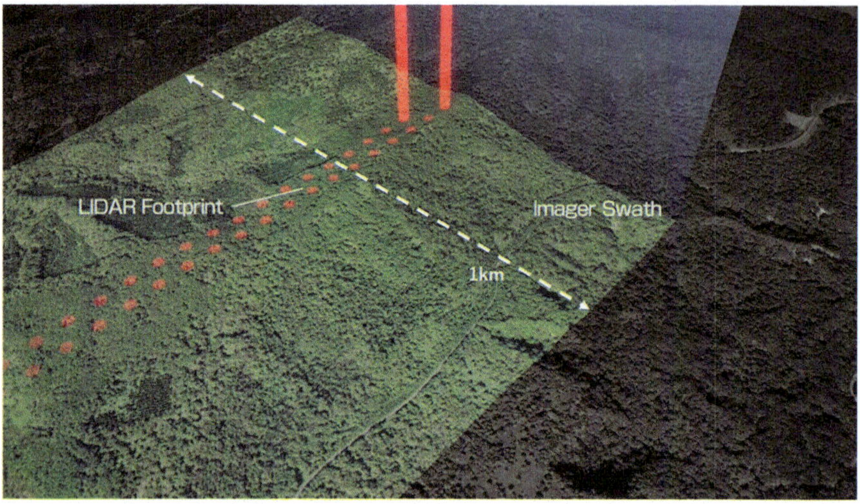

Fig. 13.5 Simultaneous observations of the Lidar instrument and imager

13.2.4 Product

Table 13.2 shows the products that the MOLI project plans to offer. These products will be free of charge from G-Portal (https://gportal.jaxa.jp/gpr/?lang=en).

13.2.5 Observation Experiment Using Airborne Lidar

We conducted airborne measurements to evaluate the validity of the observation method using multi-footprint data. We installed a multifootprint Lidar system on an aircraft (King Air 200T) with a CCD camera (CANON EOS 5D Mark III). The aircraft flew over laurel and conifer forests at an altitude of 5500–7500 m. One month after the experiment, we validated the canopy height data directly to verify the observation results obtained at the two sites.

Figures 13.6 and 13.7 are two verification sites. Figure 13.8 shows the Lidar-observed waveform #1_1 point in Muroto, corresponding to a laurel forest (Fig. 13.6). The relative heights were calculated by the integrated waveform energy from the ground peak to the signal start. Therefore, RH100 represents the maximum tree height (Sun et al. 2008). Each footprint's maximum canopy height correlates to RH100 (Chen 2010). We confirmed that each verification site had a uniform canopy height and assumed that most verification site footprints overlapped with the maximum canopy height indicated in the land vegetation observations. Therefore, we compared the maximum canopy height at the verification sites with the canopy height from RH100. We also corrected for the ground elevation in the footprint using the terrain slope calculated from multifootprint sampling. We used the Sawada

Table 13.2 MOLI mission product list

Level	Product
L1 (Standard)	Lidar footprint and imager products
	• Geolocated waveforms Footprint position accuracy ≤15 m • Geolocated image 1-km swath, 3 bands
L2 (Standard)	Lidar footprint products
	• Terrain height Accuracy ±3 m (terrain slope ≤ 30° and canopy cover > 95%) ±5 m (terrain slope > 30° and canopy cover < 95%) • Canopy heights (Φ25-m footprint) Accuracy ±3 m (canopy height ≤ 15 m) ±20% (canopy height > 15 m) • Forest biomass (Φ25-m footprint) Accuracy ±25 t/ha (biomass density ≤ 100 t/ha) ±25% (biomass density > 100 t/ha)
L3 (Research)	Integrated products with Lidar and imager
	• Canopy heights (1-km swath) Target accuracy ±~5 m (canopy height ≤ 15 m) ±~40% (canopy height > 15 m) • Forest biomass (1-km swath) Target accuracy ±40 t/ha (biomass density ≤ 100 t/ha) ±40% (biomass density > 100 t/ha)
L4 (Research)	Wall-to-wall map products (with GCOM-C/SGLI Data)
	• Canopy height map • Forest biomass map

et al. (2015) equation for slope correction. Figures 13.9 and 13.10 show the results from two verification sites. These histograms show that the differences between the canopy heights obtained by the correction and the maximum canopy heights measured at the verification sites (RMSE) are smaller than 2 m (Muroto, 1.54 m and 1.08 m; Gero, 1.18 m). These results show the validity of MOLI's system design for the mission requirements.

In addition, integrated products containing Lidar and imager data (L3) and wall-to-wall map products (L4) will be produced as research products. The L3 and L4

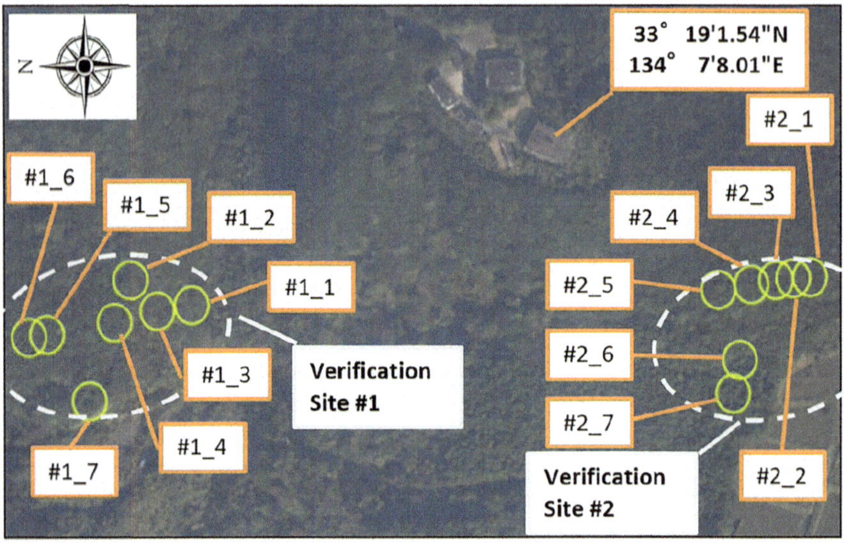

Fig. 13.6 The picture from the aircraft and annualized footprint (Muroto, laurel forest)

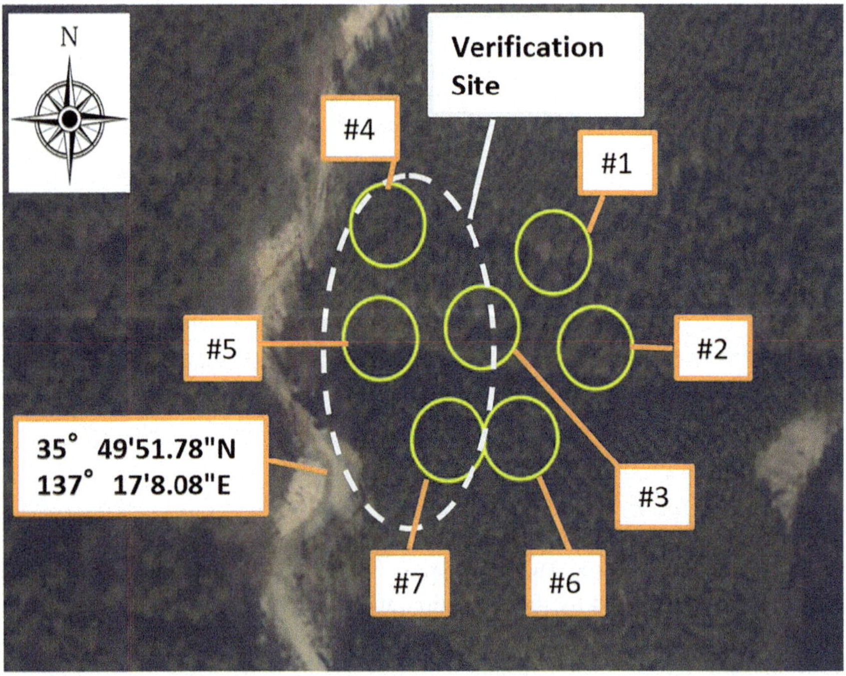

Fig. 13.7 The picture from an aircraft and annualized footprint (Gero, conifer forest)

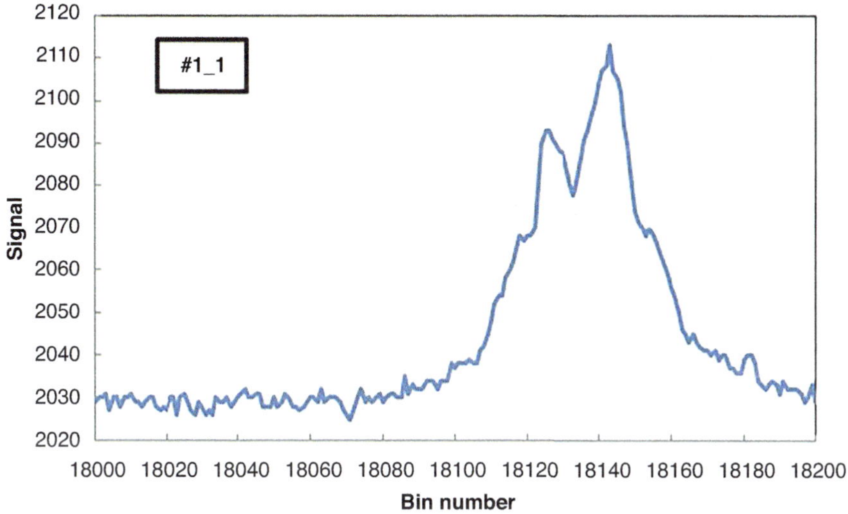

Fig. 13.8 Observed Lidar wave

products were prototyped using deep learning to simulate the data acquired in this airborne experiment. Figure 13.11a shows a prototype of the L3 product created using the Lidar and image data from the airborne experiment. Figure 13.11b shows a prototype of the L4 product, which combines the Lidar data from the aircraft experiment with Landsat-8 imagery from the same period.

13.3 MOLI Observations in Peatlands

Since MOLI is aboard the ISS, it can make observations only between 51°N and 51° S. However, it can observe tropical peatlands, which are essential in the carbon cycle. Although actual MOLI observation data do not yet exist, we combined the GEDI product, which is available and has similar data specifications as MOLI, with Norway's International Climate and Forests Initiative (NICFI) Planet tropical base maps ("Norway's International Climate and Forest Initiative (NICFI)" n.d.) to create simulated data representing a peatland observed by MOLI.

Figure 13.12 shows the coordinate information corresponding to two periods of GEDI L1B products (7/13/2019 and 6/3/2021) observed close to each other in West Kalimantan. The MOLI observations are simulated using these two periods of GEDI data and the two corresponding Planet tropical base maps (a half-year composite for June–November 2019 and monthly composite for June 2021, hereafter referred to as the Planet data). Figure 13.13 shows a simulated dataset. This path is 19.6 km long and is shown with the horizontal axis in the along-track direction. The upper-row GEDI product data are plotted for each waveform value by distance, with the

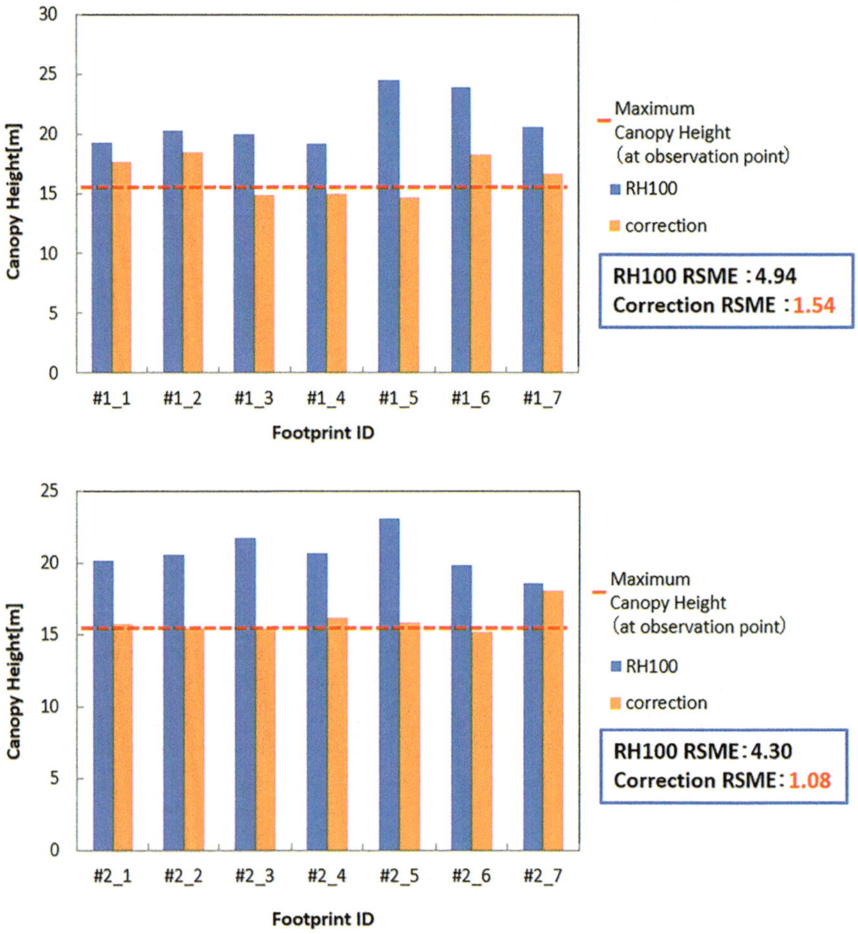

Fig. 13.9 Comparison of the RH100 and corrected results (Muroto, laurel forest)

amplitude plotted in the third dimension in shades of green. Each point represents the TanDEM-X value stored in the GEDI L1B product (Krieger et al. 2007), the lowest ground surface mode stored in the GEDI L2A product (Hofton et al. 2019), and the AW3D30 data (Tadono et al. 2014). The GEDI footprint locations are plotted in the bottom row in the simulated imager data. The images are shown 1-km wide as in the MOLI specifications, with the red, NIR, and green bands of the Planet data assigned to RGB. The spatial resolution of the Planet data is 4.77 m, roughly that of MOLI's Imager, making it the best choice for simulating MOLI imager data.

In Figs. 13.14 and 13.15, the observed characteristic topography shown in Fig. 13.13 is divided into areas A and B. Figure 13.14 shows an enlarged view of area A in Fig. 13.15. Area A includes data observed by GEDI or Planet data where logging from GEDI data may have occurred. The Planet data confirm logging

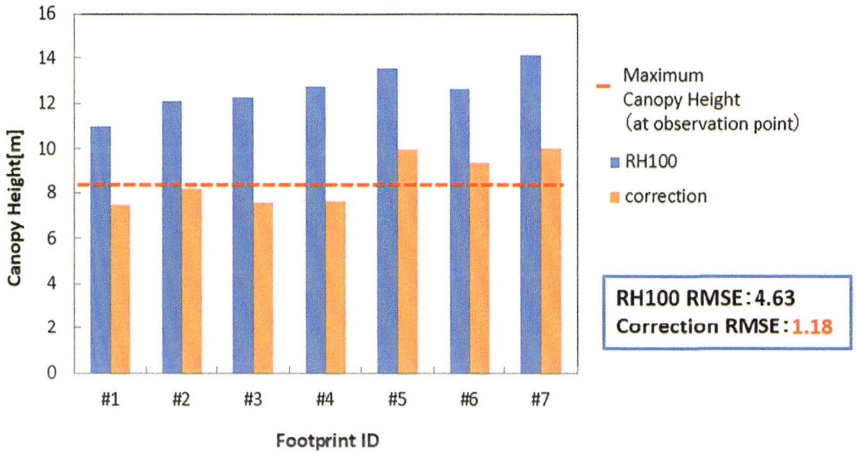

Fig. 13.10 Comparison of the RH100 and corrected results (Gero, conifer forest)

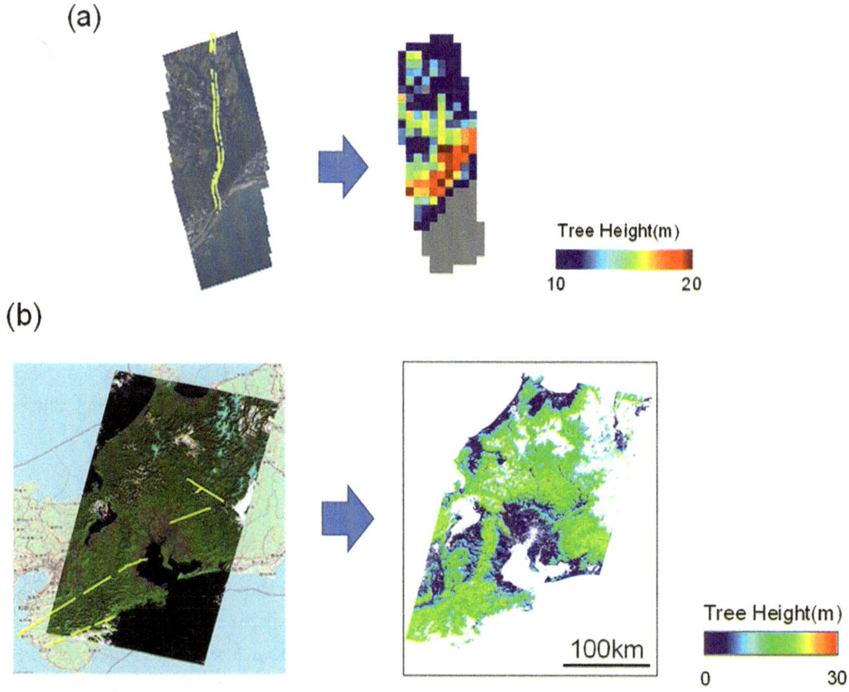

Fig. 13.11 Prototypes of the MOLI L3 and L4 products. (**a**) L3 product: left, georeferenced aerial 5-m-resolution image camera with Lidar footprints (yellow dots); right, 40-m gridded tree height map (RMSE = 3.7 m). (**b**) L4 product: left, Landsat-8 images with Lidar footprints (yellow dots); right, 90-m gridded tree height map (RMSE = 4.6 m)

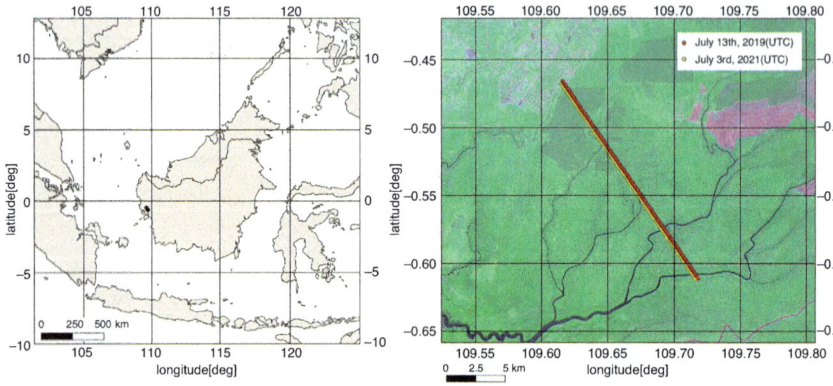

Fig. 13.12 Coordinates of GEDI observation data collected in West Kalimantan, Indonesia, used to simulate the MOLI data. Left: Overall view of Kalimantan Island. Right: Enlarged view of the GEDI data used for visualization

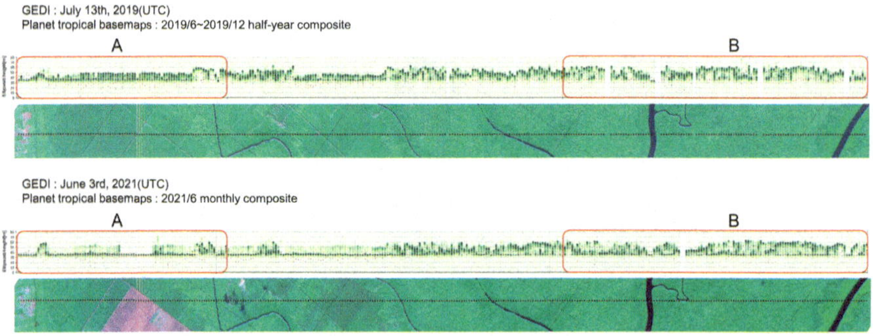

Fig. 13.13 GEDI products (upper panel) and Planet tropical base maps (lower panel) used to simulate MOLI observations on July 13, 2019, and June 3, 2021. For the upper plots: red, GEDI L2A elevation lowest mode; blue, AW3D_30; and black, TanDEM-X elevation (including the GEDI L1B product). The extent of the image is the same for both periods

(clear-cutting) in area (2). In area (2), the canopy peak in the GEDI waveform almost disappeared between 2019 and 2021. Comparing the Lidar waveform data representing 2019 and 2021 in region (1), the peak position of the Lidar waveform moves closer to the ground surface, while the starting position of the received Lidar signal increases. This waveform pattern may be due to thinning associated with growth between the two observation periods. Satellite images also show thinning in the vicinity of the footprint. Simultaneous observations such as MOLI are desirable for analyzing such combinations of areal image data and vertical profiles because of the possibility of inconsistencies occurring between data sources due to different observation periods.

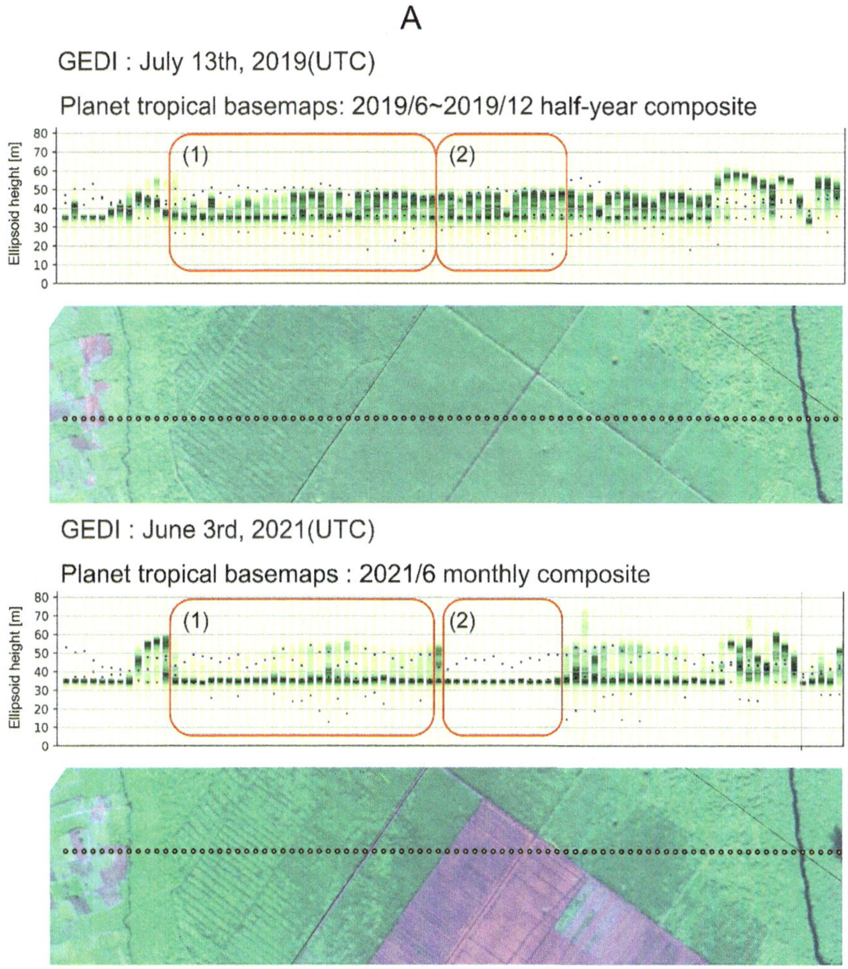

Fig. 13.14 Enlarged view of area A in Fig. 13.12. (1) The points where there is no apparent change between the two periods of Planet data, but there is a change in the GEDI (Lidar) data. (2) The points where changes occur in both the Planet and GEDI (Lidar) data between the two time periods

This result also demonstrates that spaceborne Lidar can identify changes in the three-dimensional structure of forests that cannot be identified in satellite imagery. The GFOI report notes that existing remote sensing data may underestimate forest degradation (Espejo et al. 2020). Spaceborne Lidar has the disadvantage of not being able to observe exact locations at certain times. However, it has been suggested that those data be identified statistically after zoning, such as at the landscape level or by watershed. Considering that changes in vegetation recovery due to differences in clear-cutting and thinning in peatlands (Hamberg et al. 2019) and their effects on hydrology (Heikurainen 1970) have been confirmed, it is important to be able to

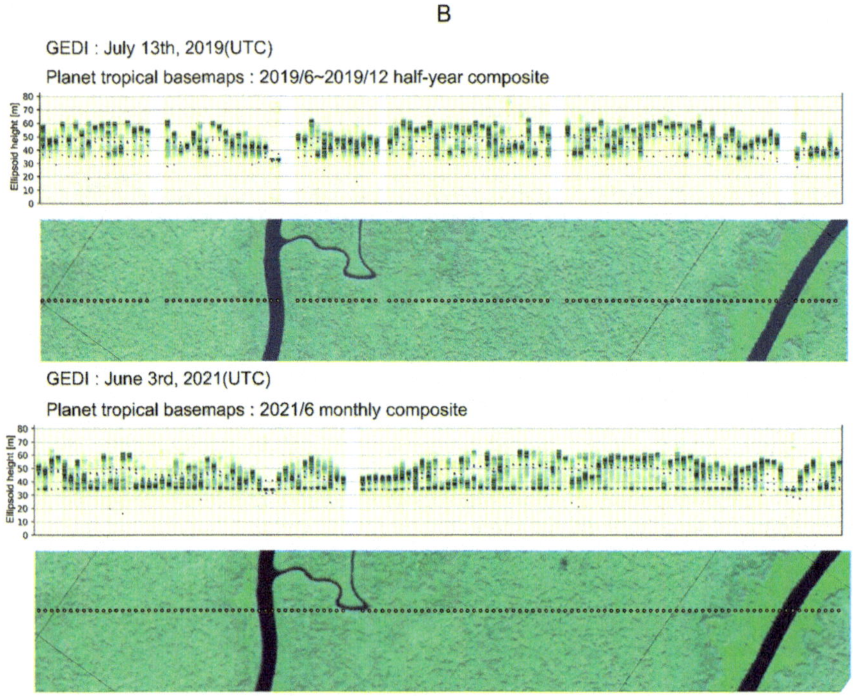

Fig. 13.15 Enlarged view of area B in Fig. 13.13

acquire three-dimensional information, including land cover and ground-surface cover materials, using two sensors, a high-resolution imager, and a Lidar instrument like MOLI, to observe the extent of logging in more detail than can be obtained from conventional satellite images.

Furthermore, it is clear from the GEDI data that AW3D_30 and TanDEM-X in Fig. 13.13 are not digital terrain models (DTMs) or digital surface models (DSMs). Highly accurate DTMs correlate well with the field-measured peat thickness (Vernimmen et al. 2020) and should be helpful for mapping deep peat with a high potential for future carbon emissions from degradation and fire. Spaceborne Lidars, such as GEDI and MOLI, can accurately observe the height of the Earth's surface at each observation time. While the observations are scattered along the spacecraft's orbit, these instruments should still improve the situation.

Figure 13.15 shows an enlargement of area B in Fig. 13.13. Unlike area A, this area is not a plantation but is instead an untouched natural forest, and the GEDI data show that the waveforms indicating the canopy height and vertical structure are different at each location. The data measured by full-wave Lidars such as GEDI and MOLI can estimate the canopy height and foliage height diversity (FHD) from the shape of recorded waveforms stored in the GEDI L2B product (Hao and John 2019). The FHD has been reported to be related to the bird diversity (MacArthur and MacArthur 1961), and a study performed with airborne Lidar (LVIS) data simulating spaceborne Lidar observations revealed relationships between the bird species diversity and parameters representing the forest structural complexity in a temperate forest in Maryland, USA (Goetz et al. 2007). Hansen et al. (2019) combined forest canopy height maps created by ICESat/GLAS and Landsat-7 and -8 (Goetz et al. 2007) with tree cover and loss year data to create the Forest Structural Condition Index (SCI), which was reported to be in close agreement with the FHD measured by ALS (Hansen et al. 2016, 2019). Supposing MOLI can obtain the FHD using the Lidar and develop an accurate, areally extended dataset with high-spatial-resolution imager data, such a dataset could be valuable in identifying hot spots, such as tropical forests, that require immediate protection.

13.4 Summary

We introduced the space Lidar mission (MOLI) currently under development and described its potential. Because the MOLI mission simultaneously observes high-resolution images and Lidar data that can be used to acquire vertical information, it is expected to significantly contribute to forest monitoring and topographic information acquisitions, including in peat forest research. The MOLI mission plans to develop and release a data structure for each latitudinal and longitudinal degree to allow the simultaneous confirmation of vertical information indicated by the Lidar waveform data and areal information provided by the imager, as shown in Fig. 13.13.

MOLI is also Japan's first Earth-observing Lidar mission and is meant to establish the basic technology for future spaceborne Lidar development. We plan to acquire technology that will contribute to Lidar missions that perform areal Lidar observations such as LIST (Yong et al. 2022), which NASA is considering, and that will perform wall-to-wall 3-D mapping over multiple periods, which has been challenging thus far. Another development direction is the use of multiwavelength Lidar to detect more accurate ground surface information. While obtaining satellite forest observations with imagers is now standard, several missions have been conducted with spaceborne Lidars since the launch of ICESat/GLAS in 2003. The MOLI mission will enhance these existing Earth observations with simultaneous Lidar and imager observations and their fused data products.

Acknowledgments The authors thank the members of the MOLI Science Team and Mr. Tsuyoshi Kato of the Sumitomo Forestry Co. for his valuable suggestions and discussions with us.

References

Avitabile V, Baccini A, Friedl MA, Schmullius C (2012) Capabilities and limitations of Landsat and land cover data for above-ground woody biomass estimation of Uganda. Remote Sens Environ 117:366–380. https://doi.org/10.1016/j.rse.2011.10.012

Baccini A, Laporte N, Goetz SJ, Sun M, Dong H (2008) A first map of tropical Africa's above-ground biomass derived from satellite imagery. Environ Res Lett 3(4):045011. https://doi.org/10.1088/1748-9326/3/4/045011

Ballhorn U, Jubanski J, Siegert F (2011) ICESat/GLAS data as a measurement tool for peatland topography and peat swamp forest biomass in Kalimantan, Indonesia. Remote Sens 3(9): 1957–1982. https://doi.org/10.3390/rs3091957

Beck J, Wirt B, Armston J, Hofton M, Luthcke S, Tang H (2021) Global Ecosystem Dynamics Investigation (GEDI) level 02 user guide. Doc version, vol 2

Chen Q (2010) Retrieving vegetation height of forests and woodlands over mountainous areas in the Pacific Coast region using satellite laser altimetry. Remote Sens Environ 114(7):1610–1627. https://doi.org/10.1016/j.rse.2010.02.016

Ciais P, Sabine C, Bala G, Peters W (2013) Carbon and other biogeochemical cycles. In: Stocker TF, Qin D, Plattner GK, Tignor M, Allen SK, Boschung J, Nauels A, Xia Y, Bex V, Midgley PM (eds) Climate change 2013: the physical science basis. Contribution of Working Group I to the fifth assessment report of the Intergovernmental Panel on Climate Change. Cambridge University Press, pp 465–570. https://doi.org/10.1017/CBO9781107415324.015

Ciais P et al (2014) Current systematic carbon-cycle observations and the need for implementing a policy-relevant carbon observing system. Biogeosciences 11(13):3547–3602. https://doi.org/10.5194/bg-11-3547-2014

Dargie GC, Simon LL, Ian TL et al (2017) Age, extent and carbon storage of the central Congo Basin peatland complex. Nature 542(7639):86–90. https://doi.org/10.1038/nature21048

Dong JR, Kaufmann RK, Myneni RB et al (2003) Remote sensing estimates of boreal and temperate forest woody biomass: carbon pools, sources, and sinks. Remote Sens Environ 84(3):393–410

Espejo AB, Feferici S, Green C et al (2020) Integrating remote-sensing and ground-based observations for estimation of emissions and removals of greenhouse gases in forests: methods and guidance from the Global Forest Observations Initiative. Edition 3.0

Fayad I, Baghdadi N, Alvares CA et al (2021) Terrain slope effect on forest height and wood volume estimation from Gedi data. Remote Sens 13(11):2136. https://doi.org/10.3390/rs13112136

Foody GM, Boyd DS, Cutler MEJ (2003) Predictive relations of tropical forest biomass from Landsat TM data and their transferability between regions. Remote Sens Environ 85(4): 463–474. https://doi.org/10.1016/S0034-4257(03)00039-7

Friedlingstein P, Jones MW, O'Sllivan M et al (2022) Global carbon budget 2021. Earth Syst Sci Data 14(4):1917–2005. https://doi.org/10.5194/essd-14-1917-2022

Friend AD et al (2014) Carbon residence time dominates uncertainty in terrestrial vegetation responses to future climate and atmospheric CO_2. Proc Natl Acad Sci U S A 111(9): 3280–3285. https://doi.org/10.1073/pnas.1222477110

Global Carbon Budget Archive (n.d.). https://www.globalcarbonproject.org/carbonbudget/archive.htm. Accessed 15 Dec 2022

Goetz S, Steinberg D, Dubayah R, Blair B (2007) Laser remote sensing of canopy habitat heterogeneity as a predictor of bird species richness in an eastern temperate forest, USA. Remote Sens Environ 108(3):254–263. https://doi.org/10.1016/j.rse.2006.11.016

Hamberg L, Hotanen JP, Nousiainen H, Nieminen TM, Ukonmaanaho L (2019) Recovery of understorey vegetation after stem-only and whole-tree harvesting in drained peatland forests. For Ecol Manag 442:124–134. https://doi.org/10.1016/j.foreco.2019.04.002

Hancock S, Armston J, Hofton M et al (2019) The GEDI simulator: a large-footprint waveform lidar simulator for calibration and validation of spaceborne missions. Earth Space Sci 6:294–310. https://doi.org/10.1029/2018ea000506

Hansen MC, Potapov PV, Geotz SJ et al (2016) Mapping tree height distributions in Sub-Saharan Africa using Landsat 7 and 8 data. Remote Sens Environ 185:221–232. https://doi.org/10.1016/j.rse.2016.02.023

Hansen A, Barnett K, Jantz P et al (2019) Global humid tropics forest structural condition and forest structural integrity maps. Sci Data 6(1):232. https://doi.org/10.1038/s41597-019-0214-3

Hao T, John A (2019) Algorithm Theoretical Basis Document (ATBD) for GEDI L2B footprint canopy cover and vertical profile metrics. Goddard Space Flight Center, Greenbelt, MD

Hayashi M, Saigusa N, Oguma H, Yamagata Y (2013) Forest canopy height estimation using ICESat/GLAS data and error factor analysis in Hokkaido, Japan. ISPRS J Photogramm Remote Sens 81:12–18. https://doi.org/10.1016/j.isprsjprs.2013.04.004

Healey SP, Patterson PL, Saatchi S, Lefsky MA, Lister AJ, Freeman EA (2012) A sample design for globally consistent biomass estimation using lidar data from the Geoscience Laser Altimeter System (GLAS). Carbon Balance Manag 7(1):10. https://doi.org/10.1186/1750-0680-7-10

Heikurainen L (1970) The effect of thinning, clear cutting and fertilization on the hydrology of peatland drained for forestry, 104th edn. Suomen Metsätieteellinen Seura

Hofton M, Blair B, Story S, Yi D (2019) Algorithm Theoretical Basis Document (ATBD) for GEDI transmit and receive waveform processing for L1 and L2 products. Version 1.0. Goddard Space Flight Center, Greenbelt, MD

JAXA and NTT DATA Launch Joint Research on space-borne laser altimeter to create advanced 3D map (n.d.). https://global.jaxa.jp/press/2021/03/20210326-1_e.html. Accessed 15 Dec 2022

Joosten H, Clarke D (2002) Wise use of mires and peatlands. International Mire Conservation Group, International Peat Society, p 304

Kondo M, Ichii K, Takagi H, Sasakawa M (2015) Comparison of the data-driven top-down and bottom-up global terrestrial CO_2 exchanges: GOSAT CO_2 inversion and empirical eddy flux upscaling. J Geophys Res Biogeosci 120(7):1226–1245. https://doi.org/10.1002/2014JG002866

Krieger G, Moreira A, Fiedler H et al (2007) TanDEM-X: a satellite formation for high-resolution SAR interferometry. IEEE Trans Geosci Remote Sens 45(11):3317–3341. https://doi.org/10.1109/TGRS.2007.900693

Lang N, Kalischek N, Armston J, Schindler K, Dubayah R, Wegner JD (2022) Global canopy height regression and uncertainty estimation from GEDI LIDAR waveforms with deep ensembles. Remote Sens Environ 268:112760. https://doi.org/10.1016/j.rse.2021.112760

Le Toan T, Beaudoin A, Riom J, Guyon D (1992) Relating forest biomass to SAR data. IEEE Trans Geosci Remote Sens 30(2):403–411. https://doi.org/10.1109/36.134089

MacArthur RH, MacArthur JW (1961) On bird species diversity. Ecology 42(3):594–598. https://doi.org/10.2307/1932254

Miettinen J, Shi C, Liew SC (2016) Land cover distribution in the peatlands of Peninsular Malaysia, Sumatra and Borneo in 2015 with changes since 1990. Glob Ecol Conserv 6:67–78. https://doi.org/10.1016/j.gecco.2016.02.004

MOLI Multi-footprint Observation LIDAR and Imager mission (n.d.). https://www.kenkai.jaxa.jp/research/moli/moli-index.html. Accessed 15 Dec 2022

Neuenschwander AL, Magruder LA (2016) The potential impact of vertical sampling uncertainty on ICESat-2/ATLAS terrain and canopy height retrievals for multiple ecosystems. Remote Sens 8(12):1039. https://doi.org/10.3390/rs8121039

Norway's International Climate and Forest Initiative (NICFI) (n.d.). https://www.nicfi.no/. Accessed 15 Dec 2022

Page S, Hoscilo A, Wösten H et al (2008) Restoration ecology of lowland tropical peatlands in Southeast Asia: current knowledge and future research directions. Ecosystems 12:888–905. https://doi.org/10.1007/s10021-008-9216-2

Penman J, Gytarsky M, Hiraishi T et al (2003) Good practice guidance for land use, land-use change and forestry. Institute for Global Environmental Strategies IGES, Japan. http://inis.iaea.org/search/search.aspx?orig_q=RN:36103537

Popescu SC, Zhao K, Neuenschwander A, Lin C (2011) Satellite lidar vs. small footprint airborne lidar: comparing the accuracy of above-ground biomass estimates and forest structure metrics at footprint level. Remote Sens Environ 115(11):2786–2797. https://doi.org/10.1016/j.rse.2011.01.026

Potapov P, Xinyuan L, Hernandez-Serna A et al (2021) Mapping global forest canopy height through integration of GEDI and Landsat data. Remote Sens Environ 253:112165. https://doi.org/10.1016/j.rse.2020.112165

Roy DP, Kashongwe HB, Armston J (2021) The impact of geolocation uncertainty on GEDI tropical forest canopy height estimation and change monitoring. Sci Remote Sens 4:100024. https://doi.org/10.1016/j.srs.2021.100024

Saatchi SS, Harris NL, Brown S et al (2011) Benchmark map of forest carbon stocks in tropical regions across three continents. Proc Natl Acad Sci U S A 108(24):9899–9904. https://doi.org/10.1073/pnas.1019576108

Sakaizawa D, Mitsuhahi R, Murooka J, Imai T, Kimura T, Asai K (2018) Current status of the ISS-vegetation lidar Mission-MOLI. In: IGARSS 2018—2018 IEEE international geoscience and remote sensing symposium, pp 1861–1864. https://doi.org/10.1109/IGARSS.2018.8518379

Santoro M, Cartus O (2018) Research pathways of forest above-ground biomass estimation based on SAR backscatter and interferometric SAR observations. Remote Sens 10(4):608. https://doi.org/10.3390/rs10040608

Sarker LR, Nichol JE (2011) Improved forest biomass estimates using ALOS AVNIR-2 texture indices. Remote Sens Environ 115(4):968–977. https://doi.org/10.1016/j.rse.2010.11.010

Sawada Y, Suwa R, Jindo K et al (2015) A new 500-m resolution map of canopy height for Amazon forest using spaceborne LiDAR and cloud-free MODIS imagery. Int J Appl Earth Obs Geoinf 43:92–101. https://doi.org/10.1016/j.jag.2015.04.003

Setyawati W, Suwarsono S (2018) Carbon emission from peat fire in 2015. IOP Conf Ser Earth Environ Sci 166:12041. https://doi.org/10.1088/1755-1315/166/1/012041

Shugart HH, Saatchi S, Hall FG (2010) Importance of structure and its measurement in quantifying function of forest ecosystems. J Geophys Res Biogeosci 115(4). https://doi.org/10.1029/2009JG000993

Sun G, Ranson KJ, Kimes DS, Blair JB, Kovacs K (2008) Forest vertical structure from GLAS: an evaluation using LVIS and SRTM data. Remote Sens Environ 112(1):107–117. https://doi.org/10.1016/j.rse.2006.09.036

Tadono T, Ishida H, Oda F, Naito S, Minakawa K, Iwamoto H (2014) Precise global DEM generation by ALOS PRISM. ISPRS Ann Photogramm Remote Sens Spat Inf Sci II–4 (May):71–76. https://doi.org/10.5194/isprsannals-ii-4-71-2014

Ueyama M, Ichii K, Iwata H et al (2013) Upscaling terrestrial carbon dioxide fluxes in Alaska with satellite remote sensing and support vector regression. J Geophys Res Biogeo 118:1266–1281. https://doi.org/10.1002/jgrg.20095

Vernimmen R, Hooijer A, Akmalia R et al (2020) Mapping deep peat carbon stock from a LiDAR based DTM and field measurements, with application to eastern Sumatra. Carbon Balance Manag 15(1):4. https://doi.org/10.1186/s13021-020-00139-2

Yang F, Ichii K, White MA et al (2007) Developing a continental-scale measure of gross primary production by combining MODIS and AmeriFlux data through Support Vector Machine approach. Remote Sens Environ 110:109–122. https://doi.org/10.1016/j.rse.2007.02.016

Yong F, Li Z, Hui G, Bincai C, Li G, Haiyan H (2022) Spaceborne LiDAR surveying and mapping, Ch. 4. In: Hong DLT, Thi D (eds) LiDAR technology—from surveying to digital twins. IntechOpen, Rijeka. https://doi.org/10.5772/intechopen.108177

Zhao P, Lu D, Wang G, Wu C, Huang Y, Yu S (2016) Examining spectral reflectance saturation in landsat imagery and corresponding solutions to improve forest above-ground biomass estimation. Remote Sens 8(6):469. https://doi.org/10.3390/rs8060469

SPRINGER NATURE

GPSR Compliance

The European Union's (EU) General Product Safety Regulation (GPSR) is a set of rules that requires consumer products to be safe and our obligations to ensure this.

If you have any concerns about our products, you can contact us on ProductSafety@springernature.com

In case Publisher is established outside the EU, the EU authorized representative is:

Springer Nature Customer Service Center GmbH
Europaplatz 3
69115 Heidelberg, Germany

The manufacturer's authorised representative in the EU is Springer
Nature Customer Service Centre GmbH, Europaplatz 3, 69115 Heidelberg,
Germany. If you have any concerns regarding our products, please
contact ProductSafety@springernature.com

Printed and bound by CPI Group (UK) Ltd, Croydon, CR0 4YY
01/10/2026
02225866-0010